1990

Proceedings of
the Sixth
International
Congress on
Mathematical
Education

EDITED BY

Ann & Keith Hirst

ICMI SECRETARIAT

JÁNOS BOLYAI MATHEMATICAL SOCIETY

Editors:
Ann & Keith Hirst
Department of Mathematics
The University Southampton,
SO9 5NH United Kingdom

This book was typeset by TYPOTEX GT at the Mathematical Institute of
the Hungarian Academy of Sciences
P.O. Box 127 H–1367 Budapest, Hungary

János Bolyai Mathematical Society
P.O. Box 240
H-1368 Budapest, Hungary

PRINTED by MALÉV in Hungary

ISBN 963-8022-48-5

ERRATA to the PROCEEDINGS of ICME - 6

Inside front page, middle: Insert the name of the other VICE
   PRESIDENT: "Emilio Lluis, Cincinnati 23, Ciudad de
   los Deportes, 03710, D.F. Mexico"

pp.31 - 47: The running head should read as "FRAMEWORKS
   AND FACTS" instead of "FRAMEWORKS AND
   FACTS IN THE PME"

p.33, line 3, "(and algorithms)" instead of "(et algorithms)"

p.37, line 2 should start as: "   partition   quotition"
   line 8, "mathematical, physical, or" instead of "mathematical physical or"
   last but one line: "fractional quantity" instead of "fractionalquantity"

p.38, line 10, "Rogalski" instead of "Rogarski"

p.46, in the reference Kieran - Wagner: "International" instead
   of "Internation"

p.47, item 4, "Rogalski" instead of "Rogarski"

p.101: Country of L.P.Steffe is **USA** and not Italy

p.102, line 3, "panel. Knowledge" instead of "panel, knowledge"

p.114, line 30, "Neil" instead of "Niel"

p.115, line 7, "Piattelli - Palmarini" instead of "Piateli - Palmarine"
   line 18, "Watzlawick" instead of "Watylawick"

p.191, lines 20: Insert "subplenary 1 on 'New Technology'
   (chair/reporter: R. Strässer) and" after "subplenaries
   were sheduled:"

p.201, last-but-one reference: "Pengelly" instead of "Pengellerey"

1

p.219, line 1, "($<$12 years old)" instead of "(12 years old)

p.221, line 3 in Session 1: "Finzi" instead of "Finz", The main title should read: "The Dynamics of the Classroom: considering the older students ($>$12 years old)"

p.222, in Session 2, twice: "Nachmias" instead of "Nachias"

p.223, line 8, "A. Arcavi/R. Nachmias" instead of "A. Arcavi"

line 9, "Feghali" instead of "Feghavi"

line 10: Insert 4 more contributors: J. Cesar (France), I. Capuzzo Dolcetta et al (Italy), H. Johnson (USA), A. Rubin (USA)

p.224, line 13, "to face new" instead of "to face with new"

p.227, line 3 from below, "In the discussion questions were raised" instead of "In the discussion was raised"

p.230, line 3 in the paragraph on the ATM: correct version: "set a problem and then observe the children's responses"

p.232, line 11 in Strand 2: Insert 7 lines between "LOGO" and " was concerned": "to Dissection Motion Geometry, in which geometrical figures are dissected, and their parts are transformed and re-assembled to construct new figures of equal area. *Chronis Kynigos (UK/Greece)* has used H. Loethe's POST/DIRECTION/DISTANCE microword as a "bridge" between the intrinsic turtle geometry and Euclidean geometry.

As far as linking school geometry and LOGO"

p.233, line 17, "grade 11 and 12" instead of "grade and 12"

p.254, line 26, "streaming and teaching of students in" should read as "streaming and tracking students into"

line 27, "track" instead of "learn"

p.256, line 5, "students' abilities" instead of "students, abilities"

line 31, "students' achievement" instead of "students, achievement"

p.258, line 13, "students' learning" instead of "students, learning"

2

p.259, line 11, "students' success" instead of "students, success"

p.260, line 18, "students'" instead of "students,"

p.262, line 6, "pupils'" instead of "pupil,s"

p.264, middle: The sentence "*What can teachers . . .* " was intended to be a bold face subtitle like the one on p.266 next paragraph: "his/her teaching" instead of "their teaching"

p.266, line 2: Use the plural "theories" instead of "theory", also on p. 267, lines 26, 28, 34, 37.

p.271, line 25, "the knowledge" instead of "knowledge"

p.272, first word: "A"

p.274. line 12, last word: "the" instead of "this"

p.305, line 9, "Mike Price" instead of "Mike Pierce"

p.306, line 23, "remuneration" instead of "renumeration"

p.309, line 24, should read: "This summary is prepared by, and therefore the responsibility of, the Theme organizers."

p.311: Add: "Christine Keitel is the Chief Editor of this report. Hungarian Coordinator: László Máté"

p.312, last line, "Schubring" insted of "Schdubring"

p.316, line 7 in section 2.7: "Diana" instead of "Diane"

p.317, line 4 in section 2.10: "members" instead of "member"
line 6 in section 2.10: "Souviney" instead of "Souvviney"
line 9 in section 2.10: "ethnomathematical" (one word)

p.318, line 3 in section 3.1: "Mari" instead of "Marie"
line 12 in section 3.1: "video - taped" instead of "video - tape"
line 5 in section 3.2: "problem" instead of "programme"

p.320, line 5, "functions" instead of "function"

p.321, line 5 in section 3.9: "Yackel" instead of "Jackel"

p.322, line 3: The first Questioner: "Christine Keitel (West - Berlin)

p.324, line 5 from below: "colonial" instead of "colonical"

3

p.331, lines 11 -12 should read as "In this session three movies were shown: *"Flatland"*, *"Symmetry and Space: M. C. Escher"*, *"Geometries and Impossible Worlds: M. C. Escher"*, all produced

p.331, lines 16,19: "Peitgen" instead of "Peitigen"

p.332, The country of K.Clements: Australia

p.342, line 12, "Whitman" instead of "Witman"
line 14, "imparted" instead of "imported"
line 4 from below: Insert closing parenthesis at end of sentence.

p.343, line 8, "lead" instead of "had"
line 20, "writers" instead of "Writes"
line 27: Insert "in British Columbia" after "major change"

p.344, line 8, first word "The" should read "each"

p.349: Delete starting hyphen under heading "Section 5.4"

p.351: Insert missing line between lines 7 and 8: "alternative view, that mathematical formalism is primarily a means for"

p.355, line 8, "rethorical" instead of "ethorical"

p.374, 2nd paragraph, the last two lines should read: "ics and physics tradition. D. Siemon reported about studies on the role of metacognition in the teaching of mathematics as compared to the teaching of sciences."

4

CONTENTS

FOREWORD

The *Sixth International Congress on Mathematical Education* was held in Budapest, Hungary, from July 27 – August 3, 1988. There were 2414 registered participants from 74 countries, accompanied by more than 500 additional visitors. The following countries were represented by more than 10 delegates: Australia (113), Belgium (13), Brazil (18), Bulgaria (61), Canada (47), Czechoslovakia (35), Denmark (11), Finland (26), France (112), FRG (75), GDR (11), Greece (17), Hungary (306), Israel (43), Italy (81), Ivory Coast (13), Japan (228), The Netherlands (65), New Zealand (12), Norway (21), Poland (28), Portugal (22), South Africa (25), Spain (80), Sweden (62), Switzerland (15), UK (274), USA (380), USSR (90).

These congresses are now held every four years, having previously taken place in Lyon (France), Exeter (United Kingdom), Karlsruhe (Federal Republic of Germany), Berkeley (United States of America) and Adelaide (Australia). They are held under the auspices of the International Commission on Mathematical Instruction, and on this occasion the host organization was the János Bolyai Mathematical Society.

As on previous occasions, the *Executive Committee of ICMI* set up a planning structure based on an *International Programme Committee*. This committee was responsible for inviting the plenary speakers, and for making the major planning decisions, which were executed through a Hungarian Organizing Committee.

It was decided that the general pattern used at Adelaide would be followed. Accordingly, *Action Groups, Theme Groups* and *Topic Areas* were established. For the first two of these, in addition to a Chief Organizer and a Hungarian Coordinator, an international panel was appointed. This was to ensure as wide a range as possible of presentations and discussions, with contributors from many parts of the world.

The reports of these groups, prepared by chief organizers with the cooperation of their panel members and others, form the major part of these Proceedings. The variety of forms of organization of these reports is a reflection of the diversity of the work which took place. The work of the groups was facilitated by the production of a comprehensive book of *Programme Statements* which group organizers prepared prior to the Congress, and which participants were given on arrival in Budapest. A *list of participants* was also published. An additional aspect of the groups' work was the provision, in a number of cases, of a related Survey Lecture.

One of the most interesting features of these congresses is to learn from the experiences of other countries. Much of this is achieved by individual conversation as well as presentations within the groups. *National Presentations* add to this experience, and at this Congress contributions under this heading were invited from *Argentina, Bulgaria, Malawi and Spain.* Some other countries mounted exhibitions of their work as part of the displays, which also included a selection of curriculum development projects, and material from some individual institutions.

The place of Mathematics and its Education in Society is a major theme of growing importance. This was recognized by a whole day being devoted to it. A separate publication will result from that day's activities, and in fact also some other groups plan to produce publications.

For individual participants it is important to be able to exchange ideas, both by presenting their own work and learning about the work of others. To maximize this opportunity, time was given for short oral communications, and space was devoted to poster presentations. Around 200 of each were accepted, and *two books of abstracts were prepared* for Congress participants. As in all Congress activities the range of topics in these presentations was very wide, covering national and regional issues, teaching of particular topics, the use of computers, including software demonstrations, presentation of teaching aids and apparatus, epistemological issues and professional matters. The use of video materials was represented, but little emphasis was given to calculators.

As with any large congress, its success relies also on the support and efforts of many organizations and individuals. Although for reasons of space it is not possible to thank everybody here, their contributions are highly valued and are recorded in a number of the Congress documents.

The main venue for the Congress was the Technical University. Plenary sessions and the opening and closing ceremonies took place at the Budapest Convention Centre, a facility of which any city in the world would be proud. The hospitality and friendliness of the Hungarian people greatly appreciated by all, and deserve special mention.

The achievements of the Congress depended ultimately on the participants, but these could not have been realized without the hard work before, during and after the conference by all those involved in organizing the groups' activities, mounting the projects and exhibitions, and providing the essential organizational support invested for the smooth running of the Congress.

Much of the preparation of these Proceedings took place during a week of intensive writing in Esztergom immediately after the Congress. We are most grateful to all those who made this possible, including those chief organizers who attended, and also to Margit Gémes, Gabriella Köves, Katalin Lesnyik and Jenő

Székely. The final typesetting was done by Dezső Miklós at TYPOTEX GT using the TEX text-processing system.

Finally, the greatest thanks must go to Dr. Tibor Nemetz, whose boundless energy and unfailing good humour in the face of many problems played such a major part in the whole work of the Congress.

We have accepted the standpoint of the organizers that the Proceeding should reach the participants of ICME–6 as soon as possible. Consequently, the deadlines of submitting the reports and eventual corrections must have been kept. Therefore we could not attend the last version in a few cases of late submissions.

We have been greatly honoured to be asked by the Congress organizers to act as editors of these Proceedings. We hope that they will act as a reminder of a productive conference, a stimulus to further research in mathematics education, and as a source of memories of a happy visit to the beautiful city of Budapest.

Budapest, October 1, 1988

ANN HIRST and KEITH HIRST, editors.
Department of Mathematics
University of Southampton,
SOUTHAMPTON, S09 5NH, UK.

PLENARY PRESENTATIONS

PLENARY ADDRESS:

SCHOOL MATHEMATICS IN THE 1990's: RECENT TRENDS AND THE CHALLENGE TO DEVELOPING COUNTRIES

Bienvenido F. Nebres (Philippines)

Background of the Talk

I would like to begin with the geographical and cultural context in which I have worked both in mathematics and in mathematics education. That is, East Asia, especially Southeast Asia. The Southeast Asian Mathematical Society was established in 1972 and has since been very active in the development both of mathematics and mathematics education. In 1972, all our countries with the exception of Japan would have been considered developing in the economic sense. But it was clear that the development of mathematics and mathematics education was quite different in each country. We realized quite early that the categories of economic development or underdevelopment were inadequate to explain important differences. It could not explain the strong mathematics being done in Hanoi even during the worst years of the Vietnam War. Nor could it explain the great number of students doing pure mathematics in Singapore, when careers in pure mathematics were extremely limited in that island republic. The economic viewpoint does not explain either why some countries handled the new mathematics with care and took only those parts that could be integrated into their system and others created major problems by the poor handling of the introduction of the new mathematics.

I would like to dedicate this talk to the memory of my student, Doris Capistrano. I learned immediately after delivering this plenary address that she had died in a tragic shooting incident. This address and her life and dreams will always remain intertwined in my memory.

It was quite clear after some discussion that much of it had to do with culture, a tradition of learning, and values that uphold learning.

A regional conference on mathematics education held in Tokyo, Japan in October 1983, allowed me to articulate some of my thoughts on the question of mathematics education and the societies of Southeast Asia [1]. In that talk I classified the problems of mathematics education into two types:

Micro problems, or problems internal to mathematics education, such as curriculum, teacher training, textbooks, the use of calculators, problem-solving and the like.

Macro problems, or problems affecting mathematics education because of *external pressures* from other sectors of society: economy, politics, culture, language.

I argued that in many developing countries, the problems that merit most thought and research are those due to pressures from outside society. The purpose of study regarding these pressures is to provide some scope and freedom for the educational system so that it can attend to the internal problems of mathematics education. Since 1983, I have been led to isolate the area of culture and the values that support learning as the key variable to study and understand in developing mathematics and mathematics education in a developing country like the Philippines.

At the Adelaide Congress in 1984, a theme group on "Mathematics For All" was formed and zeroed in on the cultural history of the canonical school mathematics curriculum. Discussions led to a clearer understanding of the problems created by the lack of fit when a curriculum developed for an elite in western Europe is transplanted into every classroom in a developing country [2].

It is from this context and these concerns that this talk dialogues with the ICMI study series monograph "School Mathematics in the 1990's" [3]. I was privileged to participate in the first draft of this monograph. For this talk, I have tried to read the monograph from the point of view of a country like the Philippines, a developing country with a young tradition of learning. In the monograph itself, it is clear that the challenge for school mathematics is how to face a world transformed by the changing demands of society and culture and by a major revolution in technology. This paper concentrates on the first challenge, namely the demands of society and culture, and only takes up briefly the challenge of technology. It is not that the challenge of technology is less important for developing countries, but experience has shown that making progress in facing society and culture is necessary before the challenge of technology can be rationally met.

I. The Content of the School Mathematics Curriculum

Some Surprising Facts

The ICMI study reminds us of some well-known, but still surprising, facts. *The school mathematics curriculum is remarkably uniform throughout the world.* We also hear the same remarkably uniform reasons as to why mathematics should be taught to all and why the mathematics curriculum should be as it is. These reasons usually focus on the importance of mathematics for teaching us how to think, on the pleasure that it can give (at least to a few), and the usefulness of mathematics in a growing technological world. This uniformity is there despite extremely different school conditions.

To give one example of this variety of socioeconomic circumstances, consider the school enrollment patterns of Mexico and of Japan. In practice, only about 60% of Mexican children who start primary school continue beyond the first year. Roughly 10% start secondary education, and a mere 3% complete it. By contrast, practically all Japanese children complete lower secondary education, and about 95% stay in full-time education until they are at least 18. Yet the mathematics syllabuses year by year of the two countries are very similar. Both are based on syllabuses developed elsewhere, for pupils whose circumstances resemble those of students neither in Japan nor in Mexico [4].

The uniformity of the reasons given for the *usefulness* of mathematics is also quite surprising considering the diverse employment opportunities in differing countries and in different regions of a given country.

But of course we know that the uniformity of the school mathematics curriculum is not due to the internal structure of mathematics or to uniformity of situations.

The familiar school mathematics curriculum was developed in a particular historical and cultural context, that of western Europe in the aftermath of the industrial revolution. Those who designed it only had a minority of society in mind, for at that time only a small elite sector had access to a substantial number of years of schooling. In recent decades, what was once provided for a few has now been made available to — indeed, forced upon — all [5].

As the discussions at the Adelaide Congress of the theme group "Mathematics For All" showed, this problem of a curriculum made for an elite but given to all is a problem for all countries. But the problems in developing countries are particularly severe, because in many of them the tradition of schooling is still not fully rooted. There is a dual society separating an elite from the majority. Resources are limited and the values of society do not uphold the role and status of teachers. It is extremely crucial then that developing countries with a relatively young tradition of schooling be more critical of experience derived from elsewhere and pay more attention to their own circumstances and needs. *It is these countries that I have mainly in mind in this talk.*

What should be the *content* of the school math curriculum, especially for developing countries? Consider the different "curricula".

The intended curriculum — the curriculum which we find in official syllabuses and in the official textbooks. As noted above, this is remarkably uniform throughout the world.

The implemented curriculum — the curriculum which is actually taught in the schools.

The achieved curriculum — the curriculum that is actually mastered and learned by the students.

The kind of unintended *gaps* that can occur in developing countries may be illustrated by the experience of the Philippines with the new mathematics in the middle 1960's [6]. The Department of Education invited Peace Corps volunteers to train teachers in the new math. As was natural, they emphasized the new aspects, namely numeration systems, sets and set operations, the laws that govern number operations. Moreover, the curriculum followed what is called the spiral method, that is, the sequence of topics is the same every year, beginning with sets, set operations, numeration systems, commutative and associative laws, with actual operations on numbers coming only in the second half of the curriculum. Not surprisingly, the teachers thought that the new mathematics replaced the old and taught mainly intersections and unions and different kinds of numeration systems. Moreover, since it is normal that the first half of the curriculum is covered more fully than the second half, we had a generation of students who thought that school mathematics was basically sets and set operations, commutative and associative laws and never learned how to add, subtract, divide or multiply.

Closing the Curriculum Gap

What efforts are being made to close the gap between the intended, the implemented, and the achieved curriculum? I cite three examples, which also represent different models or approaches.

1. Total Reorganization.

In my 1983 Tokyo paper, I cited the effort of Japan in the second half of the 1970's as reported by Shokichi Iyanaga, former President of ICMI [7]. The gap problem was diagnosed to be due to an intended and implemented curriculum which was too difficult for too many schoolchildren. Because of the values of uniform development in Japanese society, the decision was to revise the intended curriculum downwards to lessen the number of students who could not keep up. The high discipline and efficiency of the Japanese school system assured that this intended curriculum was actually implemented. A recent report of Hiroshi Fujita shows that the gap problem cannot easily be fully solved. "We appreciate the fact that the 1978 curriculum has improved the situation ... but again the change was too limited ... the curriculum is still too wide ranging in scope and too hard to follow for the majority of students, while at the same time stifling the development of brighter students" [8].

2. Organic Transformation.

In a symposium held at the University of Chicago in 1985, Tamás Varga reports an effort begun in the 1960's. An experimental curriculum was started in two first grade classes by two teachers. Through a patient organic process, the number of classes grew to two hundred in ten years. Then social and political pressure led to a speeding up of the process and a molding of a "milder version of the curriculum"

[9]. By 1984–85, "100% of the seventh and about 80% of the eighth graders *enjoy the fruits of our work*" [10]. Not exactly all, continues Varga, and he then recalls the problems: the pace became too quick, teachers followed not because they believed or understood, but because they had to, etc.

3. Ethnomathematics.

A different approach and philosophy was presented by Ubiratan d'Ambrosio, when he proposed ethnomathematics. In his plenary address at the Adelaide Congress [11], he spoke of the mathematics contained in the experience of boat construction among the Amazon Indians, the example of sieves and the geometry that is present in their weaving patterns. His approach and that of others working on ethnomathematics is to change the content of the intended curriculum from the canonical school mathematics curriculum to one which arises from and is closely related to the experience of mathematics in a given culture. If we were to follow this approach, then we would no longer have a canonical school mathematics curriculum, at least in the early years, but would have different curricula in the different cultural contexts of the world.

Different Curricular Tracks

The ICMI Study reminds us that it is helpful to look back to the different curricula in the history of school mathematics. Figure 1 shows the *learning track of arithmetic*. This emphasizes a very practical form of mathematics and may be called the *vocational track*. Figure 2 shows the learning track of *traditional pure and applied mathematics*. This approach along more traditional mathematics emphasizes the arithmetic and computational aspects at primary and early secondary levels. Finally, Figure 3 shows the learning track of *modern mathematics*. This learning track has brought the approach of modern mathematics down to the elementary level.

Today we are being asked to consider three types of mathematics:

(i) ethnomathematics
(ii) school mathematics
(iii) higher pure mathematics.

It can be argued that the new school mathematics reforms of the 1960's sought to link (ii) and (iii) more closely: to transplant the aims, methods and structures of (iii) to (ii). Now, greater efforts are being made to link (i) and (ii). It is good at this point to remember that in the first two learning tracks, elementary school mathematics was always more vocational and practical [12]. In that form it may be more easily linked with ethnomathematics. The challenge would then be to link these more traditional learning tracks with the mathematics in a given culture, and at the later stage, to link it to higher mathematics.

There is an added question for developing countries, namely the problem of *dropouts*. There is a need to round off the curriculum, for practical purposes probably after four years and after six years. This rounding off would allow students who finish only four years (or only six years) to have some closure in their learning of mathematics. It may be that returning to the more traditional learning tracks

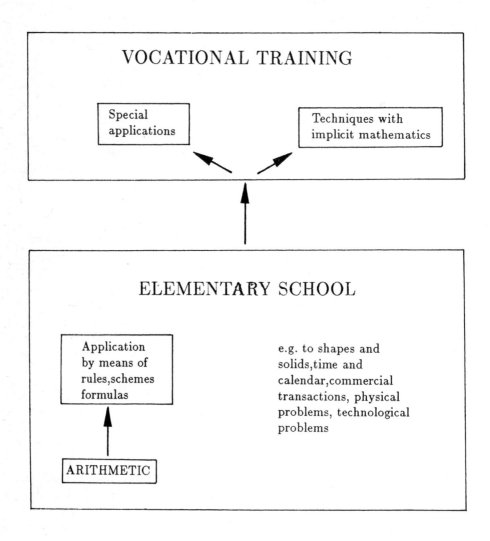

Fig. 1: Learning Track of arithmetic

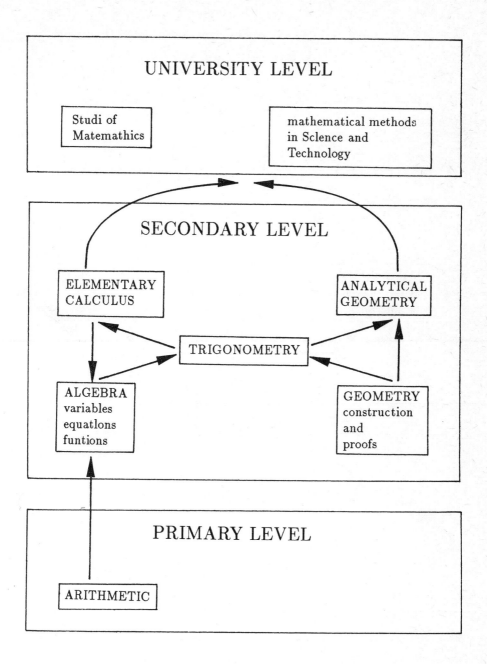

Fig. 2: Learning Track of Traditional Pure Mathematics

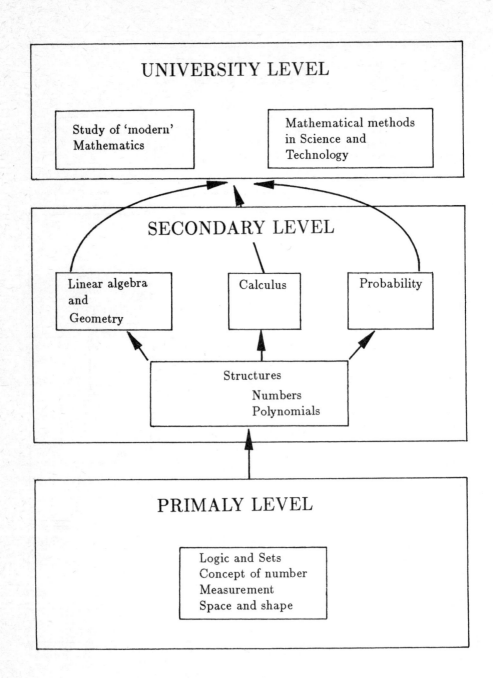

Fig. 3: Learning Track of "Modern" Mathematics

may make this job of rounding off curricula for dropouts easier for developing countries.

II. The Impact of Technology on School Mathematics

The ICMI Study says, "Perhaps the major concern relating to the content of school mathematics curriculum in the 1990's is the extent to which it will/should be affected by new technology and, in particular, by the hand-held electronic calculator and the microcomputer" [13]. This statement underlines the perceived *importance* of the impact of technology on school mathematics.

Importance

The importance of the impact of technology on school mathematics is beyond doubt. One monograph in the ICMI Study series has been dedicated to this question: "The Influence of Computers and Informatics on Mathematics and Its Teaching" [14]. A whole issue of the *Arithmetic Teacher* in the United States (Feb. 1987) was dedicated to the role of calculators in school mathematics. The discussion on technology and school mathematics has attracted some of the biggest numbers of participants in the last two international congresses. This question is important as well for developing countries. Even for poorer countries, calculators can be made readily available for the majority of schoolchildren. The microcomputer is important for a small but influential segment of the population. While it may take some time for developing countries to make microcomputers available to the larger population, they also must pay attention to it in a very serious way.

But what exactly shall we do?

Having underlined the importance of the impact of technology on school mathematics, however, we still do not know how to proceed. A serious discussion on the role of the calculator on mathematics education was held at ICME 3 (Karlsruhe, 1976). ICME 4 at Berkeley in 1980 and the *Agenda for Action* of the *National Council of Teachers of Mathematics* of the United States emphasized the importance of "taking full advantage of the power of calculators and computers at all grade levels" [15]. But as the *Arithmetic Teacher* issue cited above states, more than ten years later we are still far from developing a calculator-integrated curriculum. While more and more of our students are becoming familiar with the microcomputer (especially at home), the place of the microcomputer in the classroom and in the mathematics curriculum is still unclear.

It is quite likely that developing countries will have to depend on advances in the developed world in these two areas of a calculator-integrated and/or microcomputer-integrated curriculum. We do not have the human and financial resources to go into the intense research, development, testing and evaluation necessary to develop such new curricula. Having said this, I will still propose an agenda for developing countries with respect to the impact of technology on the school mathematics curriculum. I suggest that we look on it as the classic case of technology transfer. The success of the newly industrializing countries in my part of the world in absorbing new technology has followed a pattern: *importing the technology, imitating it, making small but important modifications, eventually mastering the technology and reconstructing it according to their own needs and purposes.* The *Southeast Asian Mathematical Society* has followed a similar path in our effort to build up mathematics research in our part of the world.

My agenda then for the developing world, and here I have greatly in mind Southeast Asia, is that we pool our resources together and organize our efforts over the next decade. I believe that a pattern following the paradigms above of technology transfer or the experience of the Southeast Asian Mathematical Society would be helpful: listening to and learning from experts abroad, beginning similar efforts and experiments in our own home countries, bringing together these efforts and experiments in workshops where we can learn from one another and develop our own technology-integrated curriculum.

III. The Macro Environment: The Importance of Culture, Beliefs, and Values for Mathematics Education

In E. G. Begle's landmark review "Critical Variables in Mathematics Education: Findings from a Survey of the Empirical Literature", there are very few articles regarding the role of cultural values in their mathematics education. But discussions of mathematics education in Third World countries have come to focus more and more on the question of cultural variables. In the Asian experience, this is probably because the sharp contrast of traditions and cultures in East Asia makes us aware of the importance of cultural variables. For example, we have realized that questions like the status and salaries of teachers are not simply a function of the economy but also of cultural values. Similarly the rise of what we might call the internal force of genius, wherein mathematics talent springs up even under difficult economic circumstances, seems to be fostered by the cultural environment.

These intuitive perceptions regarding the importance of cultural variables have been helped immensely by a series of cross - cultural studies done by H. W. Stevenson and others at the University of Michigan [17]. They have made recent studies comparing mathematics education in Japan (Sendai), Taiwan (Taipei), U.S.A. (Minneapolis). The levels studied were students in first grade and fifth grade. It is widely known that the achievement of students from Japan and China has been notably superior to those of American children. The studies corroborate these earlier findings. But what is notable in the studies is that the variables that explain the differences seem to be clearly in the area of culture, beliefs, and values. It is not possible to do a sufficient summary of the studies, but we can highlight some aspects that help clarify our discussion.

(1) Belief in the Importance of Mathematics

Parents, teachers, and pupils in Japan and Taiwan are clearly convinced of the importance of mathematics for their education. It is seen as crucial and at least on the same level as reading. In the United States, however, achievement in reading is seen as crucial for the education of the child, but mathematics much less so. In fact, most American parents seem to see no serious problem with the mathematics education of their children and are quite satisfied with it and with the achievement of their children. In contrast, in Japan and China, parents worry a lot about the mathematics education and achievement of their children.

(2) What Makes a Student Do Well in Mathematics?

Teachers and parents in the United States usually attribute achievement in mathematics to the innate abilities of the pupils. In the Asian tradition, however, there is a de-emphasis on differences of talent among students and a belief that achievement in mathematics is primarily due to hard work and to giving the proper time to it. If a child's rate of learning is slower than others, it means only that the child must study harder. They quote a Chinese proverb, "The slow bird has to start out early".

(3) The Status of Teachers

In general, it does seem that teachers are accorded higher status in Japan and in Taiwan than in the United States. This is both in terms of their social standing in society and in terms of the remuneration that they receive. For example, in Japan, the government pays special attention to the salaries of grade school teachers and they are normally better paid than high school teachers or university faculty, except at the highest ranks of the university.

There are some clearly observable effects of these differences of culture, beliefs, and values:

(i) First, more time and effort is given to mathematics in the Japanese or Chinese classroom than in the American classroom. This is equally true with respect to time for homework, time given by the teacher, parents, and pupils in the study of mathematics.

(ii) Teachers in Japan and China continue to work with the whole class including the slow-learners. There is effort to help the whole class learn the lesson and to have them keep up with the pace. Thus the problem of children dropping out of the learning process is lessened in these classrooms.

(iii) The expectations of school and of society of the teacher provide him or her with more time to do teaching duties in the Asian classroom than in the American classroom. Careful measurements show that more is given to teaching and to teaching-related activities by the Asian teacher than by the American teacher.

IV. An Organic Model for Mathematics Education in Developing Countries

The importance of the cultural matrix (culture, values, beliefs) for mathematics education leads us to an image of the development of mathematics in a developing country along the lines of an *organic model*. This way of thinking is less prevalent in modern western thought. It may be, as Fritjof Capra argues in "The Turning Point", because of the dominance of the Cartesian and Newtonian mechanistic view of reality [18]. But it is a familiar framework in eastern thought. For example, a recent article I read on the introduction of western cosmology to Japan in the 17th century, speaks as follows: "The soil in which the flowers of modernization bloomed so profusely had been enriched over centuries. It was due to this rich soil that Japan was so successful in modernizing itself swiftly after the Meiji restoration. Many called the modernization a miracle, but in fact, it was nothing of the kind: it was the outgrowth of several centuries of effort" [19]. In my discussions with mathematician friends in Japan and China, the importance of the human and social aspects of the development of mathematics, the importance also of the affective and non-cognitional aspects, have come up over and over again. In this vision then, the canonical school mathematics curriculum or some modification of it has to be seen as an organic growth being planted in foreign soil. It is important then to understand the cultural matrix, the soil into which this transplant is placed. My own reflection on our efforts in Southeast Asia and, in particular, the Philippines presents the following important areas of concern if we are to handle this transplant well.

(1) Pace

In the article from the 1985 Chicago symposium cited above, Tamás Varga discusses the process of implementation of the pilot work which he had developed in the 1960's in Budapest. He writes,

> In a decade the number of classes grew to about two hundred, 0.5% of the classes in the eighth grade all over the country. The number of classes from year to year grew in a geometric progression. The quotient of the progression was $\sqrt[10]{\frac{200}{2}} \approx 1.58$. This is the pattern of an organic growth. The number of classes joining in each year was proportional to the number of classes within the scheme. As with contagious diseases, when the sources of infection increased (the increased possibility of visiting classes), the chances of "catching the disease by personal contact" increased proportionately [20].

He goes on, "Attempts to speed up a natural, organic growth rarely give satisfactory results, however, and unfortunately such attempts were made" [21]. He then goes on to record the problems that occurred because of the effort to speed up the growth too much. Pace, therefore, says that the development of school mathematics especially in a developing country, has to follow the laws of organic growth. It is foolish to try to speed things up too much as it will simply lead to a withering or distortion of the growth.

(2) Aspects

There may be certain parts and aspects of mathematics that should be given emphasis in a particular country and culture. Ethnomathematics points out many such examples, normally mathematics associated with house-building, boat-building, geometrical patterns, with which a particular culture is familiar. My own belief in this matter is that what is important is that children "own" the mathematics they learn. It is important, therefore, to emphasize and to highlight areas which children can relate to and claim as their own. It may be because of familiar patterns as in the efforts of ethnomathematics. It may simply be that the particular problem that is given to them is one which students in their own culture had solved in a mathematics olympiad some years back. Whatever it may be, at least some aspects of mathematics emerge from the unfamiliar, impersonal mass and come up to the child as a possible friend.

(3) Style

Just as the canonical school curriculum has developed all over the world, there is a tendency also towards canonical methods of teaching mathematics in countries and cultures. Let me note two areas of style where balance is needed. The first is the balance between methods which emphasize *insight* and *understanding*, (*understanding why*) and *mastery of skill (knowing how)*. Without wanting to over-simplify philosophies of learning, one can say that western philosophy places most emphasis on insight and understanding. If we can get the child to understand why, it will be easy for him or her to learn how. But eastern philosophy places a premium as well on the right way of doing things. There is thus a place for memorization, for group chanting, for methodologies that the West would characterize as learning by rote. There is a belief that mastering the right way of doing things will also lead to insight and understanding [22]. I believe that each culture should reflect on the balance and the style of teaching and learning most suitable for it.

Secondly, Western school tradition focuses on the individual. Teaching style emphasizes individual learning and effort. But Asian self-consciousness places emphasis on the group and one's identity and reality is largely determined by the group. We also need a teaching style that balances focus on the individual and focus on the group.

(4) Continuity and Growth

Organic development is continuous and dynamic. It may start, of course, with a singularity, as with the first fertilized cell at the beginning of life. This consciousness of the continuity of growth leads one to the importance of the preparation of the cultural matrix (the soil), care for initial growth (when the transplant is rather fragile), and making sure that the growth process does not outstrip internal capabilities and resources. It also leads one to realize that like all life, it has a dynamic cycle of youth, maturity, and need for re-invigoration.

There is also need for continuity throughout the school system, from weakest schools to strongest ones. One measure that we might develop regarding this continuity would be to divide schools into different categories a, b, c, d, and check that there is a continuity up through the ladder. In many developing countries, where dual societies are prevalent, one will often see big breaks in this educational ladder.

(5) Non-Cognitional Factors

Most mathematics education studies of learning focus on cognitional factors. In fact, one of the forthcoming studies in the ICMI study series is on mathematics and cognition. But again if culture, values, beliefs affect mathematics education in an important way, then non-cognitional factors such as the cultural environment or affective factors must be understood as well. This has come up quite a bit in my discussions with mathematician friends from Japan. They have emphasized the importance of the affective dimension in doing mathematics. They have pointed out that it is important to do mathematics research in one's own language, because research involves the whole person, cognition and feelings, head and heart. To do mathematics in a foreign language is like boxing with one hand tied behind one's back. One uses one's head, but not one's heart [23]. I think we understand as teachers the importance of affective factors in learning and doing mathematics. Our experience with Ph.D. students shows that on the level of the dissertation, it is not so much intelligence and talent that count, but motivation and emotional factors that differentiate the students who succeed and those who give up. If this is so, it is important for the development of school mathematics that we be much more aware of non-cognitional factors and take them into account in the development of school mathematics in our countries.

V. Research and the Process of Change

The above discussion, that development of school mathematics in developing countries has to give much attention to culture, values, and beliefs, seems to place development in the world of the private and the internal and beyond the reach of planning and organized change. However, my own experience of values and value-change says that the area to focus on is something one can touch, namely *choices and decisions*. Our values and beliefs of course shape our choices and decisions. But our values and beliefs are also shaped in great part by the choices that we make. There are certain crucial forks in our lives and our future is determined by which fork we take. Similar holds true for nations and cultures.

When one stands before the Great Wall in Beijing, where for centuries the results of the imperial examinations were read and the new mandarins were determined, one cannot help but think of the influence of this structure chosen by the Chinese emperors. The high value placed on learning, even seemingly useless learning like Chinese classics or mathematics, may have been largely shaped by the fact that Chinese emperors had determined that these were the path to high position in government. The special attention paid in Japan to the salaries of grade-school teachers is expressive, of course, of the values of Japanese society, but it also assures the flow of talent to grade schools which strengthens and reinforces these values. The attention of media and funding agencies to the needs of science and technology in the United States in the post-sputnik era opened attractive forks in the road for young people to pursue careers in science and technology.

If the leadership of a nation believes in the importance of reform in mathematics education, then its task is to create the structures (the channels and pathways) that may shift the choices and decisions of its citizens to these efforts and values. This would include a higher value to be placed on *the learning of mathematics*, an emphasis on the *necessity and importance of hard work, higher status and recognition of teachers.*

(1) The Process of Change

I suggest that the model of organic transformation, exemplified by the experience of Varga, may be most suitable for developing countries. Our experience in the Philippines of building up graduate and research programs in mathematics, physics, and chemistry verifies the laws of growth mentioned in his paper. Similarly, my own longer experience of developing programs for social consciousness and responsibility of students in the Philippines follows this paradigm [24]. It began with a small group of faculty and students with alternative values and choices (what we eventually called a counterculture), but which gradually began to "infect" a larger portion of the academic community. There was also the experience of being pushed to move too fast and eventually the time when many of these values were accepted by the whole academic community.

In this light, we can look again at the essential elements presented in the ICMI Study for a successful process of change for school mathematics in the 1990's:

a) A group or groups to develop a plan for change.

b) A measure of the "capital" available for change. "Capital includes finance, talent and goodwill".

c) Developing awareness of the problems of educational change in key figures in the society.

d) Well-planned intervention strategies and evaluations of these strategies.

e) Identifying the elements involved in bringing about change and determining the interrelationships of these elements.

I would focus the study of the process of change on what the ICMI Study calls *intervention strategies*. These could be the so- called *magnet schools* or *lighthouse schools*. The planning for and understanding of these intervention strategies would require a serious measure and understanding of the "capital" available to the society and culture. Secondly, the planners would have to monitor with great care the pace of growth and development. If the national leadership is seriously intent on change, then one of the difficult pressures from them would be to move too widely, too soon. Planners may have to agree on a two-track strategy. One which would carry the original vision and which may move more slowly and a second one which would meet the more immediate demands of the national leadership.

(2) Research

If we look at the program for research presented in the ICMI Study (Fig. 4), it becomes clear that the areas of research which I believe are most important for developing countries are those to the right. That is, *processes of change, learning context, theoretical constructs*.

Under *theoretical constructs*, I would be particularly interested in case studies and a deeper theoretical understanding of organic models for change. In my own reflection on this matter, I have found myself turning instinctively to the language and the conceptual framework of catastrophe theory (singularities, unfolding, potential). I do not know if such a conceptual framework will prove fruitful.

136,205

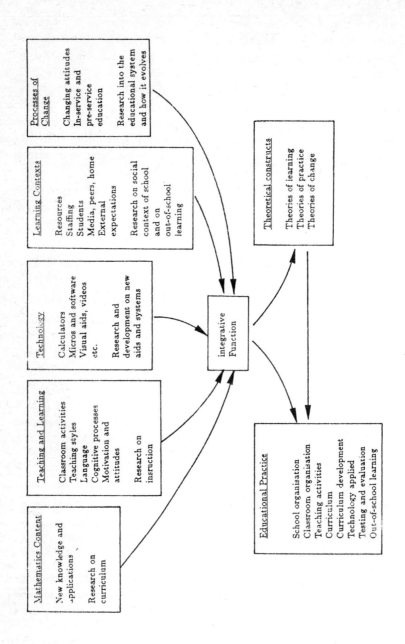

Fig. 4

Under *processes of change*, I have been much helped by studies and by conversations on the development of mathematics and mathematics education in Japan. More careful studies of the history and development of school mathematics systems in different countries and the socio-cultural context within which they have developed would probably be helpful for future planners.

Finally, under *learning context*, the Stevenson studies show clearly how important are the attitudes and values of parents, the attitudes and values of teachers, the structure of time and activities in school and out of school. My impression is that cross-cultural studies are particularly helpful in this area. They help different cultures to use one another as a mirror within which they can see more clearly the lights and shadows in their own system.

References

[1] B. F. Nebres, Problems of Mathematical Education in and for Changing Societies – Problems in Southeast Asian Countries, *Proceedings of ICMI-JSME Regional Conference on Mathematical Education*, Tokyo, Japan, 1983, 10–19.

[2] M. Carss, ed., Theme Group 1: Mathematics for All, *Proceedings of the Fifth International Congress on Mathematical Education*, Birkhäuser, 1986, 133–145.

P. Damerow et al, eds., *Mathematics for All*, UNESCO Science and Technology Education Document Series No. 20. 109.

[3] A.G. Howson and B. Wilson, eds., *School Mathematics in the 1990's*, ICMI Study Series, Cambridge University Press, 1986, 104 Hereafter referred to as *ICMI Study*.

[4] ICMI Study, 19.

[5] ICMI Study, 8.

[6] B. F. Nebres, Major Trends from ICME IV: A Southeast Asian Perspective, *Teaching Teachers, Teaching Students*, Steen & Albers, eds., Birkhäuser, 1981, 59–60.

[7] B. F. Nebres, Problems of Mathematical Education , 11–12.

[8] H. Fujita, The Present State and Current Problems of Mathematics Education at the Senior Secondary Level in Japan, *Developments in School Mathematics Education Around the World*, I. Wirszup & R. Streit, eds., National Council of Teachers of Mathematics, 1987, 195–96.

[9] T. Varga, Post–'New Math' Since 1963: Its Implementation as a Historical Process, *Development in School Mathematics Around the World*, I. Wirszup & R. Streit, eds., National Council of Teachers of Mathematics, 1987, 242.

[10] *Ibid.*

[11] U. d'Ambrosio, Socio-Cultural Bases for Mathematical Education, *Proceedings of the Fifth International Congress on Mathematical Education*. M. Carss, ed., Birkhäuser 1986, 61–66.

[12] I would like to thank Jeremy Kilpatrick for calling my attention to the fact that there have been two cultures in elementary school mathematics, one more vocational and practical, the other more linked to higher math.

[13] ICMI Study, 66.

[14] Published by Cambridge University Press, 1986. 155.

[15] *An Agenda For Action, Recommendations for School Mathematics of the 1980's*, NCTM 1980, 1.

[16] E.G. Begle, *Critical Variables in Mathematical Education: Findings from a Survey of the Empirical Literature*, MAA and NCTM, 1979, 165.

[17] S.Y. Lee, V. Ichikawa, and H.W. Stevenson, Beliefs and Achievement in Mathematics and Reading: A Cross-National Study of Chinese, Japanese and American Children and Their Mothers, *Advances in Motivation and Achievement Enhancing Motivation*, Vol. 5, JAI Press Inc., 1987, 149–179.

   H.W. Stevenson, S.Y. Lee, J. W. Stigler, Mathematics Achievement of Chinese, Japanes, and American Children, *Science*, Vol. 231, February 14, 1986, 693–699.

   H.W. Stevenson, America's Math Problems, *Educational Leadership*, October 1987, 4–10.

[18] F. Capra, *The Turning Point: Science, Society, and The Rising Culture*, Bantam Books, 1983, 53–74.

[19] S. Ito, The Introduction of Western Cosmology to Japan in the Seventeenth Century, *The East*, Vol. 13, no. 3 (Aug. 1987), 55.

[20] T. Varga, *op. cit.*, 241.

[21] *Ibid.*, 242

[22] Note that the Stevenson studies point out that the Asian children did equally well in test items that emphasized rote learning as in those that emphasized insight and understanding.

[23] I would like to thank S. Mizohata of Kyoto University and M. Morimoto of Sophia University for these insights.

[24] B.F. Nebres, The Task of A Jesuit University in the Philippines, *Philippine Studies*, Vol. 27, 1979, 82–92.

PLENARY ADDRESS:

THEORETICAL FRAMEWORKS AND EMPIRICAL FACTS IN THE PSYCHOLOGY OF MATHEMATICS EDUCATION

Gérard Vergnaud (France)

Psychology of mathematics is not really a new field of research, as several psychologists started to investigate that field a long time ago. But it has developed and changed a lot during the last 15 years. The international group "Psychology of Mathematical Education", which is a working group of the International Congress on Mathematical Education, has played an important part in gathering ideas and new results together. Let me mention the name of Ephraim Fischbein who launched that working group and was its first president.

The aim of my lecture is to present in a synthetic fashion some of the results that have been obtained and some theoretical ideas which, I think, are essential to understand how students learn mathematics and develop their own ideas, and how teachers can improve their teaching, by taking account of the way students learn, or fail to learn.

Many mathematicians think that primary school mathematics is not mathematics, because there is nothing like proof in it, nor does it include any sophisticated concepts like those of function, continuity, or algebraic structure. This is a wrong view, and my first task will be to give examples of the conceptual importance, from a mathematical point of view, of children's first acquisitions.

Another task will be to show that mathematical concepts and procedures are learned and developed over a long period of students' growth, more than 10 years for additive structures or for multiplicative structures.

My next purpose will be to analyse the role of symbols and language in concept formation and problem solving, and to illustrate the specific, and sometimes difficult, operations of thinking that are required by the reading and the use of mathematical symbolisms like graphs and algebra.

And finally I will stress the need for a better interaction of epistemology of mathematics, cognitive, developmental and social psychology, and didactics.

What is a mathematical behaviour?

There are many kinds and many levels of mathematical behaviour, even in school mathematics. It is not easy to compare such behaviours as counting a set of discrete objects, measuring the length of a room, or the volume of a container, perform a difficult subtraction, choose the data and the operations that are necessary to solve a double proportion problem, a linear equation or a linear system, analyse the variation of a function, show the equality of two angles in a complex geometrical figure.

And yet it is important to consider that all these behaviours are tied to mathematical concepts, and would not be possible if such concepts as those of one-to-one correspondence, cardinal, additivity of measures, place-value notation, isomorphism of measures, linearity and n-linearity, function and variable, did not exist.

The fact that such concepts exist does not mean that students are fully aware of the relationship between these ideas and the way they behave. Most often mathematical ideas are only implicit in students' behaviour. Let me take two examples:

The first one is inspired by the work of Gelman and Gallistel (1978), Fuson and Hall (1982), Steffe et al. (1983), Chichignoud (1986) and others. When counting a set of six elements, most five or six year-olds count: one, two, three, four, five, six... six! Not only do they have to establish one-to-one correspondences between the objects to be counted, the gestures of the finger, the movements of the eyes and the number-words, but also they feel the need to say the word "six" twice. The first utterance refers to the sixth element of the set, the second utterance refers to the cardinal of the set: this double utterance means that the concept of cardinal has been recognized.

Older children may not repeat "six", but only stress it differently from the other number-words. But there are children who do not repeat "six" and when asked how many objects there are, are unable to answer, and start counting again.

For children who can cardinalize, "six" summarizes the information on the set gathered by the counting procedure. This is not the case for children who do not cardinalize. One can therefore infer the existence and the non-existence of the concept of cardinal from children's behaviour. Actually rather than a concept it is an invariant property of discrete sets relying upon an invariant organization of the counting behaviour. Let us call such invariant organization of behaviour, for a certain class of situations, a "scheme".

One can see from this example that the scheme of counting a set is not only made of rules of production but also of implicit strong mathematical ideas, namely those of one-to-one correspondence and cardinal. Let us call these ideas "operational invariants". Children progressively strengthen and extend the scope of validity of such invariants: cardinals do not depend on the spatial distribution of the elements, nor on the size of objects... And the counting procedure is altogether the same, even if it is more difficult to count a flock of moving sheep than a pile of plates on the table.

My second example will be taken from algebra. When dealing with equations like

$$4x + 25 = 53 \quad \text{or} \quad 41 = 3t + 26$$

many 13–14 year olds in France can apply the same scheme (invariant organization of the action sequence), for instance:

$4x + 25 - 25 = 53 - 25$	$41 - 26 = 3t + 26 - 26$
$4x = 28$	$15 = 3t$
$4x/4 = 28/4$	$15/3 = 3t/3$
$x = 7$	$5 = t$

or simplified versions like:

$4x = 53 - 25$	$41 - 26 = 3t$
$4x = 28$	$15 = 3t$
$x = 28/4$	$t = 15/3 = 5$

The same scheme applies to all equations of the shape $ax + b = c$ whatever the name and the place of the unknown may be, provided a, b and c are positive, b is smaller than c, and a is a small whole number, smaller than $c - b$. Why these restrictions? Because for many students, the scheme cannot be fully extended to negative numbers and to difficult divisions.

Yet the procedure has some generality and it relies on such mathematical ideas as the conservation of equality when the same number is subtracted from both sides, or when both sides are divided by the same number. For many students these ideas are only implicit theorems. Let us call them theorems-in-action. They are operational invariants, like the ideas of cardinal and one-to-one correspondence.

I can give quickly a few more examples of implicit concepts and implicit theorems that can be traced in the emergence of new competences in children.

It is now well known for instance that when they have to count a set of children after having counted 4 boys and 3 girls, many five or six year-old children count the whole set again. It is a big step to be able to say $4 + 3 = 7$ or even to start from four, and count three steps forward. It shortens the counting-all procedure. This discovery (it is a fresh discovery for students as nobody usually teaches it to children) can be considered as the spontaneous recognition of the fundamental axiom of the theory of measure:

$$\text{card}(A \cup B) = \text{card}(A) + \text{card}(B)$$
provided A and B have no common part.

Don't count the whole set again; just add the cardinals of the subsets.

It is also well known (Carpenter and Moser, 1983; Fuson, 1983) that young students tend to simulate as closely as possible the structure of problems. For instance, when they have to find the result of winning three marbles when the winner had 6 marbles before playing, they would proceed by starting from 6 and count 3 steps on. If the situation is a start of 3 and a win of 6, the natural tendency is to start from 3 and count 6 steps on. This is not so easy and the risk of going wrong is bigger; another step for 5 to 7 year olds, is to start from the greater number 6 and count 3 steps on. One can consider this discovery as commutativity in action:

$$3 + 6 = 6 + 3$$

$$a + b = b + a$$

The realm of validity of such a theorem-in-action is not as large as that of the real theorem. Moreover, children cannot usually explain clearly why it is possible to do $6 + 3$ instead of $3 + 6$, but the mathematical idea is nevertheless present.

My last example will be the spontaneous solution given by some students to double proportion problems. In a verbal problem students had to calculate the quantity of sugar necessary for 50 children going to a vacation camp for 28 days. They had found in a book that the quantity needed for 10 children for 1 week was 3.5 kg. Some students said that 50 children is 5 times more than 10, and 28 days 4 times longer than a week; therefore the consumption of sugar should be 20 times bigger. The non trivial theorem-in-action revealed by this procedure can be written:

$$\text{Consumption } (5 \times 10, 4 \times 7) = 5 \times 4\times \text{ Consumption } (10, 7)$$

which is a particular case of the general property of bilinear function:

$$f(\lambda_1 x_1, \lambda_2 x_2) = \lambda_1 \lambda_2 f(x_1, x_2)$$

Of course the easy numerical values made it possible for 11 year-olds to extract ratios 5 and 4; and there is no conceptual difficulty for them in recognizing the double proportion of consumption to the number of persons and to the length of time. Therefore, there is a gap between the theorem-in-action they used and the general theorem. But again the mathematical idea was there.

In summary a mathematical behaviour is a behaviour that relies upon some mathematical idea. A mathematical behaviour consists of an invariant organization of behaviour called a scheme. Some schemes are algorithms, but not all mathematical schemes are algorithms; and even when students have been taught an algorithm, they don't always follow the rules they have been taught, and replace them by more meaningful schemes or by memorized shortcuts.

Schemes (et algorithms) are not made of rules only; they are also made of goals and expectations, inferences and operational invariants. I have given a few examples of such operational invariants: cardinal, additivity axiom, commutativity, conservation of equality, bilinear theorem. The implicit knowledge contained in schemes can be analysed as made of concepts and theorems that are used in action, without being clearly identified and worded as objects, properties and relationships.

Situations and conceptual fields

A scheme is associated with a class of situations. There are many different schemes because there are many different classes of situations. For instance it is not the same conceptual problem and not the same scheme to add 7 and 5 in the three following problems:

1. Peter had 5 marbles. He plays a game of marbles with John and wins 7 marbles. How many marbles does he have now?
2. Robert has just played a game of marbles with Celia. He has lost 7 marbles. He has now 5 marbles. How many marbles did he have before playing?
3. John has played two games of marbles. He has lost 7 marbles in the second game, but he does not remember what happened in the first game. When he counts his marbles in the end, he finds that he has altogether won 5 marbles. What happened in the first game?

Problem 2 (Robert) is usually solved with a delay of one year and a half after Problem 1 (Peter) is solved. The main reason for this lies in the structure of the problems. Whereas Problem 1 (Peter) consists of searching the final state knowing the initial state and the transformation, Problem 2 (Robert) consists of searching the initial state knowing the final state and the transformation: this requires either the inversion of the transformation, or a hypothetical reasoning on the initial state such as: had he 10 marbles in the beginning, he would be left with 3; he needs 2 more; Robert had 12 marbles.

Let me represent both problems:

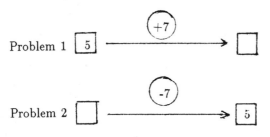

The addition 5+7 does not have the same meaning in both cases. In problem 1, it fits perfectly with the primitive conception of addition as an increase (Gelman and Gallistel, 1978). In problem 2, it goes against that primitive conception (you don't add when you have lost marbles), and it requires the inversion of the transformation -7, from the initial to the final state, into +7, from the final to the initial state.

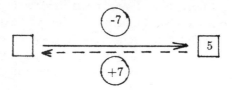

Problem 3 (John) is the most difficult case; 75% of 12 year old students fail to solve that problem. Addition 5 + 7 is now totally counterintuitive, as you need to add a part and the whole to find the other part.

Problem 3

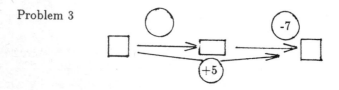

As a matter of fact the two games are considered as parts of the combined transformation which connects the initial and the final states. Had John lost 15 marbles in all, the answer would have been easy and natural. But problem 3 requires an addition 5 + 7, which is actually a subtraction of two directed numbers:

$$x + (-7) = (+5)$$
$$x = (+5) - (-7) = 5 + 7 = +12$$

The research work that has been achieved by Carpenter and Moser (1983), Riley, Greeno and Heller (1982), Vergnaud and Durand (1976), Nesher (1982) and others shows consistent results and substantial agreement about the classification of addition and subtraction tasks.

Whereas addition is usually conceived by mathematicians and teachers as the binary commutative combination of two parts into a whole, and subtraction as the search for one part knowing the whole and the other part, the psychological classification of cognitive tasks reveals that, beside the binary combination of two parts into a whole, there is also the unary operation of a transformation of the initial state. There are essentially six different tasks related to the initial-state — transformation — final-state relationship, among which two are solved by adding and four by subtracting. A similar situation exists for comparison problems, when you deal with three-term relationships: reference-state — comparison — compared-state, for instance:

John has 7 sweets, he has 2 sweets less than Janet. How many sweets does Janet have?

The binary commutative combination cannot model these relationships; one rather needs a unary non-commutative operation, as can been seen in the above arrow-diagrams.

Let me also stress the fact that it is not wise to keep directed numbers away from children at the primary school level, as they do meet situations involving transformations and relationships which should be modelled by directed numbers. I do not find as much support as I would like in favour of this idea, probably because some operations with negative numbers are still difficult for many 15 or 16 year-old students. But the fact that some tasks with negative numbers are difficult for the majority of 15-year olds must not hide that other fact that most 9 year-olds can understand what a negative transformation is (I have eaten four sweets), or what a negative relationship is (I owe you 3 dollars), and can also understand that negative and positive transformations are inverses of one another.

In summary, if one calls "additive structures" the set of situations that involve the addition or subtraction of two numbers, one sees some primitive competences and conceptions emerging in 3 or 4 year-old children; and still some problems requiring just one addition are failed by a majority of 15 year-olds. Between these two periods of their cognitive development, children discover or learn how to solve a great variety of problems.

Researchers are now able to give a differentiated picture of the variety of cognitive tasks which children meet, of the mental "revolutions" they have to achieve, of the main obstacles on which some students keep failing for a long time.

Several important concepts are involved in additive structures: cardinal, measure, state, transformation, comparison, difference, inversion and directed number are all essential in the conceptualizing process undertaken by students. Some of these concepts remain implicit for them. Understanding them sometimes requires the teacher to provide explicit wording, symbolizing and explaining.

The place of language and symbols is certainly an important issue in mathematics education. I will come to this point later.

But before I do this, I would like to take other examples of the long term development of mathematical competences and conceptions. My next example will be "multiplicative structures", i.e. the set of situations that involve the multiplication or the division of two numbers, or a combination of such operations. Most of these situations are in fact simple-proportion or multiple-proportion problems, in which two variables are proportional to each other (simple proportion), or one variable is proportional to several other independent variables (multiple proportion).

Children are faced with proportion problems as soon as they have to find the cost of several identical objects, share a number of sweets, or find how many pastries they can buy with a certain amount of money.

In simple proportion problems, there are two kinds of ratio: scalar ratios (between magnitudes of the same kind) and function ratios or rates (between magnitudes of different kinds), sometimes called intensive quantities (Schwarz, 1988).

Take for instance a multiplication problem like the following: Miniature cars cost 5 dollars each. How much do you have to pay if you buy 4 of them?

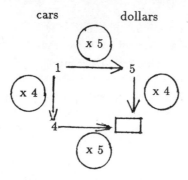

The above diagram shows that you can either use the invariant scalar ratio (vertical):

$$\frac{\text{cost of 4 cars}}{\text{cost of 1 car}} = \frac{\text{4 cars}}{\text{1 car}}$$

or the invariant function ratio or rate (horizontal):

$$\frac{\text{cost of 4 cars}}{\text{4 cars}} = \frac{\text{cost of 1 car}}{\text{1 car}}$$

The first one is a scalar and has no dimension; it will be expressed, in natural language, by such expression as "4 times more". The second one is a quotient of two magnitudes "dollars per car" and it relates the two different variables; this is the reason why I call it a function-ratio.

If one introduces multiplication as repeated addition of the same number, it is meaningful to do this with the scalar operator in mind:

4 times 5 dollars ↔ 5 dollars + 5 dollars + 5 dollars + 5 dollars

but not with the function operator, as 5 times 4 cars cannot give dollars.

The correct analysis would be:

$$4 \text{ cars} \times \frac{5 \text{ dollars}}{1 \text{ car}} = \boxed{} \text{ dollars}$$

Thus, students meet dimensional analysis at the primary school level.

Many results have been collected on the comparative difficulty of multiplication and division problems in the simple proportion case. One may vary the structure of the problem, the numerical values and the domain of experience to which such problems refer. See for instance Bell, Fischbein and Greer (1984), Vergnaud (1983).

multiplication	type I division case partition	type II division quotient	general case rule of three
1 b	1 ☐	1 b	a b
a ☐	a c	☐ c	c ☐

a, b, c can be taken among small or large whole numbers, among decimals, larger or smaller than 1, among fractions.

And the problems may refer to familiar or unfamiliar domains of experience, to easy or difficult mathematical physical or technical domains.

- Type II division is, on average, rather more difficult than type I division.
- Multiplying by a decimal is a striking obstacle, probably because it moves the child away from the primitive conception of multiplication as iterated addition.
- It is even more difficult to divide by a decimal.
- In children's primitive conception, multiplication is supposed to make bigger and division smaller. Therefore multiplying or dividing by numbers smaller than 1 gives counterintuitive results.
- It is also more natural for students to divide the larger number by the smaller. Many errors appear when the correct division is smaller by larger.
- The difficulty of rule-of-three problems is altogether tied to the numerical values. The procedures used by students favor the use of the isomorphic properties of the linear function

$$f(x + x') = f(x) + f(x')$$
$$f(\lambda x) = \lambda f(x)$$
$$f(\lambda x + \lambda' x') = \lambda f(x) + \lambda' f(x')$$

rather than the constant function coefficient:

$$f(x) = ax$$
$$x = \frac{1}{a} f(x)$$

This means that students consider scalar ratios, between magnitudes of the same kind, and linear combinations of such magnitudes, rather than function-ratios, between magnitudes of different kinds. Nevertheless the best students can shift easily from one point of view to the other, whereas the weak students keep using the same stereotyped strategy. These results are important for teachers, as some taught procedures are practically incomprehensible for most children, like the cross-product for rule-of-three problems.

Several researchers (Noelting, 1980; Karplus et al, 1983; Behr et al, 1983; Kieren, 1988) have studied extensively the difficult development of the concept of ratio. The synthesis into the concept of rational number of such different ideas as those of fractionalquantity, scalar ratio and function-ratio, is a long-term product of mathematical education.

But the most important conceptual problem in multiplicative structures, is probably the multiple proportion structure, which is involved in the measure of space (area and volume), in different domains of physics (quantity of movement, heat and energy, electricity...) and in many other domains like those of production and consumption, when some variable is proportional to two or more independent variables.

Let us take the example of the concept of volume. There is some understanding of that concept by 6 or 7 year-old students, when they have to compare the capacity of different containers, or when they have to measure, and even estimate the capacity of one container with the help of a unit of volume. Rogarski and I call unidimensional, the conception of volume that is sufficient to accomplish such tasks, and tridimensionnal the conception that is involved in the understanding of the concept of volume as the product of three lengths, or the coordination of ratios concerning lengths, areas and volumes.

In a task concerning the comparison of the capacity of two fish tanks (the large one was twice as long, three times as wide and twice as deep as the small one), we could observe several different procedures used by students, correct or incorrect (Vergnaud, 1983). I will take the example of two correct procedures:

1) $2 \times 3 \times 2 = 12$ times larger.
2) mentally pave the large tank with the small one: two in the length, three in the width; this makes 6; two in the height; 6 and 6 make 12.

The first procedure is tridimensional; it can be generalized to non-whole number ratios. The second procedure is unidimensional; it fails with ratios like 1.5, 0.8, 1.2, as one cannot easily pave the large tank with the small one.

One can also make different uses of the formula for the area of regular prisms:

$$V = A \cdot H$$

1) To calculate the volume, you must know the basic area and the height, and multiply.
2) You can also calculate the height (or the area) when you know the volume and the area (or the height). This reading of the formula is not contained in the first reading.
3) The volume is proportional to the area when the height is held constant, and proportional to the height when the area is held constant. This reading of the formula is actually the best way to understand it, and the true reason for it. But it is very rarely provided by schoolbooks, at least in France.
4) If you take units ten times smaller for lengths, the measure of the volume will be 1000 times bigger.

Students may be able to read the formula at a certain level and be unable to understand the tridimensional nature of the concept of volume, which is yet essential. And still they know something about the concept of volume.

It is for this very reason that I have developed the framework of conceptual fields: we need a framework to understand connections and jumps, in the development and the learning of competences and conceptions. The main organizer of such fields is the content of knowledge, and not such abstractions as the logical structure, the linguistic structure, or the level of complexity as measured by information theory. The complex conception of volume comes from the enrichment of former conceptions of volume, together with the interaction of these conceptions with new situations which require students to take account of new relationships, both spatial and multiplicative.

Additive structures are a conceptual field. Multiplicative structures are another conceptual field, not totally independent from the former one, but sufficiently independent to be studied separately. The conceptual field of multiplicative structures is made of such concepts as those of linear and n-linear functions, ratio, and rational number, dimensional analysis, vector-space... It involves situations of different kinds, taken in different domains, that can be analysed by simple and multiple proportion structures. It also involves different words and natural language expressions, different symbolizations like tables, graphs and formulas. It finally involves the implicit or explicit recognition by students of a variety of operational invariants like those mentioned above: scalar ratio, function ratio, independence and dependence of variables, theorems-in-action such as the isomorphic properties of the linear and the n-linear functions.

This framework is a critical means to understand how students learn: it claims that concepts are rooted in situations, consist of invariants of different kinds and levels, and need to be represented by linguistic and non-linguistic symbolic elements.

Algebra

I do not have the time to explain in detail the numerous problems raised by the learning and teaching of algebra. Convergent results have been collected on the way students shift, or have difficulties in shifting, from arithmetic to algebra, also on the errors they make in reading and transforming algebraic expressions, on the conceptual difficulties raised by the concepts of function and variable, and on the use that can be made of calculators and computers : Filloy, Rojano (1984, 1985), Booth (1984), Vergnaud et al. (1987), Kieran (1988).

I would also like to stress the important theoretical problem of the relationship between symbols and syntax on the one hand, mathematical knowledge and schemes on the other hand.

Algebra is not an independent conceptual field, that could be taught and learned independently of additive and multiplicative structures. Even if teachers and schoolbooks take it for granted that algebraic expressions are just relationships between numbers, students do not: they keep considering numbers as magnitudes (cardinals, lengths, areas, money, physical quantities...), and as relationships between magnitudes. Moreover, as they read expressions from left to right, they often consider that algebraic expressions model situations in which time is considered as going from an initial state on the left, to a final state on the right. In such a conception, the equality sign does not mean the symmetric and transitive equality relationship that is required to understand equations, but rather a relationship between a production process and an outcome.

Therefore the teaching of algebra, especially the introduction of algebra to 12 to 15 year-old students, requires a careful epistemological and cognitive analysis.

As algebra makes important use of syntactic rules, (you must do this, you may not do that), it is important to clarify the relationship between symbols and symbolic operations on the one hand, and the mathematical magnitudes, relationships and operations that are represented by those symbols on the other hand. Let me take four examples:

1) $3x + 12 = 6x - 3$
2) $3x^2 + 12x = 0$
   $3x(x + 4) = 0$
3) $(n - 1) + (n + 1) = 2n$
4) $(a + b)(a - b) = a^2 - b^2$

The meaning of example 1 is usually: find the value of x so as both functions $3x + 12$ and $6x - 3$ have the same value. The equality concerns numbers not functions. It is true only for x correctly chosen. The meaning of example 2 is somewhat different. It is aimed at showing that 4 and 0 are solutions. The meaning of example 3 is again different. It demonstrates that the sum of the predecessor and the successor of any whole number is an even number. The equality concerns predecessor and successor as functions of any given number, and the equality is always true. This is also the case in example 4. The meanings of elementary algebra are manifold.

Algebra is usually introduced as a way to solve arithmetic problems. This is usually viewed by students as more complex than an arithmetic solution, because in most simple equations $a + x = b$, $ax = b$ and $ax + b = c$, the algebraic solution actually depends upon the arithmetic solution and does not offer any obvious benefit.

To appear as a profitable tool, algebra must be seen as a way to solve arithmetical problems that cannot be solved easily by purely arithmetical means. Researchers have studied problems of this type. They have usually arrived at problems that can be expressed either as

$$ax + b = cx + d$$

unknown on both sides or as

$$ax + by = c$$
$$a'x + b'y = c'$$

two unknowns.

To solve these equations, students must accept operating upon the unknowns. This is just what some of them don't understand. How can you operate on something or with something you don't know? (Collis, 1975)

Another difficult cognitive problem concerns negative numbers. Students' most frequent conception is that numbers are magnitudes; they cannot be negative. The minus sign means subtraction of a positive quantity, not the inversion of a transformation or a directed number, nor a difference between two transformations or two relationships. With such conceptions what does it mean to find a negative solution?

Here again one needs to find problems that make negative solutions meaningful: problems in which unknowns are transformations, relationships, or coordinates.

When one observes students solving equations, one can usually identify organized and standardized patterns of behaviour. Automaticity is a powerful property of algebra. But automatic algorithms are only the visible part of the iceberg. The profound ideas of algebra need to be clarified and this cannot be done without the identification of the concepts of function and variable, of directed number, and so on. Not only must algebra be a useful tool, it consists also of new objects, the epistemological status of which is not clearly seen by students and even by teachers.

The meaning of the concept of function and variable is not conveyed by equations only, and not even essentially, as letters are conceived of as unknowns rather than variables. Therefore computers and programmable pocket calculators, graphic curves and other devices are essential ways for students to experience functions and variables and identify them as mathematical objects, having names, properties, and relationships to other objects.

Symbols and mathematics

The role of symbols can therefore be clarified. Symbols are necessary to identify mathematical objects and make explicit their properties and their relationships to other objects. Whereas in the arithmetic solution of a problem, one very often leaves implicit the choice of the relevant data and operations, the algebraic solution consists of making the relationships explicit and summarized in a laconic expression; then there are algorithmic or quasi-algorithmic ways of dealing with these expressions.

Actually the role of symbols in thinking is a very old psychological and even philosophical problem. Vygotski developed strong ideas about this problem more than 50 years ago, when he studied the relationship between language and thinking.

Research on mathematics education makes use of and is in tune with some of his ideas. Natural language and mathematical symbols of all kinds (tables, diagrams, graphs, algebra...) play an important part in the process of conceptualizing, also in the control and regulation of schemes et algorithms, also in the solving of new problems, and in reasoning about them, i.e. combining and transforming relationships, planning, choosing data and operations.

Algebra is certainly the most obvious case in school mathematics, of the help of symbols in thinking, but there are many earlier cases, at the elementary school level with diagrams and tables, and even at the kindergarten level, when children start counting with words, making a specific use of natural language in counting, in counting on and down, in counting up to a certain number from a given number, in adding, subtracting and comparing, also in expressing spatial relationships and movements.

If I may choose just another example of the help of symbols in thinking, I will take the case of the above-mentioned bilinear reasoning (consumption of sugar).

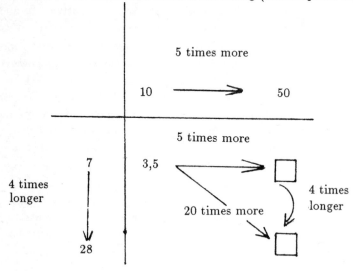

5 times more children and 4 times more time make 20 times more consumption of sugar.

But graphs and diagrams also help thinking a lot. For instance graphs are a powerful tool in algebra, to represent variables, functions and solutions of systems during the introductory phase of algebra.

But a new symbolic tool like graphs does not go without difficulties. The reading, understanding and use of such a tool raises specific conceptual difficulties.

For instance, it is not obvious that numbers can be represented by dots on a line. It requires the understanding of the concept of origin and the identification of the succession of dots to the inclusive succession of segments separating these dots from the origin (Vergnaud and Errecalde, 1980). Some 14–15 year-old students keep seeing the dots as ordered whole numbers 1, 2, 3, 4... from left to right, without being able to give any meaning to the interval between 2 and 3. When the origin cannot be represented, or when students have to change the origin of the reference system, many of them just fail. This conceptual difficulty is especially important in the use of graphic computer aids, when the origin and the scale are changed to focus some part of the graph (Nadot, 1988).

Linguistic and non-linguistic symbols are both a help and a problem for students. They help students in identifying the relevant mathematical objects and relationships, but they also raise problems of reading and understanding.

Psychology, epistemology and didactics

Research on Mathematics Education is a complex story. Psychology is only part of it, but has an essential part to play in it, from a theoretical and from a methodological point of view; not only to understand the different steps of the development of mathematical ideas and competences in students' minds, and the difficulties and the errors (as I have tried to show); but also to understand the process of learning in the classroom and the process of teaching.

Teaching and learning in the classroom is a social process, that depends on some macrofeatures of the educational and social system, and on microphenomena that take place in the interaction of students with mathematical situations and with the other actors (teacher and other students). The curriculum, the school books, the social environment, the system of training for teachers are macrophenomena. Chevallard (1985) has analysed the social process of transposition that governs the transformation (and deformation) of the scientific knowledge of mathematicians into the knowledge to be taught, and the knowledge actually taught. Among the factors of that process, teachers' representations, abilities, and difficulties play a very important part.

Three kinds of representations appear to be essential

- representation of the knowledge to be taught; .
- representation of the students' competences and conceptions;
- representation of the learning process.

Brousseau has analysed different aspects of the erroneous representations of teachers concerning mathematics and the learning process. He has also developed the theory of didactic situations that concerns the invention, the choice and the management of situations to be offered to students. The theory of situations stresses the epistemological ground of the concept to be taught and proposes systematic ways of organizing students' activity and cooperation. He has for instance made the important distinction between situations aimed at producing an action, or a message or a proof. Brousseau has also developed the concept of "didactical contract" which implicitly rules the interaction of the students and the teacher.

One of the most important parameters of teachers' decisions in the choice of situations and in the teaching process is their representation of students' competences and conceptions. This is why, as a cognitive psychologist, I have stressed that point in this address. The emergence of competences can and must be described in mathematical terms, as "theorems-in-action". But the analysis of these competences requires more than the stabilized formal description that is usually offered by mathematics.

I have shown for instance that the analysis of additive structures requires change over time to be taken into account, also the use of a unary-operation model, also of children's own untaught procedures and errors. Mathematics does not usually take change over time into consideration, and sees addition as an internal binary law of combination. Actually students have to do with both a unary and a binary conception of addition, depending on the situations they have to master. There lies the conceptual problem.

Along the same idea, the analysis of multiplicative structures requires dimensional analysis to be taken into account, even at the primary school level. This means that the learning, and therefore the teaching of mathematics are not concerned only with pure numbers but also with magnitudes and relationships between magnitudes including quotients and products of dimensions. This analysis of additive and multiplicative structures might have been made from an *a priori* point of view. It just happens that it is the difficulties of students, their errors and procedures that have compelled us to revise and complete our views, *a posteriori*.

Cognitive psychology is essential to epistemology of mathematics as much as epistemology of mathematics is essential to cognitive psychology.

If epistemology is concerned with the relationship of concepts and procedures with the practical and theoretical problems to be solved, then epistemology is important not only to understand the initial competences and conceptions of children in a conceptual field, and the situations which shape these competences and conceptions, but also to understand how these competences and conceptions are enriched, widened, restricted and sometimes profoundly changed during the process of learning and during the cognitive development of students. This process covers many years. Some aspects of the concept of volume are grasped by 6 or 7 year olds: comparing the quantity of orange juice in glasses, and measuring and comparing the capacity of containers. But it is another story to grasp a tridimensional conception of volume; most 13 or 14 year olds do not satisfy the criterion of combining ratios on lengths, areas and volumes, as I have shown above.

My final conclusion is that didactics, psychology and epistemology must be strongly tied together in mathematics education research.

Epistemology is essential to imagine didactic situations and to characterize in mathematical terms students' competences and conceptions.

Psychology is essential to analyse carefully the short term process taking place in the classroom both from a cognitive point of view and from the point of view of social interaction; also to analyse the long term process of development.

Didactics includes epistemology and psychology, but it also takes it as a burden to theorize about the conception of situations and activities that should take place in the classroom and the way to manage the ensuing process.

A mathematical concept, if one looks at it developing in students' minds, is a triplet of three sets:

— the set of situations that make the concept meaningful in a variety of aspects;

- the set of operational invariants (properties, relationships, objects, theorems-in-action...) that are progressively grasped by students, in a hierarchical fashion;
- the set of linguistic and non-linguistic symbols that represent those invariants and are used to point at them, to communicate and discuss about them, and therefore to represent situations and procedures.

The symbolic dimension of mathematical concepts is essential, but we must never forget that mathematics is not a language but knowledge; solving practical and theoretical problems is both the source and the criterion of mathematical knowledge for students as much as for mathematicians.

References and complementary bibliography:

Balacheff, N. (1988): *Une étude des processus de preuve en mathématique chez des élèves de collège.* Thèse. Université de Grenoble 1.

Behr, M.J. – Lesh, R. – Post T.R. – Silver, E. (1983): Rational Number Concepts. In Lesh, R. – Landau, M. (Eds.): *Acquisition of Mathematics Concepts and Processes.* New York, Academic Press, 91–126.

Bell, A. – Fischbein, E. – Greer, B. (1984): Choice of operation in verbal arithmetic problems: the effects of number size, problem structure and context. *Educational Studies in Mathematics,* 15, 129–147.

Booth, L.R. (1984): *Algebra: children's strategies and errors.* Windsor, NFER-Nelson.

Brousseau, G. (1981): Problèmes de didactique des décimaux. *Recherches en Didactique des Mathématiques,* 2, 37–127.

Brousseau, G. (1986): Fondements et méthode de la didactique des mathématiques. *Recherches en Didactique des Mathématiques,* 7, 33–115.

Carpenter, T.P. – Moser, J.M. – Romberg, T.P. (1982): *Addition and Substraction. A cognitive perspective.* Hillsdale, Erlbaum.

Carpenter, T.P. – Moser, J.M. (1983): The acquisition of addition and substraction concepts. In Lesh, R. – Landau, M. (Eds.): *Acquisition of Mathematics Concepts and Processes,* New York, Academic Press, 7–44.

Chevallard, Y. (1982): *Balisage d'un champ de recherche: l'enseignement de l'algèbre au premier cycle.* Seconde Ecole d'Eté de didactique des mathématiques. Orléans.

Chevallard, Y. (1985): *La transposition didactique.* Grenoble. La Pensée Sauvage.

Chichignoud, M.P. (1986): *Le concept de nombre. Etude des structures additives et soustractives en relation avec la suite numérique chez des enfants d'âge préscolaire.* Ecole des Hautes Etudes en Sciences Sociales. Thèse de troisième cycle.

Collis, K. (1975): *Cognitive development and mathematics learning.* Chelsea College. Psychology of Mathematics Education Workshop.

Douady, R. (1986): Jeux de cadres et dialectique outil-objet. *Recherches en Didactique des Mathématiques,* 7, 5–31.

Filloy, E. – Rojano, T. (1984): From an arithmetical to an algebraic thought (a clinical study with 12–13 year olds). *Proceedings of the sixth annual conference of the International Group for the Psychology of Mathematics Education,* North- American Chapter. Madison, 51–56.

Filloy, E. – Rojano, T. (1985): Operating the unknown and models of teaching. *Proceedings of the Seventh Conference for the Psychology of Mathematics.* North-American Chapter. Columbus, 75–79.

Fischbein, E. (1987): *Intuition in Science and Mathematics: an educational approach.* Dordrecht, Reidel.

Fuson, K.C. – Hall, J.W. (1983): The acquisition of Early Number Word Meanings: a conceptual analysis and Review. In Ginsburg, M.P. (Ed.): *The Development of Mathematical Thinking,* New York, Academic Press, 49–107.

Gelman, R. – Gallistel, R. (1978): *The child's understanding of number.* Cambridge, Harward University Press.

Greer, B. (1987): Non conservation of multiplication and division involving decimals. *Journal for Research in Mathematics Education,* 18, 37–45.

Hart, K. (1981): *Children's understanding of mathematics.* London, Murray

Herscovics, N. – Bergeron, J. – Bergeron, A. (1987): Kindergartner's knowledge of numbers: a longitudinal case study; part I Intuitive and procedural understanding; part II Abstraction and Formalization. *Proceedings of the eleventh International Conference. Psychology of Mathematics Education.* Montréal, 2, 344–360.

Janvier, C. (1982): *Approaches to the notion of function in relation to set theory.* Université du Québec, Montréal

Karplus, R. – Pulas, S. – Stage, E. (1983): Proportional Reasoning of Early Adolescents. In Lesh, R. – Landau, M. (Eds.): *Acquisition of Mathematics Concepts and Processes.* New York, Academic Press, 45–90.

Kieran, C. – Wagner, S. (1987): RAP Conference on the teaching and learning of algebra. *Proceedings of the eleventh Internation Conference, Psychology of Mathematics Education.* Montreal, III, 425–435.

Kieran, C. (1988): Two Different Approaches among Algebra Learners. In Coxford, A.F. – Shulte, A.P. (Eds.): *National Council of Teachers of Mathematics, 1988 Yearbook. The Ideas of Algebra K-12,* 91–96.

Kieren, T. (1988): Personal knowledge of Rational Number: its intuitive and formal development. In Behr, M.J. – Hiebert, J. (Eds.): *National Council of Teachers of Mathematics 1988 Research Agenda. Number concepts and operations in the middle grades.*

Laborde, C. (1982): *Langue naturelle et écriture symbolique: deux codes en interaction dans l'enseignement mathématique.* Thèse. Université de Grenoble I.

Nadot, S. (1988): Oral communication.

Nesher, P. – Katriel, T. (1977): A semantic analysis of addition and substraction word problems in arithmetic. *Educational Studies in Mathematics,* 6, 41–51.

Nesher, P. (1982): Levels of description in the analysis of addition and substraction word problems. In Carpenter, T.P. – Moser, J.M. – Romberg, T.A.: *Addition and Substraction. A cognitive perspective,* Hillsdale, Erlbaum, 25–38.

Noelting, G. (1980): The development of proportional reasoning and the ratio concept. Part I. Differentiation of stages. *Educational Studies in Mathematics,* 11, 217–253.

Riley, M.S. – Greeno, J. G. – Heller, J.I. (1983): Development of children's problem solving ability in arithmetics. In Ginsburg, H. (Ed.): *The development of mathematical thinking,* New York, Academic Press, 153–196.

Rogarski, J. (1982): Acquisition de notions relatives à la dimensionalité des mesures spatiales (longueur-surface). *Recherches en Didactique des Mathématiques,* 3, 343 –396.

Schwarz, J. (1988): Intensive Quality and Referent-Transforming Arithmetic Operations. In Behr, M.J. – Hiebert, J. (Eds.): *National Council of Teachers of Mathematics 1988 Research Agenda. Number concepts and operations in the middle grades.*

Steffe, L.P. – Von Glasersfeld, E. – Richards, J. – Cobb, P. (1983): *Children's counting types: philosophy, theory and application,* New York, Praeger

Vergnaud, G. – Durand, C. (1976): Structures additives et complexité psychogénétique. *Revue française de Pédagogie,* 36, 28–43.

Vergnaud, G. – Errecalde, P. (1980): Some steps in the understanding and the use of scales and axis by 10–13 year-old students. *Proceedings of the fourth international Conference for the Psychology of Mathematics Education,* Berkeley, 285–291.

Vergnaud, G. (1981): *L'enfant, la mathématique et la réalité.* Bern, Peter Lang

Vergnaud, G. (1983): Multiplicative Structures. In Lesh, R. – Landau, M. (Eds.): *Acquisition of Mathematics Concepts and Processes,* Academic Press, 127–174.

Vergnaud, G. (1983): Didactique et acquisition du concept de volume. *Recherches en Didactique des Mathématiques,* 1, 5–133.

Vergnaud, G. – Cortes, A. – Favre-Artigue, P. (1987): Introduction de l'Algèbre auprès de débutants faibles. Problèmes épistémologiques et didactiques. In Vergnaud, G. – Brousseau, G. – Hulin, M. (Eds.): *Didactique et Acquisition des Connaissances Scientifiques,* Grenoble, La Pensée Sauvage, 259–279.

SPECIAL IPC-INVITED LECTURE:

COMPUTERIZATION OF SCHOOLS AND MATHEMATICAL EDUCATION

Andrei Ershov (USSR)

I. Introduction

It is not accidental that the Organizing Committee has chosen a computer terminal as an emblem of our Congress.

But computerization is not an easy path. A very fresh and pertinent experience was the computerized registration procedure last Tuesday.

Computerization puts the teacher in a vulnerable and unstable position.

Nevertheless, computerization is like a tidal wave or a warming up of the climate: you can neither hide nor prevent it but only cope with it.

So I am going to speak about relations between Education, Mathematics and Informatics.

Professionally I am neither a teacher, nor a mathematician. So I especially appreciate the honourable invitation from the International Program Committee to deliver this lecture. But I must explain why I took the risk of accepting.

Two personal reasons and one general.

The first reason is that working professionally in Informatics, essentially in research and development, I was more or less constantly engaged in what is called a mathematical and educational practice.

The second reason is that, getting a mathematical education and having the happy privilege to be for 4 years a member of the Mathematical Division of the Soviet Academy, I felt and, possibly, possessed a flow of powerful mathematical and educational ideas from many great authors and personalities. I can't avoid mentioning some of them: Poincare and Lusin, Courant and Kolmogorov, Kleene and Markov, von Neumann and Bourbaki, Vigotsky and Piaget.

To speak about sources, one recollection. For me, this Congress is a private celebration of the 30th anniversary of my first trip abroad, namely to Hungary, in 1958. I had just been a student — seven years of study in the mathematical faculty of Moscow University.

You know the Russian mathematical literature never suffered an inferiority complex. Nevertheless my bookshelf was for all those years well populated by Hungarian authors. Only among the basic books I had

Pólya and Szegő for Calculus

Szőkefalvi-Nagy for Functional Analysis

Rózsa Péter for Recursive Functions

László Kalmár for Logic, and not forgetting

John von Neumann's classic on computers.

Really, there is no better place to conduct this Congress on Mathematical Education.

As to the third, general reason, I want to say that Mathematics is not just a material science for Informatics. There has been a steady tendency towards the mathematization of Informatics in the course of its formation and maturation . On the other hand, there is more and more evidence that some methods of Informatics, some information technologies penetrate into Mathematics, influencing its style, technique, and content.

My elaboration of this theme will be put into two global contexts.

The first context is the State Program of Introducing Computers and Informatics into school education in the USSR.

The first three years of rather intensive work in this direction have allowed us to accumulate some first experiences and gain an irreversible momentum. Moreover, this work has resulted in the development of a conception of school computerization as a long-term process with explicitly stated intermediate and final goals. I hope, some facts on the Program and some formulations from the Conception will be interesting to this audience.

The second context is the very active process of studying and discussing computers in mathematical education [1–4] and my conclusions rely heavily on what can be found in the corresponding material.

And I am happy to observe that computers in mathematics is one of the prevailing activities at this Congress.

So, my plan for the main part of the lecture is as follows:

Some evidence about the Computerization Program in the USSR.

Some formulations on Computers in Education from the Conception.

Then I plan to mention the major problems of mathematical education. In the spirit of our philosophy they will be formulated as a list of contradictions or dilemmas. In this context the so-called Kolmogorov Reform of School Mathematics in the USSR will be discussed.

The positive part of my presentation will consist of a list of theses on the impact of computers and Informatics on Mathematics and its teaching.

And I will conclude with a statement about Language, Mathematics and Informatics as three basic major subjects in School Education.

II. The program

Now, about the National Program of Computerization.

The Program was launched by adopting a governmental decision to introduce promptly and frontally in all schools a 70 hour course in grades 9 and 10 on "Fundamentals of Informatics and Computers".

In a sense this decision symbolized the beginning of Perestroika: in February 1985 Mikhail Sergeevich Gorbachev submitted the decision to the Politbureau as Chairman of the Commission on School Reform and on March 11 he was elected Secretary General of the Party and started to steer the country towards revolutionary changes.

This strong decision had many scientific and educational motivations behind it (see, for example, [7]) but essentially, it was a political decision. It triggered all the machinery in the educational system that stands behind any mandatory discipline in the curriculum. So in the following year much more was done than in all the previous years of partial experiments with computers in school.

Some visual evidence

Our first text-book in Informatics and Computers [8–11] has been published in 15 languages. I have participated in writing and editing the book and it was a frightening, I would say a Lutheranian, experience (after Martin Luther who translated the Bible into plain German) to write a book that must be read by millions of readers.

A selection of the educational literature that has flourished throughout the country in local pedagogical institutes and regional teachers' advancement centers. [5,6]

The newly established magazine "Informatics and Education" is now circulating with over 100000 copies per issue.

The first Soviet school computer "Agat" is of Apple-II architecture. The Japanese "Yamaha" computer of MSX architecture imported in several thousand units is extremely popular among teachers.

Two kinds of computers, "Korvet" of INTEL 8080 architecture and "UKNTs" of PDP–11/LSI architecture, have been especially designed for mass production.

The state plan is to deliver by the 1990/1991 academic year about 400000 school computers — the first priority among various public sectors of the use of personal computers. We expect about 3000 classes to be equipped with 12–15 computers installed in schools by the next academic year.

From the very beginning the new course in Informatics has had many ties with Mathematics. First of all, teachers of mathematics dominate among those who teach Informatics. A very rough breakdown is: 55% from Mathematics, 20% from Physics, 20% from Informatics professionals, 5% other. This situation has influenced the didactical style of the text-book.

The book as a whole supports a more or less established approach to the computer literacy course (as compared, say, with that of the ACM curriculum[12]) based on the structured style of algorithm and program development with an extensive use of the ordinary mathematical lexicon.

Quite recently, new versions of text-books on Informatics have appeared that continue the line of connection with mathematics [14,15].

The highly probable absence of computers is compensated for by a general shift from computer manipulations to concepts from informatics supported by a reasonable variety of mathematics-oriented problems [13]

Certainly where computers are available the course is supported by an intensive 40-hour computer practice. If BASIC is the only programming language available it is taught not as a programming ABC but rather as a tool for program coding of previously designed algorithmic specifications expressed in *if-then-else* and *while-do* procedural notation.

These, more or less, superficial relations are based on a much more fundamental connection between Mathematics and Informatics.

algebraic structure .... processor
carrier .... environment
carrier element .... environment state
operation .... action
predicate .... probe
signature .... instruction set
free semigroup .... expected behaviour
word .... behaviour
precondition predicate .... problem condition
postcondition predicate .... problem goal
admissible word .... problem solution
algorithm .... program
theory .... object domain

Fig. 1. Some important relations between concepts of Math-
ematic and Informatics

Fig. 1. shows parallel counterparts of important concepts from mathematics and informatics that form a base for the mathematization of informatics.

Certainly, this interconnection is rather intensively explored and elaborated in the special literature. Education, however, stays behind and only in some university courses are these ties systematically exploited.

What about the current state of computerization? Of course, like any process that is regulated essentially by external forces, it is a mixture of success and compromise, enthusiasm and frustration, getting involved and staying aside.

Nevertheless, the community of teachers as a whole is positive towards computerization, essentially because it is a backbone, moving force and symbol of the school reform. And, of course, when the computer is available and reliable, children's enthusiasm can greatly stimulate even the old teacher. Let us repeat that as a whole computerization in the USSR has gained an irreversible momentum.

What about its efficiency? The feed-back is not yet properly developed so the indicators are not complete. The percentage of those who demonstrate a satisfactory grasp of main concepts and techniques varies from an optimistic 40–50% to a more realistic 15–20% which is highly correlated with corresponding data on Mathematics. Nevertheless, the first wave of computer awareness has already reached the doors of higher education.

What can be said about the next seven years? I am, personally, skeptical about having 400000 computers installed by the 1990/91 academic year but firmly believe that by 1995 each secondary school will have a computer class.

The basic course in Informatics will move into grades VII and VIII of our 10 year school plan, thus leaving the last two grades for the advanced use of computers in pre-college and pre-professional work.

As a possible outcome of intensive scientific and pedagogical work a so-called integrated grade IV–VII course in informatics can emerge in which the study of informatics concepts and the possession of abilities to work with the computer will be tightly connected with everyday school practice heavily relying on an intensive use of the computer as an indispensable tool for intellectual work.

There are many proponents of an amalgamated course in Mathematics and Informatics. I don't think that it will happen soon but we all expect many positive results from their studies and experiments.

III. The conception

Now I turn to the main statements of the overall conception of school computerization with emphasis on pedagogical aspects [17].

I shall allow myself to be not too systematic.

First of all we consider school computerization as a premise and a component of a much more global process which in our literature is unfortunately getting the rather ugly name of "Informatization"

We understand informatization as an overall and indispensable period in the development of human civilization, a period that embraces roughly a 100 years from the mid fifties up to the mid 21st century.

This process has as a target a full use of trustworthy, exhaustive and timely knowledge in all socially meaningful kinds of human activity. Information becomes a strategic resource of mankind like material resources.

Computers and means of communication as a global network form a kind of central neural system of the organism of human society as a whole, thus giving to this organism an unseen integrity, flexibility and foresight.

Epistemologically informatization consists of the formation, separation, and maintenance of a coherent and integrated information model of the world, a model that enables society to perform a predictive, dynamic regulation of its development at all levels of activity, from individuals' behaviour to world-wide institutional operations.

As a process of the development of productive forces, informatization consists of bringing an adequate information and computer capacity to every working site, reducing the manpower requirements of every production process, a radical rise in human performance in other areas of activity, in a transition from the PRODUCTION era (what to do) to the TECHNOLOGICAL era (how to do it).

In our literature [13,18] we treat Informatics as an emerging fundamental natural science, studying laws and methods of the acquiring, transmitting and processing of information in natural, artificial and social systems. Its constituent disciplines are

. theoretical foundations of computing and communication
. algorithmics
. programming
. artificial intelligence
. theory of cognitive processes
. computational experimentation
. informology.

We distinguish Informatics as a science from various kinds of applications of informatical knowledge. Applied informatics transfers into the society so-called information technologies — stable and commonly available procedures of systematic or automated information processing, starting from such milestones of human civilization as

the creative arts
writing
book printing
the postal service
typewriting
the telegraph
the telephone
radio
television
audio technology
to such novelties as
xerography
personal computing

video

electronic mail

desk-top publication systems

optical discs.

Now I would like to go into more details of the pedagogical use of computers.

Many educators demonstrate a restrictive view of the computer. Moreover, a proclamation of such a restrictive view is sometimes considered to be a good style of thinking, a certificate of an educated approach.

Not wishing to offend the dignity of such scholars, I think that it is more productive to overcome such a restrictive view.

Indeed, the computer is just a machine, a human tool. But it is a "magic" tool whose potential is restricted only by the limits of our imagination.

Forty years of computer use allows us to give a reliable formulation of the computer as a list of its most productive features:

availability

performance

universality

programmability

adaptability

extendability

globality

ideality.

In more pedagogical terms the most pertinent expectations from the computer are as follows:

1. The computer is a most adequate technical educational tool which supports an activistic approach to the learning process in all its stages: need–motivation–goal–condition–means–operation–action.

2. Being an active partner in the dynamic balance between challenge and assistance, the computer highly stimulates the learner.

3. Programmability together with dynamic adaptability facilitates the individuality of the learning process without destroying its integrity.

4. Flexibility and a variety of user interfaces make the computer an ideal vehicle for the training phase of the learning process.

5. The formal character of computer work, strictness in observing the "rules of the game", and the explicit character of these rules raise the level of consciousness and self-control of the learning process.

6. The ability of the computer to visualize its work greatly improves the capacity of the information channels of the learning process.

7. The computer brings into the learning process new and powerful cognitive means such as computational experimentation problem solving with the help of expert systems, algorithm construction, and the enrichment of the knowledge base.

8. Being a leading and commonly used tool in technical and scientific progress the computer as an indispensable component of the educational environment brings education closer to the real world.

9. And last but not least, because of its universality and programmability, the computer allows us to make many things in school more cheaply and quickly than by any other means.

A few words about the changing role of the teacher in the computerized classroom.

There is a mistaken opinion of the computer as an obstacle that destroys the contact between teacher and pupil. Some educators view a computer lesson as a prolonged state when everybody looks at the display and keyboard not seeing each other, and the teacher, besides controlling the class, must bear a heavy burden of manipulation with his or her own computer.

Actually, such a view is just a result of initial incompetence and too straight-forward an approach to the computerized lesson.

The computer can provide a teacher with a good deal of support and can even make his or her relation with a pupil more human than before.

Firstly, the computer takes on much of the feed-back control. All pupils failures are mercilessly registered by the computer, but this becomes a more private affair than a subject of the teacher's acknowledgement and dissatisfaction. The teacher becomes freer and more positive in his attitude towards pupils.

Secondly, the computer, while maintaining partnership relations with a pupil, reduces the teacher's efforts aimed at supporting the tempo and energy of the pupils' activity. Many more events in the class happen just by "themselves" so the teacher has more time to observe the overall situation in the class-room or to pay more attention to a single pupil.

Of course all this becomes possible only when the lesson has an adequate methodological support, the computers work smoothly, the software is friendly and the teacher is fluent enough with the computer.

One more piece of evidence of the restrictive approach to the computer is viewing computerization essentially as just a logistic process of delivering computers to schools, putting the subject of Informatics into the curriculum, and attracting the teacher to the computer somehow, with a blind hope that the consequent process will be self-establishing.

Actually, school informatics is a rich conglomeration of many kinds of activity arranged in two major groups: 1) core or front-end activity which is for the most part, occurring at the school, and 2) supporting activity carried on outside the school.

Figure 2. displays the list of main front-end and supporting activities arranged in a matrix form.

I hope all the identifiers are self-explanatory; the main idea is that school informatics is a penetrating and global activity comparable with the educational system as a whole, requiring a great deal of policy, planning and resources.

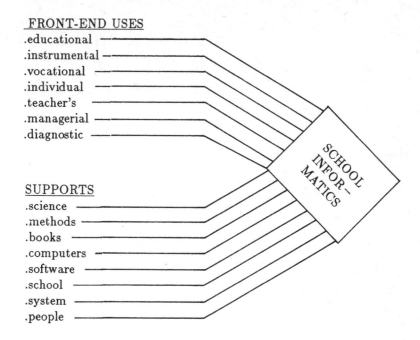

Fig. 2. The matrix of school informatics

IV. Mathematics

Now, let me speak about mathematics and mathematical education. Here, I am approaching the limits of my competence so please, take my account just as personal observations rather than a comprehensive and fully objective treatment.

It is almost customary to speak about mathematical education as something that is in a crisis. Many ICMI as well as various national studies support such a critical view [19,20].

But what is interesting, is that such a critical view of educational mathematics coexists with a literally triumphant development of professional mathematics.

Let me mention the most important and striking achievements; at least those that were more or less comprehensible to me.

Firstly, the great achievements in the theory of proofs, formation of mathematics, new proof techniques and methods, deeper understanding of limits of mathematical knowledge and mathematical methods of reasoning.

Secondly, the establishing of tight connections between theory of proofs and the theory of exhaustive search, deeper understanding of the concepts of finiteness and infinity, formation of the theory of complexity of systematic procedures or mathematical structures.

Thirdly, the deep breakthrough into nonlinear mathematics that greatly reinforced the role and power of mathematics as an instrument of study of the physical world.

Fourthly, the formidable intellectual achievement of the synthesis of algebra, geometry, topology and analysis that allowed unification of the main mathematical structures, to unite in harmony the discrete and continuous worlds of mathematical objects.

If in the beginning of the century the famous Hilbert program was viewed by a mere one thousand mathematicians as a roomy space to explore. Now it is just a well attended but memorial hall in the enormous temple of Mathematics inhabited by tens of thousands of talented and exploring people.

So, what do we actually have? Certainly, all of us observe a great distance between mathematics as a school discipline and mathematics as a living science. There is a constant gap between the needs in mathematical practice and our ability to exercise it. Our mathematical and educational universe has expanded explosively and all non-homogenities and singularities become visible.

I would call the situation not a crisis but one of growing contradictions in the goals of mathematical education. Before listing them, I have to say that many of these contradictions are of a general character and not specific to mathematics.

So, what are these main contradictions or dilemmas?

Scientific knowledge or common sense? This conflict is as old as life itself and reflects the eternal opposition between ordinary and scientific consciousness. In more technical terms, the dispute is about what kind of concepts — logically perfect but abstract or intuitively comprehensible but vaguer — should be laid down at the fundamentals of a school course.

Elitist or egalitarian education? This dilemma, which is common for education, is particularly dramatic for mathematics. Every attempt to master an integrated and complete mathematical education in school is immediately compromised by the pupils' capabilities or teachers' perfection. On the other hand, every attempt to formulate a common minimum of mathematical knowledge, makes everybody unhappy because of "important omissions".

Variability or uniformity? Perhaps this dilemma is specific to the Soviet Union with its emphasis on the uniformity and similarity of the learning process. But even if you adopt variability, difficulties caused by the previous dilemma still remain.

Higher education or common knowledge? What is the main function of secondary school: to prepare better for a college or university or to give some basic knowledge that will never be extended into future life? In fact this dilemma has led in our school reform to the separation of a basic secondary education (grades VIII–IX) and an extended secondary education (grades X–XI).

One specific instance of the previous dilemma is: to teach mathematics as a science or as a practical subject. Academician Andrei Nikolayevich Tikhonov, while establishing in Moscow University a new Faculty of Computational Mathematics and Cybernetics, tirelessly repeated his thesis of a difference between the outlook and methodology of pure and applied mathematicians though being agreeable about the unity of mathematics itself. If this university dilemma is a real one then it has a projection into the school course.

And finally *tradition or modernization?* Perhaps, this dilemma is for the pedagogical institute course rather than for the school one. What should be the basis of teacher's outlook: keeping on historical values or permanent enthusiasm about current events? The gap between tradition and novelty is well perceived.

V. The Kolmogorov reform

Now, let me speak about an important chain of events that has happened in the last 25 years in the Soviet Secondary Schools and acquired the name of the Kolmogorov Reform.

Three features make the story especially dramatic: (1) it is tightly connected with the name of one of the most distinguished mathematicians of our time, (2) its scale and extension, (3) the prevailing opinion that this undertaking was a failure.

In outline the sequence of events was as follows. In the mid-60's, as the implementation of a governmental decision about raising the scientific level and enrichment of the content of school curricula, an activity emerged aimed at developing a new curriculum for school mathematics. Formally and factually, it was a collective, committee enterprise embracing specialists from the Academy of Sciences, the Academy of Pedagogical Sciences, and universities with some participation of school teachers. But very soon this activity got its personal colour because of direct involvement of Acad. Kolmogorov in this work, his great scientific authority and, especially, because of its overall and integrated approach to the school course of mathematics. He worked at all levels: from global planning to routine lessons in the classrooms of the University Mathematical Boarding School. The leading Soviet magazine "Mathematics in School" had published more than 70 of Kolmogorov's methodological papers and notes at that period [21].

The new curriculum had been put into action promptly and frontally with extraordinary support of preliminary methodological instructions published in "Mathematics in School" and ministerial circular letters. Short-term advancement courses for teachers were conducted throughout the country.

In a few years new text-books emerged. Their use, however, caused considerable difficulties for teachers. Mathematics became harder for children. Interest in mathematics diminished and the level of university applicants fell considerably. Teachers found themselves disoriented.

A critique of the reform gradually emerged. First, about particular deficiencies in the text-book. Then, some complaints concerning methodological issues. Finally, opposition to the basic ideas of the reform appeared. Some alternative text-books had been written.

This critical mood has gradually matured into a powerful counter-reforming movement in which a traditionalist view of school mathematics took over. The counter-reformation was not so monolithic, many compromises were achieved, some so-called Kolmogorov text-books survived, but on the whole, the newly adopted mathematical curriculum was considered by many as rather conservative.

All this was accompanied by very heated discussions and personal attacks, so that the last years of Kolmogorov's life were saddened not only by his serious illness.

Certainly, it is a very crude picture. Actually, a million events took place, many of which are absolutely positive and the current situation is far from what we had 20 years ago.

First, a new generation of successful mathematicians has grown from the soil of the Kolmogorov curriculum, a generation which dominates as the best examples of our mathematical thought and practice. Secondly, teachers, in spite of all the difficulties experienced by them, felt a taste of new and exciting ideas and thus raised their level of consciousness. The "Kvant" magazine had appeared and its brilliant series "Library of Kvant". Kolmogorov's own activity greatly stimulated his colleagues, which resulted in considerable enrichment of the literature on school mathematics.

I think that it is impossible to evaluate the essence, role and fate of the Kolmogorov reform staying within its scientific and methodological content. Its fate cannot be separated from the evolution throughout the two decades which our journalists call delicately "stagnation period".

I would say, that if the Kolmogorov reform as an action was a failure then this failure is just a projection of the failure of another much more global action undertaken by Brezhnev and his silent assistants — I mean the transition to mandatory complete secondary education while preserving all the rigidity, uniformity and authoritarian character of the educational system and an attempt to do so with "zero" expense, without proper support and resources.

Thinking about the dramatic fate of the Kolmogorov reform and its leader I cannot resist drawing a parallel with the fate of another great Russian personality who belongs to the same generation. I mean Boris Pasternak and his chief work "Doctor Zhivago".

The same measure of talent, high professionalism and ability in ordinary work.

The same distance from and incompatibility with many realities of contemporary life and environment.

The same intimate connection with Culture and Nature.

The same deadly jealousy and unquietness from some of their guild-brothers.

The same high feeling of devotion and uncompromised will in performing a mission.

You may ask me why I put the Kolmogorov reform in the context of my lecture. There are three reasons for that.

Firstly, for me personally, this story is an inexhaustible source of experience, inspiration and warning in our surgical work on annotation of the computer into school education.

Secondly, the Kolmogorov reform itself focuses on and is real evidence of all the above mentioned problems of mathematical education.

And thirdly, it is my strong opinion that the computer can and must become a serious assistant and partner to mathematical education in solving also those problems at which the Kolmogorov reform had been aimed.

VI. Informatics and mathematics

Now, about the impact of computerization on mathematical education.

As a preliminary remark, let me say that mathematicians are also human beings and the computer can help them as all other people. I see, quite a lot of mathematicians regardless of their age and academic level get started on their computer work with the enthusiasm of teenagers. Mathematicians think well and with such a capacity feel themselves quite comfortable without Artificial Intelligence, knowledge bases and expert systems. But the computer gives them the number and the image and that helps greatly.

My first thesis is on mathematical practice. The overall use of computers, the construction of the informatization model of the world have expanded the volume and variety of mathematical practice enormously. Many techniques and methods of mathematical work become literally a general commodity. Abstract design in various signatures, schematization of concrete objects by selecting their attributes and relations, model construction, deduction and reduction, recursive thinking, separation and maintenance of abstraction levels, behaviour prediction, analysis of laws, regulations and rules for consistency, then mass design of algorithms and their evaluation — all this becomes an instrument of modern intellect, a skeleton of information culture. Thus, computerization is both a tool and an expression of the expansion of mathematical knowledge, and this world-wide process cannot remain unnoticed by mathematicians.

My second thesis is on the nomenclature of mathematical knowledge. The computer imitates, or rather, reproduces a human behaviour. To this end, programming and information model construction introduce into mathematics abstractions of human activity, various properties of artificial, biological, technical and social systems. All this puts on the stage discrete mathematics. The shift in physics in the direction of quantum properties of matter also facilitates this process. Some sections of discrete analysis emerge which are parallel to corresponding sections of classical analysis. The interconnection of discrete and continuous becomes the subject of a thorough study, as for example, in synergetics and catastrophe theory. New methods of mathematical work appear, for example, proving by computer. Such new approaches to mathematical proof, like the famous computer solution of the well-known 4-colour problem still require philosophical analysis.

The next thesis will be on the system role of a mathematical theory. The concept of theory belongs to mathematics. On the other hand, informatics possesses an extremely important notion of an environment. An environment is a closed world model that is incorporated into computer memory and is an operational space of a robot or other executor to be programmed. Since all instances of the executor's behaviour must be predicted we need in practice a complete knowledge of the environment and understanding also of the limits of that knowledge in the real world. All this knowledge must precede any concrete construction. Obtaining this knowledge is a duty of systems analysis and construction of a complete theory is its outcome. Systems analysis is a new kind of mass human practice and mathematical methods are its extremely important tools.

My fourth thesis will be about computational experimentation with a mathematical model. Its great importance for engineering practice is widely acknowledged. Its practicality in the learning process has also many pedagogical confirmations. I would only add that recently mathematical experiment has become more and more a source of purely mathematical inspiration and discoveries. To this end, I would refer to an interesting recent study undertaken by Dr. Hazewinkel from Amsterdam Center of Mathematical and Computational Studies [22].

My fifth thesis will be on visualization of abstractions. How to make a thought visible has been an eternal and painful problem for scientists and educators. Intellectual graphics has a millenium of history from stone age fresco to Escher engravings. And again, the computer, with its ability to synthetize images, helps greatly. Look at those magnificent pictures from "Scientific American". Now a third of its illustrations are images generated by the cooperative mind of a scientist, a programmer and a computer.

As to education, everybody knows how much a bright, unexpected image stimulates a young mind.

My sixth thesis is about dynamics in mathematics. Mathematics is a science about invariants. You can only get a real knowledge of an invariant by comprehending the dialectics of constancy and variability of its parameter. As Karl Marx said: "Every law manifests itself only in an attempt to violate it." To observe as a logical constant all life events described by a law means to understand the law and be able to apply it. The computer, with its ability to compute and visualize allows an observer to move from the statical structure of a mathematical relation and to demonstrate various trajectories of a dynamic process in space and time, thus enriching the observer, his experience, intuition and predictive ability. All this moves the learning process to study and experiment.

My next thesis is about regularity and chaos. Among computer potentials in mathematical experiments, those which allow us to observe the formation of regular structures starting from an initial chaotic state are of a special importance educationally. In their simplest form, these are various structures which appear as the result of iteration of certain non-linear operators initially applied to random objects. Many intriguing examples of such iterative operators are shown in the pages of "Scientific American".

I must admit, however, that the educational potential of such experiments is not yet properly exposed. On the other hand, it forms an extremely powerful channel for the expansion of mathematical knowledge about a very broad class of natural events: continental drift, sea-shore formation, mountain landscapes, patterns of polar auroras, formation of leaves on trees, skin colour of animals, conflict origin and crisis escalation.

My next, eighth thesis will be about basic abilities and knowledge.

In many popular books on mathematics we may read: "This book requires no prerequisite knowledge but supposes some ability for abstract reasoning." It is a pity that such things are mentioned only in forewords and are taught only in outdoor, corridor discussions. On the other hand, may be this corridor education in mathematics, like the street education in real life, is that real basis of mathematical culture which is a main goal of school education.

Here, I am in a full agreement with Professor Seymour Papert [23] who assuringly shows that the computer with its ability for direct manipulation with visible images of abstract mathematical objects in various microworlds (like LOGO) can make this vague problem of premathematical education a pedagogically explicit goal, especially in the preschool and primary school periods.

Here we might have various logical problems, intellectual competitions, composing and observing various rules of a game, designing of microworlds, direct manipulation with mathematical objects, controlling and planning an executor and many, many other things.

And my last thesis is just a small addition to the previous one: the dynamical, visually attractive, friendly and stimulating style of the computer's behaviour combined with its universality makes it an ideal instrument for the formation of an initial interest in mathematics, its beauty, unexpectedness, predictive power and the magic connection with everything around. .

VII. Conclusion

Now, a few words in conclusion. I am taking a risk to give one more restrictive, but seemingly important definition of informatics as a science concerning the rules of purposeful activity. This definition becomes just right if we equalize informatics with computer science and agree with Church's thesis and Turing's concept of a universal machine.

If we agree with all this, we will immediately conclude that the newly born informatics becomes an inseparable sister of mathematics and language, forming a basic triangle (or trinity) of major manifestations of the human intellect, namely,

ability to communicate

ability to reason

ability to act

The discipline of action is equally necessary for a human being, like the discipline of speech. Exercising control over a computer, a human being develops an ability to control himself. Understanding how the computer solves a problem, he preserves this understanding in his memory. Observing and experiencing catastrophes in artificial worlds, he accumulates, at much less expense and losses, experience in coping with decisions having consequences.

In other words, I believe that informatics will help to overcome the distant character, reflexivity and infantile nature of the contemporary intellectual education of pupils.

References

[1] Howson, A. G. and Kahane, J.-P. (eds): The influcence of Computers and Informatics on Mathematics and its Teaching. Proc. ICMI Symposium, Strasbourg, 1985, Cambridge Univ. Press, 1986.

[2] The Influence of Computers and Informatics and its Teaching. Supporting papers for the ICMI Symposium, Strasbourg, 1985. IREM, Universitè Louis – Pasteur, 1985.

[3] Johnson, D. C., Tinsley, J. D. (eds): Informatics and Mathematics in Secondary Schools: Impacts and Relationships. IFIP TC-3 Conference, Sofia, 1987. North – Holland, 1987.

[4] Johnson, D. C. – Tovis, F. (eds): Informatics and the Teaching of Matematics. IFIP TC-3 Conference, Sofia, 1987. North-Holland, 1987, p. 172.

[5] Садовская, Н. А. (сост.): Применение ЭВМ школьном образованни. Библиографический указатель литературы. Часть I. Отечественный опыт. Новосибирск, 1984 (Вычислительный центр СО АН СССП).

[6] Юнерман, Н. А. – Кисарова, М. П. (сост): Школьная информатика. Библиографической указатель отечественной и иностранной литературы за 1981-1985 гг. Новосибирск, 1986 (Вычислителььхый центр СО АН СССП).

[7] Ершов, А. П. – Звенигородцкй, Г. А. – Первин, Ю. А. – Юнерман, Н. А.: ЭВМ в школе: опыт формулирования национальхой программы. В кн.: INFO 84, 6-10 February 1984, Dresden, Plenarvorträge, 1.S.53-63.

[8] Ершов, А. П. – Монахов, В. М. – Кузнецоб, А. А. и др.: Основы информатики и вычислительной техники. Часть II. Просвещение, 1986.

[9] Ершов, А. П. – Монахов, В. М. – Бешенков, С. А. и др.: Основы информатики и вычецлительной техники. Часть II. Просвещение, 1985.

[10] Ершов, А. П. – Монахов, В. М. – Кузнецов, А. А. и др.: Изучение основ информатики и вычеслительной техники. Часть I.: Просвещение, 1985.

[11] Ершов, А. П. – Монахов, В. М. – Битиньш, М. В. и др.: Изучение основ информатики и вычислительной техники. Часть II. Просвещение, 1986.

[12] ACM Educational Board, IEEE Computer Society Activities Educational Board: Computer Science in Secondary Schools: Curriculum and Teacher Certification – *Communcations of the ACM*, 3, No.3, 1985, pp. 269-279.

[13] Ершов, А. П.: Научнометодические основы школьного курса информатики. Вестник АН СССР, 1985. Ho.12, C.49-59.

[14] Ершов, А. П. – Кушниренко, А. Г. – Лебедев, Г. В. и др.: Основы информатики и вычислительной техники. IX–X класс. Просвещение, 1988.

[15] Каймин, В. А. – Щеголев, А. Г. – Ерохина, Е. А. – Федюшин, Д. П.: Информатика. Часть I, II. Московский Инцтитут Електроного Машиностроения 1987. с. 228.

[16] Ершов, А. П.: Школьная информатика в СССР: от грамотности к культуре. Информатика и образование, 1987, Но. 6, С.З. II.

[17] Концепция использования средств вычислитеьной техники сфере образования (Проект). – Информатика и образование, 1988, Но. 6.

[18] Программа курса "Основы информатики и вычислительной техники". – Микропроцессорные средства и сустемы, 1986, Но. 4, с. 86-89.

[19] The National Science Board, Commission on Pre-college Education in Mathematics, Science and Technology: Educating Americans for the 21st Century. Washington, National Science Foundation, 1983, p. 124.

[20] International Comission on Mathematical Instruction (ICMI): School Mathematics in the 1990-s. Centre for Mathematics Eduction, Univ. of Southampton, 1985, p. 31.

[21] Колмогоров, А. Н.: Математика – наука и профессия. Библиотечка "Квант", вып. 64. Наука, Физматлит, 1988.

[22] Hazewinkel: Experimental mathematics. Report PM – R8411. Centre for Mathematics and Computer Science Amsterdam, October 1984..

[23] Papert, Seymour: Mindstorms – Children, Computers and Powerful Ideas. New York: Basic Books, 1980.

PLENARY ADDRESS:

ALGORITHMIC MATHEMATICS: AN OLD ASPECT WITH A NEW EMPHASIS

László Lovász (Hungary)

0. Introduction

The development of computers is perhaps the single most significant technological breakthrough in this century. It is natural that it has not left untouched closely related branches of science like mathematics and its education. It is also natural that whichever fields have come into contact with it, heated debates have started and very different views, extreme and moderate, progressive and conservative, have been put forth. Is algorithmic mathematics of higher value than classical, structure-oriented, theorem-proof mathematics, or does it just hide the essence of things by making them more complicated than necessary? Does teaching of an algorithm lead to a better understanding of the underlying structure, or is it a more abstract, more elegant setting that does so? Is the algorithmic way of life best (Maurer 1985), or is applied mathematics just bad mathematics (Halmos 1981)? Should computers be introduced in elementary/secondary/college education of mathematics?

I want to start with a disclaimer: I will not attempt to give an answer to all these questions. The point I will try to make is that algorithmic mathematics (put into focus by computers, but existent and important way before their development!) is not an antithesis of the "theorem-proof" type classical mathematics. Rather, it enriches several classical branches of mathematics with new insight, new kinds of problems, and new approaches to solve these. So: not algorithmic *or* structural mathematics, but algorithmic *and* structural mathematics!

The interplay between the algorithmic and structural sides of mathematics is manifold; I will only mention the two most important lines. The design and analysis of algorithms and the study of algorithmic solvability uses deeper and deeper tools from classical structural mathematics on the one hand; and an algorithmic perspective has a more and more profound effect on the whole framework of many fields of classical mathematics on the other hand.

Let me exemplify. Perhaps the first notion of an "algorithm" that was defined clearly enough so that the question of "algorithmic solvability" could be raised was the notion of a geometric construction by ruler and compass, formulated by the Greek geometers. The mathematical interest of this notion is that there are both solvable and unsolvable construction problems. The design of construction algorithms has been stimulating in geometry for a long time, and has contributed to the development of important tools (very useful also independently of construction problems) like inversion or the golden ratio. But the proof of *unsolvability* of basic construction problems (trisecting an angle, squaring a circle, doubling a cube, constructing a regular heptagon etc.) illustrates this effect more dramatically. Such negative results were inaccessible to Greek mathematics and for quite a while later on; it required the notion of real numbers and a substantial part of modern algebra to prove them, or even to formulate them with an exactness that made them accessible to mathematical methods. In fact, modern algebra was inspired by the desire to prove such negative results (besides the non-constructability of certain configurations, the non-solvability of equations of degree at least 5 was of the same nature).

As another example, let us consider the notion of primes. These numbers were also studied by the ancient Greeks, and they proved several of their basic properties. The beautiful but very hard theory of prime numbers has been a major branch of mathematics (and a source of inspiration for many other branches) throughout the history of modern mathematics too. But only the development of computers, and even more the establishment of computational complexity theory, raised the fundamental algorithmic problem: *how to test whether a given number is prime? how to find the prime factorization if it is not?* (Mathematicians in the 18–th and 19–th century, in particular Gauss, did make extensive computations regarding primes, and developed ingenious tricks to help their work. But apparently they did not consider their algorithms as mathematical results.)

These questions have profound applications in computing, and the simple elementary procedures to solve them are far from satisfactory (practice requires the consideration of numbers up to several hundred digits). Over the last 10 years or so, more and more advanced methods from number theory have been applied to design more and more sophisticated and efficient algorithms to answer these questions.

Note that the interaction of mathematics and algorithms in these two examples is different: in the first, we have a certain notion of an "algorithm" (a construction procedure), and we want to prove that it does not suffice to solve certain problems. Such questions may be notoriously hard, and, as our example shows, the proof may require deep mathematics or even the development of entirely new fields. The theory of computing today is full of unsolved problems of a similar nature; there are very few methods to prove negative results about the algorithmic solvability of problems, in particular if limits are imposed on the time (or other resources) that the algorithm can use. In such cases, computer science is an "external" user of mathematics: it supplies hard problems, which need to be modelled and solved just like hard problems in mechanics or astronomy.

In the second example, computer science penetrates classical mathematics, putting old questions in a new perspective. Often, this is done by requiring *constructions* where "pure" existence proofs have been supplied by the classical theories. In other cases, one requires *efficient* procedures where "theoretically finite" case-analyses have been at hand (like in prime testing). In some branches of mathematics, e.g. in graph theory, this process has provided a whole new framework for the field (cf. Lovász 1986). In my talk, I would like to focus on this second development.

This new perspective on several issues in classical mathematics is, of course, a challenge to mathematics education. The introduction of computers (at whatever level) is only a very partial answer. I will give some remarks about how I think this challenge can be met; but I believe that it will take much more work — experimental and theoretical — before the contours of the answer will be clear.

1. Some old results from a new perspective

The approximation of irrational numbers by rational ones has a long history. Suppose that we have a real number α, that we want to approximate by a rational number with a small denominator. The first, trivial approach is to round: take any positive integer q, and round the product $q\alpha$ to the nearest integer p. Then p/q is a reasonably good rational approximation of α; in fact, $|\alpha - \frac{p}{q}| \leq \frac{1}{2q}$.

Can we do better, with some positive integer Q in place of the 2 here? More precisely, can we find, for a given real number α and positive integer Q, a rational number p/q such that $|\alpha - \frac{p}{q}| \leq \frac{1}{Qq}$? It is clear that in this case q cannot be chosen arbitrarily any more; but how large does it have to be?

A theorem of Dirichlet states that *for every given real number α and positive integer Q, we can find integers p and q such that $0 < q \leq Q$ and $|\alpha - \frac{p}{q}| \leq \frac{1}{Qq}$.* In other words, $|q\alpha - p| \leq \frac{1}{Q}$.

There are two basic proofs of this fundamental result, and I would like to discuss them here to illustrate the difference between the algorithmic and non-algorithmic approaches.

First proof: Consider the $Q + 1$ numbers

$$0\alpha - [0\alpha], \quad 1\alpha - [1\alpha], \quad ,\ldots, \quad Q\alpha - [Q\alpha]$$

These numbers all lie in the interval $[0, 1)$, so some two of them, say $k\alpha - [k\alpha]$ and $l\alpha - [l\alpha]$ $(0 \leq k < l \leq Q)$ are closer to each other than $1/Q$. Let $q = l - k$ and $p = [l\alpha] - [k\alpha]$. Then $q \leq Q$ and

$$|q\alpha - p| = |(l-k)\alpha - ([l\alpha] - [k\alpha])| = |(k\alpha - [k\alpha]) - (l\alpha - [l\alpha])| < \frac{1}{Q}.$$

So the rational number p/q proves Dirichlet's Theorem.

Second proof (sketch): It is well known that every real number α can be expressed as a *continued fraction*

$$\alpha = a_0 + \cfrac{1}{a_1 + \cfrac{1}{a_2 + \cfrac{1}{a_3 + \dots}}}.$$

This expansion may be finite (if α is rational) or infinite (if α is irrational). For example, we have the expansions

$$\frac{11}{8} = 1 + \frac{3}{8} = 1 + \cfrac{1}{2 + \cfrac{2}{3}} = 1 + \cfrac{1}{2 + \cfrac{1}{1 + \cfrac{1}{2}}}$$

and

$$\sqrt{2} = 1 + (\sqrt{2} - 1) = 1 + \cfrac{1}{\sqrt{2} + 1} = 1 + \cfrac{1}{2 + (\sqrt{2} - 1)} = 1 + \cfrac{1}{2 + \cfrac{1}{2 + \cfrac{1}{2 + \dots}}}.$$

Now if we stop in a continued fraction expansion at the k-th step, we get a rational number

$$\frac{p_k}{q_k} = a_0 + \cfrac{1}{a_1 + \cfrac{1}{a_2 + \dots + \cfrac{1}{a_k}}}.$$

There are a number of basic facts that can be proved about this rational number, called the k-th *convergent* of α. What we need here is the fact that p_k/q_k is a very good approximation of α:

$$\left| \alpha - \frac{p_k}{q_k} \right| \leq \frac{1}{q_k q_{k+1}}.$$

Hence, if we let k be the largest subscript for which $q_k \leq Q$ (it is known that q_k tends to infinity, in fact exponentially fast), then

$$\left| \alpha - \frac{p_k}{q_k} \right| \leq \frac{1}{q_k Q},$$

i.e. the rational number p_k/q_k satisfies Dirichlet's Theorem.

Which of these proofs is "better"? There is little doubt that the first one is much simpler: not only is it shorter but it is self-contained, while the second uses quite a bit from the theory of continued fractions.

But suppose that we also want to *find* the rational number in question. Before going into a discussion of this, we have to clarify one thing: how is α given? Since we want to have an algorithm now, we had better assume that α is represented in some finite explicit form; let us assume that it is rational, say $\alpha = a/b$. Of course, we have to assume then that $Q < b$, or else the approximation problem is trivial.

Which algorithm can we derive from the first proof? It tells us that we should form all the numbers $k\alpha - [k\alpha]$, then find two, which are close, then ... But if we want to form all these numbers anyway, we can check with the same amount of work which of them is less that $1/Q$. *So the first proof of the theorem does not provide us with any non-trivial algorithm to find the rational number whose existence it certifies;* it is a "pure existence proof" in this sense.

(It could be argued that the structural insight gained from the first proof can be developed further to obtain an efficient algorithm. This, however, takes further work and deeper insight.)

Is the second proof better from this point of view? How much work does it take to compute the continued fraction expansion? We show that not only is it easy to obtain this expansion, but also that there is nothing mystical about its use in an approximation problem.

Suppose that we want to find a good rational approximations of a number α. As a first approach, we could use the rational number $a_0/1$, where $a_0 = [\alpha]$. To obtain a better approximation, we have to approximate the error, i.e., $x_1 = \alpha - a_0$. Since this number is less than 1, it is a natural idea to replace it with its reciprocal, and approximate this by its integer part $a_1 = [1/x_1]$. Now this approximation again has an error $x_2 = (1/x_1) - a_1$. Take the reciprocal again etc. The positive integers $a_0, a_1, \ldots$ obtained this way are just the coefficients in the continued fraction expansion of α.

If $\alpha = a/b$ then we obtain a_0 by dividing a by b with remainder; then a_0 is the quotient and if r is the remainder then the error $x_1 = r/b$. So $1/x_1$ is the rational number b/r and we repeat the procedure with this in place of a/b. It is clear that this is a finite procedure, and so the continued fraction expansion of α can be obtained. The rest of the proof gives a simple recipe to obtain the approximating rational number.

(Some of my readers may have observed at this point that to obtain the continued fraction expansion of a rational number a/b, we carry out exactly the same arithmetic operations as in the euclidean algorithm used to compute the greatest common divisor of a and b.)

So the second proof is just the analysis of a very natural iterative algorithm to find better and better approximations of a number.

But is this algorithm any better than the trivial one derived from the first proof? The answer is: much better! The number of steps in the continued fraction expansion of a/b (equivalently, the number of steps in the euclidean algorithm to compute g.c.d.(a, b)), is proportional to the number of digits of a and b; the number of steps in the first trivial algorithm is proportional to Q, which may be as large as b itself. If a, b and Q are 100 digit numbers, then the first algorithm takes 10^{100} steps while the second, less than 500.

The usual way to measure the running time of an algorithm is to compare the number of bit-operations with the number of bits necessary to write down the input. So a k-bit integer (having k bits in its base 2 expansion) contributes k to the input; in the diophantine approximation problem, the size of the input is the number of bits in a, b and Q, which is essentially $\log_2 a + \log_2 b + \log_2 Q$. In this model of computation, the length of the numbers also influences the time spent on a single arithmetic operation. For example, multiplication of two k-bit integers by the method taught at school takes about k^2 bit-operations, so such an operation contributes k^2 to the running time etc.

An important distinction to make is whether the running time grows as a polynomial of the input length or faster. Polynomial algorithms tend to be mathematically interesting and usually — though not always — practically feasible. Brute force case-distinctions often lead to exponentially many cases to distinguish (all subsets of a set, etc.) and thereby to exponentially growing running times. The algorithm derived from the first proof needs exponential time; the algorithm derived from the second is polynomial.

So we see that:

– *The first proof is an existence proof; it is elegant, short, but does not give an algorithm to find the approximation. It takes more work and further ideas to develop it into an efficient algorithm.*

– *The second proof is just an analysis of an elegant, natural and efficient algorithm to construct good approximations. It takes work to analyse how good these approximations are.*

The first proof also generalizes easily to the problem of simultaneously approximating several numbers by rationals with a common denominator. The algorithmic proof (continued fractions) also has some generalizations to this case, but those are substantially less elegant and do not yield as good approximations as the generalization of the first proof. To design a polynomial time algorithm for this simultaneous diophantine approximation problem that would *find* the approximating rational numbers whose *existence* is guaranteed by the almost straightforward extension of the first proof is unsettled!

2. A glimpse of complexity theory

Most of us have met problems sounding like "Characterize those sets (numbers,...) with the property that ∗∗∗.", and also the student who gives, provocatively, the answer "A set has property ∗ ∗ ∗ if and only it has property ∗ ∗ ∗." What is wrong with this answer? And what happens if the student hides the triviality by slightly re-phrasing the property ∗ ∗ ∗? When does the answer begin to be non-trivial and thereby acceptable? Or should we simply ban this kind of problem as meaningless?

One of the great successes of the theory of computing is that *it is able to define in a mathematically exact way which characterizations are "good" and which are more-or-less just rephasing the question,* at least for a large class of structures and properties. There is no room in this talk, of course, to develop this theory. But I will try to illustrate the idea by an important example.

Consider a system of inequalities

$$
\begin{aligned}
a_{11}x_1 + a_{12}x_2 + \cdots + a_{1n}x_n &\le b_1, \\
a_{21}x_1 + a_{22}x_2 + \cdots + a_{2n}x_n &\le b_2, \\
&\ \vdots \\
a_{m1}x_1 + a_{m2}x_2 + \cdots + a_{mn}x_n &\le b_m.
\end{aligned}
\tag{1}
$$

Since we are interested in algorithmic aspects, we assume that the inequalities have rational coefficients.

When does (1) have a solution? Let us investigate the following two answers to this question:

Theorem A. *The system of inequalities (1) has a solution if and only if the system*

$$
\begin{aligned}
a_{11}(u_1 - v_1) + a_{12}(u_2 - v_2) + \cdots + a_{1n}(u_n - v_n) &\le b_1, \\
a_{21}(u_1 - v_1) + a_{22}(u_2 - v_2) + \cdots + a_{2n}(u_n - v_n) &\le b_2, \\
&\ \vdots \\
a_{m1}(u_1 - v_1) + a_{m2}(u_2 - v_2) + \cdots + a_{mn}(u_n - v_n) &\le b_m
\end{aligned}
\tag{2}
$$

has a non-negative solution.

Theorem B. *The system of inequalities (1) has a solution if and only if the system*

$$
\begin{aligned}
a_{11}y_1 + a_{21}y_2 + \cdots + a_{m1}y_m &= 0, \\
a_{12}y_1 + a_{22}y_2 + \cdots + a_{m2}y_m &= 0, \\
&\ \vdots \\
a_{1n}y_1 + a_{2n}y_2 + \cdots + a_{mn}y_m &= 0, \\
b_1 y_1 + b_2 y_2 + \cdots + b_m y_m &= -1
\end{aligned}
\tag{3}
$$

has no non-negative solution.

(Note that a solution of (3) can be viewed as a linear combination of the inequalities in (1) with non-negative multipliers — the y_i — that yields the trivially non-solvable inequality $0x_1 + \ldots + 0x_n \le -1$.)

Both theorems give necessary and sufficient conditions for the solvability of (1). But Theorem A is an essentially trivial trick to show that solvability is easily transformed into non-negative solvability, while Theorem B is an important result (called Farkas' Lemma). What makes the difference?

Assume that I want to use a concrete case of (1) in this talk as an illustration of a system of solvable linear inequalities. How can I convince you that it is indeed solvable? Easy: I just show you a solution. Suppose now that I want to use another concrete case of (1) to illustrate an unsolvable system. How can I convince you that it is indeed unsolvable? To try all possible values for the variables? There is no easy way at hand.

Do the necessary and sufficient conditions formulated in the two theorems above help? If we apply Theorem A, I have to exhibit that (2) *does not* have a non-negative solution — this is not any easier than the original task. But if we apply Theorem B, it suffices to exhibit that (3) *does* have a non-negative solution — and this I can do by showing you one. So the condition in Theorem B does have an entirely different logical structure from the original property and from the condition given in Theorem A.

It turns out that a large part of graph theory, optimization, number theory etc. has an analogous structure. Important properties of graphs, numbers etc. have the feature that if they are present, there is an easy way to exhibit this (e.g. composite numbers, 4-colorable graphs etc.) Basically, these properties are defined in terms of the *existence* of a certain object (a number is composite if it has a proper divisor; a graph is 4-colorable if it has a proper coloring with 4 colors etc.) If I want to convince you that the number

$$6175321760117257427110640691143977911706681098866281$$

is composite, all I have to do is to show you the divisor

$$7858321551080267055879091$$

(Of course you still have to verify that this is a divisor; but, as we know, this takes only polynomial time.)

Such properties are called NP-*properties* (named after a technical definition involving *N*on-deterministic *P*olynomial-time Turing-machines, whose details I don't have to give here). Sometimes (but not always!) the negation of an NP-property is again an NP-property. Theorems establishing such equivalences are often among the most important results in the field (like the Farkas Lemma above). They are sometimes called *good characterizations*.

This classification of properties is also closely related to algorithms. Most properties for which we would like to find a polynomial-time algorithm to decide them belong to NP in a natural way. If a property can be decided by a polynomial-time algorithm, then both the property and its negation are NP-properties, i.e., it has a good characterization. The converse may not be true: it is not known whether every well-characterized property can be decided in polynomial time (probably not, but as I remarked before, we do not have the means to prove such negative results in this area). But, usually, to find a good characterization is an important step towards the construction of a polynomial time algorithm. The example of the

Farkas Lemma is very illustrative: almost a century after the proof of the lemma, a polynomial time algorithm to decide if (1) has a solution was given by Khachiyan in 1978 (the famous Ellipsoid Method).

Of course, this article cannot discuss the theory of algorithms in any reasonable detail; we only sketched what was necessary to support our arguments on a possible new framework for various fields in mathematics. (For an introduction to the theory of algorithms, see e.g. Sedgewick 1983.)

3. What does this imply for mathematics education?

Whatever it implies, should be regarded with the most caution and moderation. I feel that math education should follow what happens in math research, at least to a certain extent, in particular those (rare) developments there that fundamentally change the whole framework of the subject. Algorithmic mathematics is one of these. However, the range of the penetration of an algorithmic perspective in classical mathematics is not yet clear at all, and varies very much from subject to subject (as well as from lecturer to lecturer). Graph theory and optimization, for example, have been thoroughly re-worked from a computational complexity point of view; number theory and parts of algebra are studied from such an aspect, but many basic questions are unresolved; in analysis and differential equations, such an approach may or may not be a great success; set theory does not appear to have much to do with algorithms at all.

Our experience with "New Math", the adaptation of the set-theoretic foundations of mathematics to lower level mathematics education, warns us that drastic changes may be disastrous — even if the new framework is well established in research and college mathematics. So let me just raise some ideas on the teaching of algorithms at various levels, emphasizing that they must be carefully discussed and tried out before any large scale implementation is attempted.

Basically, some algorithms and their analysis could be taught at about the same time when theorems and their proofs first occur, perhaps around the age of 14. Of course, certain algorithms (for multiplication and division etc.) occur quite early in the curriculum. But these are more recipes than algorithms; no correctness proofs are given (naturally), and the efficiency is not analysed. The children have to learn (and practice) how to carry out these simple algorithms. This is like teaching theorems (axioms) without proofs, or teaching empirical facts in the sciences without experiments: necessary but not leading to really deep understanding.

What I would consider as the beginning of learning "algorithmics" is to learn to **design**, *rather than execute, algorithms.* (For an elaboration of this idea, see e.g. Maurer 1984). The euclidean algorithm, for example, is one that can be "discovered" by students in class. In time, a collection of "algorithm design problems" will arise (just as there are large collections of problems and exercises in algebraic identities, geometric constructions or elementary proofs in geometry). Along with these concrete algorithms, the students should become familiar with basic notions of the theory of algorithms: input–output, correctness and its proof, analysis of running time and space, good characterizations read off from algorithms, algorithms motivated by good characterizations etc.

Some possible types of algorithm-design problems, suitable probably already for the high school level: enumeration problems where no closed formula exists; elementary optimization problems in graph theory (e.g. maximum independent sets in trees, shortest paths, listing of cliques, circuits etc.); sorting and searching; simple (though inefficient) methods for primality testing, factorization, and many other problems in number theory; Gaussian elimination and other manipulations in linear algebra; convex hull and other elementary plane geometry constructions.

In college, the shift to a more algorithmic presentation of the material should, and will, be easier and faster. Already now, some subjects like graph theory are taught in many colleges quite algorithmically: shortest spanning tree, maximum flow and maximum matching algorithms are standard topics in most graph theory courses. This is quite natural since, as I have remarked, computational complexity theory provides a unifying framework for many of the basic graph-theoretic results. In other fields this is not quite so at the moment; but some topics like primality testing or cryptographic protocols provide nice applications for a large part of classical number theory.

4. Computers and algorithms

At this point, I have to comment on the use of computers in the teaching of these topics. One should distinguish between an algorithm and its implementation as a computer program. The algorithm itself is a mathematical object; the program depends on the machine and/or on the programming language. It is of course necessary that the students see how an algorithm leads to a program that runs on a computer; but it is not necessary that every algorithm they learn about or they design be implemented. The situation is (again) analoguous to that of geometric constructions with ruler and compass: some constructions have to be carried out on paper, but for some more, it may be enough to give the mathematical solution (since the point is not to learn to draw but to provide a field of applications for a variety of geometric notions and results).

Let me insert a warning about the shortcomings of algorithmic language. There is no generally accepted form of presenting an algorithm, even in the research literature (and as far as I see, computer science text books for secondary schools are even less standardized and often even more extravagant in handling this problem.) The practice ranges from an entirely informal description to programs in specific programming languages. There are good arguments in favor of both solutions; I am leaning towards informality, since I feel that implementation details often cover up the mathematical essence. Let me illustrate this by two examples.

An algorithm may contain a step "Select any element of set S". In an implementation, we have to specify which element to choose, so this step necessarily becomes something like "Select the first element of set S". But there may be another algorithm, where it is important the we select the first element; turning both into programs hides this important detail. (Also, it may turn out that there is some advantage in selecting the *last* element of S. Giving an informal description leaves this option open, while turning the algorithm into a program forbids it.)

To give my second example, recall that the Fibonacci numbers are defined by the recurrence

$$\dot{F}_{k+1} = F_k + F_{k-1}$$

(and by $F_0 = 0$ and $F_1 = 1$). This recurrence provides a trivial algorithm to compute these numbers. Turning this recurrence into a program, we would get something like this: if F is the current Fibonacci number, and G is the previous, then (with an auxiliary variable H)

$$H := F, \quad F := G + F, \quad G := H.$$

We see that the program contains a number of details that do not belong *mathematically* to the procedure of computing Fibonacci numbers: it stores F_{k+1} in the same register where F_k used to sit, but to do so, it has to "salvage" F_k since its value will be needed in the next step, and to do so, we need an "auxiliary variable" etc.*

To show the other side of the coin, the main problem with the informal presentation of algorithms is that the "running time" or "number of steps" is difficult to define — as we have experienced above. Unfortunately, this depends on the details of implementation (down to a level below the programming language; it depends on the data representation and data structures used). Sometimes there is a way out by specifying which steps are counted (e.g. comparisons in a sorting algorithm, or arithmetic operations in an algebraic algorithm); but this is "cheating" in a sense since we disregard the time needed to handle the data, which may be as much as, or even more than, the time used by the "mathematically essential" steps. It should be mentioned that the polynomiality of an algorithm is "robust", i.e., it does not depend on implementation (although different implementations may have different polynomials in the bound on their running time).

The route from the mathematical idea of an algorithm to a computer program is long. It takes the careful design of the algorithm; analysis and improvements of running time and space requirements; selection of (sometimes mathematically very involved) data structures; and programming. In college, to follow this route is very instructive for the students. But even in secondary school mathematics, at least the mathematics and implementation of an algorithm should be distinguished.

An important task for mathematics educators of the near future (both in college and high school) is to develop a smooth and unified style of describing and analysing algorithms. A style that shows the mathematical ideas behind the design;

---

* I am grateful to Jack Edmonds for these examples and arguments — and regret not to have given a more detailed exposition of them.

that facilitates analysis; that is concise and elegant would also be of great help in overcoming the contempt for algorithms that is felt nowadays both from the side of the teacher and of the student.

References

P. R. Halmos (1981), Applied mathematics is bad mathematics, in *Mathematics Tomorrow* (ed. L. A. Steen), Springer, 9–20.

L. Lovász (1986), *An Algorithmic Theory of Numbers, Graphs, and Convexity*, CBMS-NSF Reg. Conf. Series in Appl. Math. **50**; SIAM.

S. Maurer (1984), Two meanings of algorithmic mathematics, *Mathematics Teacher* 430–435.

S. Maurer (1985), The algorithmic way of life is best; reflexions by R. G. Douglas, B. Korte, P. Hilton, P. Renz, C. Smorynski, J. M. Hammersley and P. R. Halmos; *College Math. Journal* **16**, 2–18.

R. Sedgewick (1983): *Algorithms*, Addison–Wesley.

L'ADRESSE DU PRÉSIDENT:

LA GRANDE FIGURE DE GEORGES PÓLYA

Jean-Pierre Kahane (France)

György Pólya est né à Budapest le 13 décembre 1887, il y a juste un peu plus de cent ans. Il est mort à Stanford en septembre 1985, il y a moins de trois ans. Entre ces deux dates, une vie extraordinairement remplie et une oeuvre monumentale.

Il est bien naturel d'évoquer cette vie et cette oeuvre ici, à Budapest, à la fin de ce sixième congrès international de l'enseignement mathématique. D'abord parce que Pólya est l'un des fleurons de l'école mathématique hongroise, qui a tant apporté au monde au cours du vingtième siècle. Ensuite parce que Pólya s'est intéressé de près, et profondément, à l'enseignement des mathématiques et que, dans le grand public, il est plus célèbre encore par son oeuvre pédagogique que par son oeuvre mathématique.

Tous ceux qui sont ici connaissent, directement ou indirectement, une partie de l'oeuvre de Pólya. Certains l'ont connu personnellement. Beaucoup se souviennent de l'hommage qui lui a été rendu au second congrès international de l'enseignement mathématique à Exeter en 1972. En 1980, à Berkeley, il était président d'honneur du quatrième congrès. Il y a longtemps que Pólya est une figure légendaire, et le voisinage de sa mort et de son centenaire a encore avivé la légende, et produit de nouveaux documents le concernant. Le plus complet vient d'être publié au Bulletin of the London Mathematical Society (1987). Il contient une excellente notice nécrologique, par Gerald L. Alexanderson et Lester H. Lange, des contributions de Kai Lai Chung, Ralph D. Boas, D. H. Lehmer, Doris Schattschneider, Ronald C. Read, Menahem M. Schiffer, et Alan H. Schoenfeld sur les différents aspects de l'oeuvre de Pólya, et la liste de ses travaux la plus complète qui existe aujourd'hui. L'essentiel de ces travaux est d'ailleurs largement accessible. Ses livres figurent dans toutes les bibliothèques. Le plus populaire d'entre eux, *How to solve it* (1945) a été publié à plus d'un million d'exemplaires, et dans dix sept langues. Ses principaux articles sont regroupés dans les quatre volumes des *Collected Papers* (1974–1984). Pólya est un merveilleux écrivain scientifique, qui connaît tous les registres de l'écriture et de la langue: il sait écrire de façon concentrée pour le spécialiste, et sait prendre tout son temps quand il le faut ; il peut avoir la limpidité du cristal, et aussi bien orner son argumentation de broderies chatoyantes et spirituelles ; il connaît — je suppose — toutes les ressources du hongrois, de l'allemand qu'il a surtout pratiqué jusqu'en 1940, et de l'anglais qui a été sa langue d'adoption depuis. Ce qu'il a écrit en français, dans la revue suisse l'Enseignement mathématique (par parenthèse, alors comme aujourd'hui organe officiel de la Commission Internationale de l'Enseignement Mathématique) ou dans les Comptes Rendus de l'Académie des Sciences, est une série de perles: sous forme à la fois claire et concentrée, on y trouve la substance de beaucoup d'écrits postérieurs plus développés.

Ainsi beaucoup d'entre vous, dans leur langue maternelle, ont pu avoir accès à la pensée de Pólya. A priori, cela rend ma tâche difficile. Si vous connaissez déjà Pólya, que vous apprendrai-je ? En vérité, la matière est si riche qu'en visitant ou revisitant l'oeuvre de Pólya, on découvre toujours du nouveau. Si je ne vous apprends rien, ce sera donc entièrement de ma faute.

Pour minimiser le risque, je vous parlerai d'abord de sa vie et de ses grandes étapes — Budapest, Zürich, Stanford — ; de sa personnalité et de son amitié de toute une vie, avec Gábor Szegő ; des influences qu'il a subies et de celle qu'il a exercées ; de ses intérêts et des orientations de sa production scientifique. Puis je vous proposerai un petit dictionnaire de termes mathématiques auxquels le nom de Pólya est attaché : courbe de Pólya, fonctions de Pólya, indicatrice de Pólya, densité de Pólya, fonctions de Pólya–Schur, dimension de Pólya–Szegő, promenade de Pólya, théorie de Pólya, sans compter tous les théorèmes qui portent son nom. Et pour terminer j'illustrerai et je commenterai quelques unes de ses idées sur l'heuristique, le raisonnement plausible, et en général sur l'enseignement mathématique.

- - - - * — ** — * * * — ** — * - - - -

Dès son enfance, György Pólya a baigné dans une ambiance de ferveur intellectuelle. Son père était juriste, employé par une compagnie d'assurances ; mais il était aussi économiste, écrivain, traducteur, linguiste. Sa traduction de Wealth of Nations, d'Adam Smith, était un classique. Son frère aîné était intéressé aux mathématiques, mais il devint chirurgien et professeur à l'Université. Sa mère était d'une vieille famille de Buda, héritière d'une tradition de grande culture.

Pólya avait dix ans quand son père mourut, laissant cinq orphelins. Ses soeurs, plus âgées, se mirent à travailler dans la même compagnie d'assurances, tandis que lui apprenait au lycée les langues vivantes et anciennes. Plus tard, il lui arrivait de citer Homère, Virgile ou Dante dans le texte. Au lycée même, il traduisait en hongrois la poésie de Heine. Sa mère le poussa vers la profession du père et, après le lycée, il s'inscrivit à l'Université pour étudier le droit. Ses études de droit ne durèrent qu'un semestre ; le droit, pour lui, était ennuyeux, alors qu'il existait tant d'autres choses intéressantes. Il se tourna donc vers des études littéraires et, au bout de deux ans, il obtint une licence d'enseignement en latin et hongrois. Il poursuivit avec la philosophie, et, comme d'usage, eut à prendre à cette occasion des cours de physique et de mathématiques. Le professeur de physique était le célèbre Loránd Eötvös, et le professeur de mathématiques le jeune Lipót Fejér. Il tomba sous le charme de Fejér, et devint mathématicien.

Il s'en expliqua plus tard de diverses manières. A l'âge de quatre vingt dix ans, il avait la passion des bons mots, et dans un entretien avec Alexanderson il ne résiste pas à la tentation: "je ne me suis pas trouvé assez bon pour la physique, et trop pour le philosophie: entre les deux, il y avait les mathématiques".Mais à d'autres occasions (1961, 1969), quand il s'interroge sur la raison pour laquelle la Hongrie, au vingtième siècle, a donné au monde tant de mathématiciens de premier plan, il met en évidence des facteurs moins personnels et de plus grande portée: la qualité de l'enseignement des mathématiques au lycée, son lien avec la recherche mathématique par l'intermédiaire du Matematikai Lapok (le journal hongrois pour les bons élèves des lycées, où les problèmes étaient posés par les meilleurs mathématiciens), et le rôle personnel de Fejér, qui avait une passion pour l'exposition, la communication et l'enseignement dont Pólya allait hériter.

Il est remarquable que, pour Pólya, l'engagement vers les mathématiques soit allé de pair avec l'ouverture à tous les problèmes du monde. A l'Université, il fut l'un des fondateurs du cercle Galileo Galilei, dont le nom seul était un symbole révolutionnaire sous l'empire des Habsbourg. A l'époque d'ailleurs, Pólya s'intéressait à Ernst Mach — le représentant le plus illustre de l'empiriocriticisme — bien plus qu'à son contradicteur, un nommé Vladimir Ilich Lénine. Il garda toute sa vie durant un intérêt très vif pour les procédures démocratiques: la représentation proportionnelle, dont il donne une excellente théorie mathématique (1919), ou le mode d'élection du président des Etats-Unis, qu'il soumet à une critique percutante (1961). Durant la guerre de 1914 — il était déjà installé à Zürich — il rallia le pacifisme de Russell, et le souci de la paix mondiale est l'un des derniers qu'il ait exprimé, en signant, juste avant sa mort, un appel international de mathématiciens pour le gel des armements nucléaires — dont, par parenthèse, le professeur Á. Császár et moi-même sommes aussi signataires.

Désormais — nous sommes en 1910 — György Pólya est mathématicien. Il passe une année à Vienne, en revient pour passer sa thèse à Budapest — sur l'interprétation des probabilités comme volumes ou rapports de volumes à n dimensions —, il visite longuement Göttingen (1912–13), où se trouvent alors Klein, Hilbert, Runge, Landau, Weyl, Hecke, Courant et Toeplitz, puis Paris (1914), et il s'installe, sur l'invitation de Hurwitz, à l'Ecole Polytechnique de Zürich (Eidgenössiche Technische Hochschule). Hurwitz devait mourir en 1919, et c'est Pólya qui édita ses oeuvres. Parmi les mathématiciens qui l'ont influencé, Pólya cite Hurwitz en première place. En particulier, c'est à Hurwitz que sont dus plusieurs exercices du fameux livre de Pólya et Szegő: *Aufgaben und Lehrsätze aus der Analysis.*

Il est temps de parler de Gábor Szegő. Il a sept ans de moins que Pólya. Quand Pólya le rencontre pour la première fois, à Budapest, Szegő vient de remporter le concours Eötvös, et Pólya vient de passer son doctorat. Dorénavant leurs vies sont inséparables. Le premier article de Szegő est la solution d'un problème de Pólya. Puis il travaillent ensemble à ce qui allait devenir la bible de plusieurs générations d'analystes, les fameux exercices de Pólya–Szegő, publiés en 1925. Ce fut toujours le livre chéri de Pólya, et il vaut la peine d'écouter ce qu'il en dit en préface à l'édition des Oeuvres de Szegő, il y a quelques années:

"It was a wonderful time ; we worked with enthusiasm and concentration. We had similar backgrounds. We were both influenced, like all other young Hungarian mathematicians of that time, by Leopold Fejér. We were both readers of the same well directed Hungarian Mathematical Journal for high school students that stressed problem solving. We were interested in the same kind of questions, in the same topics ; but one of us knew more about one topic and the other more about some other topic. It was a fine collaboration. The book PSz, the result of our cooperation, is my best work and also the best work of Gábor Szegő."

La formule est paradoxale: ni l'oeuvre multiforme de Pólya, ni l'influence profonde de Szegő sur l'analyse contemporaine ne se réduisent au Pólya–Szegő. Mais il est vrai que c'est un livre exemplaire, à la fois une somme des connaissances de l'époque et un traitement profondément original sous forme d'un enchaînement de problèmes, de sorte que c'est aujourd'hui, et pour tout l'avenir prévisible, un grand classique des mathématiques.

La collaboration avec Szegő ne s'arrête pas là. Tandis que Pólya poursuit sa carrière à Zürich, Szegő s'installe en Allemagne, à Berlin, puis Königsberg. Mais les persécutions raciales font fuir Szegő, qui émigre en Amérique et trouve un poste à Stanford. En 1940, Pólya à son tour fuit l'Europe de la guerre et du fascisme. Szegő, qui dirige alors le département de mathématiques de Stanford, l'invite à le rejoindre. C'est le début d'une nouvelle carrière pour Pólya. La collaboration avec Szegő s'était exprimée dans plusieurs articles. Elle aboutit, à Stanford, à un nouveau livre, *Isoperimetric inequalities in mathematical physics*, publié en 1951. Leur amitié est légendaire. A Stanford, même après leur retraite, on va rendre visite aux Pólya et aux Szegő. Szegő meurt le premier, le 7 août 1985, et Pólya un mois après. Reste à Stanford Madame Stella Pólya, la compagne de toute sa vie, à laquelle j'adresse d'ici et au nom du Congrès mes respectueuses et cordiales amitiés.

Je n'ai pas encore évoqué les liens de Pólya avec l'Angleterre. En 1924, il fut invité par Hardy et passa l'année à Oxford puis Cambridge. C'est là qu'il connut Littlewood et Ingham, et que fut amorcée la collaboration entre Hardy, Littlewood et Pólya qui devait mener à un autre beau livre, *Inequalities*, publié en 1934. La personnalité de Hardy était exceptionnelle et attachante. Hardy était le mathématicien pur par excellence, et aussi par réaction à l'égard de l'hégémonie traditionnelle des mathématiques appliquées en Angleterre. Pólya était sensible au charme de Hardy et il était aussi un mathématicien pur — recevant la plus grande part de son inspiration des mathématiques déjà constituées, et refusant toute autre contrainte que celles qu'il s'imposait lui-même. Mais Pólya, au contraire de Hardy, a été un mathématicien pur ouvert aux applications, et sollicitant l'inspiration du monde extérieur.

J'en prendrai seulement trois exemples — outre ce que j'ai indiqué de ses préoccupations politiques — : la promenade au hasard, la structure des molécules organiques, et la vibration des membranes.

La promenade au hasard (Irrfahrt, random walk) est une notion introduite par Pólya, sur laquelle je reviendrai dans un instant. Selon Pólya, l'idée lui en est venue après avoir rencontré inopinément, à deux reprises, l'un de ses étudiants en galante compagnie. Il se trouvait un peu dans l'embarras. Mais au lieu de s'en tenir là, il a bâti un modèle de rencontres au hasard qui est à la base de tout ce qu'on appelle aujourd'hui récurrence ou transience.

A Zürich, Pólya fréquentait ses collègues des autres disciplines et enseignait les mathématiques à des étudiants très divers: étudiants en chimie, en architecture, en ingénierie. En chimie organique, la question des isomères — c'est-à-dire des molécules constituées des mêmes atomes, mais dans différents agencements — est de nature mathématique, mais Pólya est le premier à en proposer un modèle et à fournir les solutions. C'est le point de départ de ce qu'on appelle aujourd'hui en combinatoire la *théorie de Pólya*.

A Stanford, avec Szegő, il se tourne à nouveau vers les problèmes du monde physique, et sur les rapports entre physique et géométrie. Il s'agit, sur la vibration des membranes, la résistance à la torsion, et d'autres questions, "d'estimer les quantités physiques à partir des données géométriques, d'estimer ce qui est le moins accessible à partir de ce qui l'est le plus" (je cite la préface de leur ouvrage: *Isoperimetric inequalities*). L'analyse mathématique, sur des modèles physiques déjà élaborés, apparaît là comme un moyen privilégié pour pénétrer dans la nature des choses.

Ainsi, dans l'infinie complexité du monde physique, Pólya ouvre des pistes (la promenade au hasard, la combinatoire des configurations) et creuse des sillons (les problèmes extrémaux et les inégalités du type isomérimétrique).

Mais son ouverture au monde extérieur ne s'arrête pas au monde physique. Pólya, nous l'avons vu, a eu une vocation d'enseignant avant même de devenir mathématicien. Les mathématiques sont inséparablement pour lui objet de découvertes et objet d'enseignement. A Zürich, l'enseignement aux élèves ingénieurs l'amène à une réflexion et à des méthodes pédagogiques qu'il élaborera ensuite sous la forme du *raisonnement plausible* ; en un mot, comment convaincre sans prouver ? Mais à ce propos, comment le mathématicien se convainc-t-il lui-même de la vérité d'une proposition, avant de la démontrer ? Quel est le rôle de l'observation, de la conjecture, de la vérification, dans la découverte mathématique? C'est tout le champ de l'heuristique, à laquelle il se consacrera à

Stanford et dont il reconnaît la paternité au grand Euler. Enfin, quel est le propos même de l'enseignement ? Selon Pólya, c'est d'apprendre à l'élève à penser par lui-même. Plutôt que d'exposer un sujet, il vaut donc mieux le décomposer en questions que l'élève ait lui-même à résoudre. Ces questions sont les *problèmes à résoudre*, et les *problèmes à démontrer*. Cette décomposition en problèmes, c'est le projet même des *Aufgaben* de Pólya–Szegő. Mais c'est aussi, selon Pólya, le programme de Descartes dans les *Regulae ad directionem ingenii*, qu'il appelle un classique sur la logique de la découverte. Et c'est déjà le projet d'Euclide, dans la mesure où les théorèmes , au sens d'origine, ne sont rien d'autres que des problèmes à démontrer. La plupart des ouvrages que Pólya écrit à Stanford traitent de ces thèmes: il s'agit de *How to solve it* (1945), *Mathematics and plausible reasoning* (1954), *Mathematical discovery* (1962). J'y reviendrai dans la dernière partie de cet exposé.

Pólya n'aura jamais cessé d'enseigner. Il donnait encore des cours de combinatoire à quatre-vingt-dix ans. Au cours de sa longue carrière, il a écrit sept ouvrages et plus de deux cent cinquante articles. Pour m'en tenir à la période de Zürich, voici une liste des principaux domaines qu'il a étudiés: probabilités, en particulier théorème de la limite centrale (Zentraler Grenzwertsatz: la terminologie lui est due) et promenade au hasard ; fonctions spéciales ; intégrales définies ; zéros des polynômes ou des fonctions entières ; théorie de l'approximation ; équations algébriques ; singularités et propriétés arithmétiques des fonctions analytiques ; séries de Taylor et de Dirichlet et rôle des lacunes ; fonctions entières de type exponentiel et intégrales de Fourier ; fonctionnelles linéaires sur des espaces de fonctions ; inégalités numériques ; formes bilinéaires ; capacités et dimensions, théorie du potentiel ; corps convexes ; problèmes combinatoires ; réflexions sur l'heuristique et l'enseignement. Tels sont les principaux sujets traités dans les cent cinquante articles de cette période.

Il ne sera pas possible de donner une idée de cette richesse. Avant de vous donner une sélection sous forme d'un petit dictionnaire, je voudrais conclure par une citation. L'influence de Pólya a été considérable sur beaucoup de mathématiciens de ma génération (pour moi par exemple, dans les années 1950, Pólya était déjà figure de légende). N. G. de Bruijn (cité par Alexanderson) exprime cette influence en ces termes:

"Plus peut-être que tout autre mathématicien, c'est Georges Pólya qui a orienté mon activité mathématique. Sa chaleureuse personnalité rayonne à travers toute son oeuvre. Goût merveilleux, méthodologie d'une clarté cristalline, moyens simples, résultats puissants. Si on me demandait de nommer un et un seul mathématicien que j'aurais aimé être moi-même, j'ai ma réponse prête à l'instant: Pólya".

- - - - * — ** — *** — ** — * - - - -

Le petit dictionnaire qui suit est loin de rendre justive à l'oeuvre de Pólya. Je me suis attardé sur ce qui est le plus facile à exposer. Il s'agit donc d'un balayage très superficiel, qui est à compléter par les études sérieuses déjà évoquées (commentaires des Collected Papers (1974–1984) ; notice du Bulletin de la London Mathematical Society 1987).

La *courbe de Pólya* (1913) est une variante d'une courbe de Peano, qui remplit un triangle rectangle. On part d'un tel triangle et on le divise en deux au moyen de sa hauteur: on obtient deux triangles, semblables au précédent, qu'on désigne par 0 et 1. En répétant l'opération sur ces deux triangles, on obtient quatre nouveaux triangles, 00, 01, 10, 11. On répète indéfiniment: on obtient des triangles de plus en plus petits, toujours semblables au triangle de départ, et toute suite illimitée de 0 et de 1 correspond à des triangles emboîtés dont l'intersection se réduit à un point. Ainsi, à un nombre t écrit dans le système binaire ($0 \le t \le 1$) correspond un point du triangle, et tout le triangle est balayé lorsque t parcourt l'intervalle $[0, 1]$.

Lorsque le triangle rectangle est isocèle, on retrouve une construction de Cesàro (1905), retrouvée indépendamment par Paul Lévy et par W. Sierpinski. Dans ce cas, à une étape donnée, tous les petits triangles sont égaux, ce qui a des conséquences intéressantes quand on compare les mesures et dimensions d'ensembles de valeurs de t et de leurs images. La plupart des points sont simples, mais il y a également des points doubles et quadruples.

Pólya montre qu'on peut choisir l'angle aigu α du triangle rectangle de telle façon que l'ordre de multiplicité maximum soit 3 (et on a su depuis que dans toute application continue de $\mathcal{R}^m$ sur $\mathcal{R}^{m+p}$ il y a des points de multiplicité $p + 2$ (Hurewicz 1939).

Il observe aussi qu'à l'étape n dans le cas $\alpha \ne \frac{\pi}{4}$, il y a $\binom{n}{k}$ triangles dont l'hypoténuse est $(\cos \alpha)^{n-k}(\sin \alpha)^k$ (en supposant l'hypoténuse du triangle de départ égale à 1). Poursuivons cette observation. Si $\frac{\pi}{6} < \alpha < \frac{\pi}{4}$, cela décroît moins vite que 2^{-n} (longueur des intervalles dyadiques de la n-ième étape), et on en déduit que la fonction $f(t)$ qui réalise le paramétrage de la courbe de Pólya est partout non dérivable. Si $\frac{\pi}{12} < \alpha < \frac{\pi}{6}$, la plupart des hypoténuses décroissent moins vite que 2^{-n}, mais certaines décroissent plus vite: la fonction $f(t)$ est non dérivable presque partout, mais elle admet une dérivée nulle sur un ensemble non dénombrable. Si $\alpha < \frac{\pi}{12}$, la plupart des hypoténuses décroissent plus vite que 2^{-n} : la fonction $f(t)$ admet une dérivée nulle p. p. Cette discussion, avec le détail des preuves, est due à Peter Lax (1973).

En réalité, le paramètre naturel sur la courbe de Pólya n'est pas t, mais s, le double de l'aire parcourue, qui varie de 0 à 1 quand t varie de 0 à 1 (*figure 1*). Si $F(s)$ est l'image de s, on vérifie facilement que F est partout non dérivable, et aussi que

$$(*) \qquad \parallel F(s) - F(s') \parallel^2 \le \mid s - s' \mid .$$

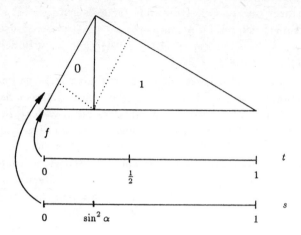

Figure 1

Montrons encore l'usage de la courbe de Pólya dans un problème essentielle-ment dû à S. Kakutani. Voici le problème: on donne un triangle rectangle de côtés a et b, et un nombre quelconque de points dans le triangle. Montrer qu'on peut les numéroter sous la forme $M_1, M_2, ..., M_n$ de telle façon que

$$\sum_{1 \le m \le n-1} (M_m M_{m+1})^2 \le a^2 + b^2.$$

Réponse: d'après (*), il suffit (dans le cas $a = \cos \alpha$, $b = \sin \alpha$, qui est typique) de numéroter les points dans l'ordre où on les voit apparaître sur la courbe de Pólya. Une variante est de considérer un rectangle de côtés a et b, un nombre quelconque de points dans le rectangle, et de montrer qu'on peut les numéroter sous la forme $M_0, M_1, ..., M_{n-1}, M_n = M_0$ de façon que

$$\sum_{0 \le m \le n-1} (M_m M_{m+1})^2 \le 2(a^2 + b^2).$$

Le problème d'origine (qu'on traite avec la courbe de Cesàro) concerne un carré $(a = b)$.

Les *fonctions de Pólya* (1949) sont les fonctions $f(x)$ $(x \in \mathcal{R})$ qui sont positives, paires, nulles à l'infini et convexes sur la demi-droite $x \ge 0$ (*figure 2*). Elles fournissent toute une classe, très utile, de fonctions caractéristiques au sens des probabilités. C'est très facile à voir en considérant les fonctions de Pólya extrémales, qui sont des fonctions triangles-isocèles.

L'*indicatrice de Pólya* (1922) d'une fonction entière de type exponentiel $f(z)$ est la fonction

$$h(\varphi) = \limsup_{r \to \infty} \frac{\log | f(re^{i\varphi}) |}{r}.$$

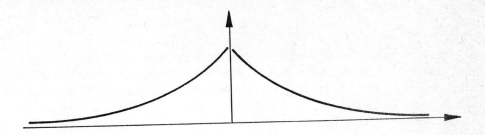

Figure 2

C'est la fonction d'appui d'un convexe fermé et borné, qui est le *diagramme de Pólya* de la fonction $f(z)$ (*figure 3*). Par exemple, le diagramme de Pólya de $f(z) = e^{az}$ est $\{a\}$, et le diagramme conjugué de $f(z) = \sum_1^N c_n e^{a_n z}$ est le plus petit convexe contenant les points a_n. Les fonctions entières de type exponentiel jouent un rôle très important en analyse de Fourier (théorème de Paley–Wiener 1934, théorème de Plancherel–Pólya 1937), et la notion a été introduite par Pólya (1919). En voici d'ailleurs une illustration.

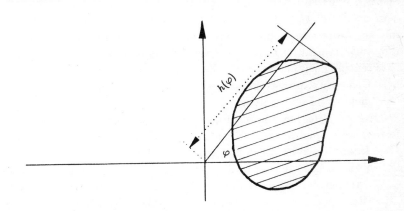

Figure 3

Les *fonctions de Pólya–Schur* (1914) sont les fonctions d'une variable complexe qui, dans un voisinage de 0, sont approchables par des polynômes à racines réelles. Ce sont nécessairement des fonctions entières de type exponentiel multipliées par une fonction de Gauss. Les travaux de Pólya sur les racines réelles ont des conséquences dans la théorie des transitions de phase, comme le montre Marc Kac dans son commentaire des *Collected Papers* (volume II).·

Les *théorèmes de Pólya* sur les fonctions à valeurs entières sur les entiers sont deux perles (1915, 1920): la plus petite fonction entière transcendante appliquant

$\mathcal{N}$ dans $\mathcal{Z}$ est 2^z, et la plus petite fonction entière transcendante appliquant $\mathcal{Z}$ dans $\mathcal{Z}$ est

$$\tfrac{1}{\sqrt{5}}\left(\left(\tfrac{3+\sqrt{5}}{2}\right)^z - \left(\tfrac{3+\sqrt{5}}{2}\right)^{-z}\right)$$

La densité d'une suite positive (λ_n) est la limite, quand elle existe, de n/λ_n quand $n \to \infty$. La *densité maximum de Pólya* d'une suite Λ positive est la borne inférieure des densités des suites densitables contenant Λ. Pólya montre que la borne inférieure est atteinte, et que la densité maximum intervient dans la distribution des singularités des séries de Dirichlet d'exposants λ_n (c'est-à-dire des séries $\sum a_n e^{-\lambda_n s}$). Le résultat le plus simple est que, si la densité est nulle, le domaine d'existence est un demi-plan (généralisation d'un théorème de Fabry), et que si la densité maximum D est strictement positive il n'en est pas nécessairement ainsi, mais qu'alors tout segment de longueur πD sur la droite de convergence contient un point singulier (1923, 1927, 1939, 1942).

La *dimension capacitaire de Pólya et Szegő* (1931) est une notion de dimension qui, au contraire de celles de Minkowski, Bouligand ou Hausdorff, n'est pas fondée sur des considérations géométriques mais sur la théorie du potentiel. Elle s'introduit aussi de façon intéressante en analyse de Fourier: δ est la dimension capacitaire d'un compact $K \subset \mathcal{R}^n$ si, pour tout $\alpha < \delta$, K porte une mesure $\mu \neq 0$ dont la transformée de Fourier $\hat{\mu}(u)$ appartient à l'espace $L^2(\mathcal{R}^n, \frac{du}{(1+|u|)^{n-\alpha}})$, et si pour aucun $\alpha > \delta$ il n'en est ainsi. Le résultat principal de la thèse de Frostman (1934) est l'égalité de la dimension de Hausdorff et de celle de Pólya– Szegő.

La *promenade au hasard* de Pólya (1921, 1938), dont j'ai déjà parlé, est le plus simple des processus: le promeneur saute au hasard d'un point de $\mathcal{Z}^n$ à un point voisin, indéfiniment, et sans privilégier aucune direction. Problème: atteint-il presque sûrement un point assigné ? La réponse est non pour $n \geq 3$, oui pour $n = 1$ ou 2. Comme l'exprime plaisamment Pólya, en dimension deux, il est vrai que tous les chemins mènent à Rome !

La *théorie de Pólya* en combinatoire est une méthode de décompte — au moyen de fonctions génératrices à plusieurs variables — pour des configurations obtenues en plaçant des objets donnés aux sommets d'un polyèdre donné. L'origine et l'application principale (encore utilisée par les chimistes) est le décompte des isomères. Le premier exposé — et l'un des meilleurs — se trouve en deux pages des Comptes–Rendus (1935). Le dernier qu'en ait donné Pólya est dans le livre de Pólya–Tarján–Woods de 1983. On y voit, par exemple, que le nombre maximum des alcools aliphatiques $C_n H_{2n+1} OH$ est respectivement de 1, 2, 4, 8, 17, 39 pour $n = 2, 3, 4, 5, 6, 7$.

La théorie de Pólya a eu beaucoup de succès en combinatoire, sans doute pour trois raisons: son intérêt propre, son application à des problèmes réels, et le renom de Pólya. Un article paru en 1927, d'un mathématicien peu connu, J. Howard Redfield, en donne pourtant les éléments principaux. Mais cet article, intitulé: *The theory of group-reduced distributions* (Amer. J. Math. 49 (1927), 433–455) est resté longtemps inaperçu. Pólya ne l'a pas remarqué, et Redfield ne semble pas avoir remarqué non plus les articles de Pólya. Redfield est mort en 1944 en ayant publié cet unique article. Le second, redécouvert, a été publié en 1984 dans le Journal of Graph Theory, et maintenant, par un juste retour des choses, la gloire de Redfield grandit à l'ombre de celle de Pólya. On peut en juger par le Pólya–Read (1987), qui groupe un texte fondamental de Pólya, datant de 1937, et le commentaire de Read sur toute l'évolution du sujet.

Je signale pour mémoire l'*heuristique de Pólya*, sur laquelle je reviendrai. Cependant je veux citer ici le *proverbe de Pólya*. Dans les différentes éditions de *How to solve it?* (17 langues, comme je l'ai dit) figure un petit dictionnaire d'heuristique, et, dans ce dictionnaire, un article: *sagesse des proverbes*. Pólya a dû beaucoup s'amuser à superviser ses diverses traductions. Il a fait traduire les proverbes anglais en vrais proverbes français, allemands, italiens, etc..., et il arrive que la traduction enrichisse l'original. Il a ajouté un proverbe de son crû, qui est beaucoup plus joli en anglais qu'en français. Vous le connaissez sans doute: *Your five best friends are What, Why, Where, When and How. You ask What, you ask Why, you ask Where, When, and How — and ask nobody else when you need advice.*

Je n'ai pas parlé de tous les honneurs que Pólya avait reçus. Mais pour terminer ce dictionnaire des termes qui portent le nom de Pólya, il est bon que je signale le *prix Pólya* de combinatoire, décerné par la Society for Industrial and Applied Mathematics, le *prix Pólya* d'exposition mathématique décerné par The Mathematical Association of America, le *prix Pólya* que vient de créer la London Mathematical Society pour honorer une oeuvre mathématique exceptionnelle dans le Royaume Uni. A Stanford, reconnaissance rare pour un mathématicien, l'un des bâtiments porte son nom: *Pólya Hall*. Son nom est ainsi gravé dans la pierre, mais il le restera plus longtemps encore dans la mémoire des hommes.

- - - - * — ** — * * * — ** — * - - - -

Dans la dernière partie de cet exposé, je me propose de développer et commenter certaines des idées de Pólya sur l'heuristique, le raisonnement plausible, le rôle des problèmes dans l'enseignement.

Les idées de Pólya sur ces sujets datent de loin. C'est en 1919 qu'il a donné, dans une revue pédagogique suisse, la primeur de ce qu'il allait développer plus tard dans Mathematical Discovery: une sorte de représentation géométrique de la façon dont la solution d'un problème se construit (*Geometrische Darstellung einer Gedankenkette*). En 1931, dans l'Enseignement mathématique, il publie, comme résumé d'une conférence faite devant la société suisse des professeurs de mathématiques, l'aide-mémoire qui allait devenir fameux dans *How to solve it ?*

1. Comprendre la question
2. Trouver un chemin qui aille de l'inconnue aux données, en passant, s'il le faut, par plusieurs problèmes intermédiaires (*analyse)*
3. Mettre en oeuvre (*synthèse)*
4. Vérifier et critiquer.

Je renvoie aux pages de garde de *How to solve it ?* pour le détail de ces rubriques. J'insisterai seulement sur l'un des conseils qu'il donne pour l'analyse, ou, si l'on veut, la conception d'un plan: pourriez-vous imaginer un problème voisin plus accessible ? un problème plus général ? un problème plus particulier ? un problème analogue ? C'est l'un des thèmes qu'il développe avec brio dans son livre sur les mathématiques et le raisonnement plausible (MRP).

Je commenterai cette démarche en considérant trois problèmes de géométrie.

Premier exemple (problème à résoudre) (MRP, chapitre 2). Démontrer le théorème de Pythagore.

L'idée naturelle (qui est chez Euclide) est de construire trois carrés ayant respectivement pour côtés les trois côtés du triangle rectangle (*figure 4*) et de montrer que l'aire du plus grand est la somme des aires des deux autres. On peut généraliser: au lieu de carrés, construire sur les trois côtés du triangle rectangles trois polygônes semblables ; leurs aires sont proportionnelles aux aires des carrés, donc il s'agit toujours de montrer que l'aire du plus grand est la somme des aires des deux autres (*figure 5*). Puis particulariser (*figure 6*): un seul tracé, celui de la hauteur, rend la solution évidente. La figure 6 est un analogue de la figure 4, qu'on obtient en généralisant puis particularisant.

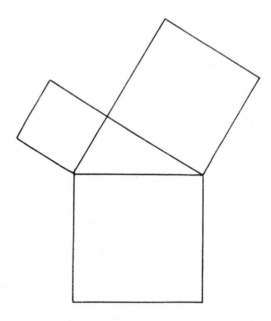

Figure 4

Il n'est pas besoin d'insister sur la parenté de la figure 6 et de la figure 1. Est-ce que cette jolie preuve du théorème de Pythagore n'est pas un sous-produit de la courbe de Pólya ?

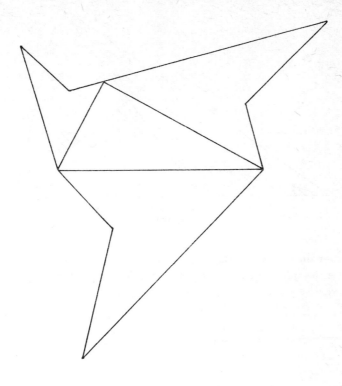

Figure 5

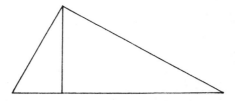

Figure 6

Deuxième exemple (problème à résoudre). On donne un triangle T et trois demi-droites, issues des sommets, qui traversent T, et délimitent avec les côtés de T, quatre triangles et trois quadrilatères. On suppose que ces quatre triangles ont des aires égales. Montrer que les trois quadrilatères ont des aires égales (souvenir pour moi d'une visite à Iraklion) (figure 7).

Il s'agit d'un problème affine. Donc tout cas particulier est équivalent au cas général. Prenons T équilatéral. Si la figure est invariante par rotation d'un tiers

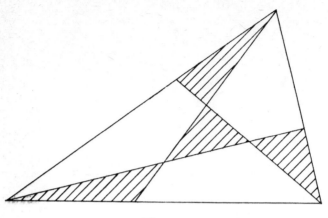

Figure 7

de tour autour du centre du triangle, les petits triangles adjacents aux côtés de T sont égaux, et les quadrilatères sont égaux. Sinon, ces petits triangles sont inégaux (*figures 8, 9*).

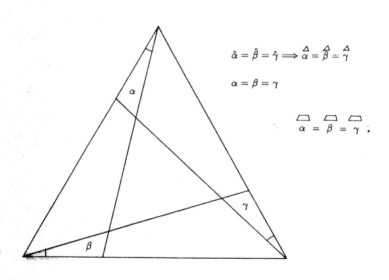

$$\hat{\alpha} = \hat{\beta} = \hat{\gamma} \Longrightarrow \overset{\triangle}{\alpha} = \overset{\triangle}{\beta} = \overset{\triangle}{\gamma}$$

$$\alpha = \beta = \gamma$$

$$\overset{\square}{\alpha} = \overset{\square}{\beta} = \overset{\square}{\gamma} \; .$$

Figure 8

La conclusion est une généralisation du problème qui s'avère plus facile à résoudre: si les trois petits triangles adjacents aux côtés de T ont des aires égales, les trois quadrilatères ont des aires égales.

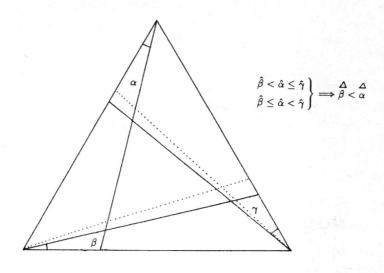

$$\left.\begin{array}{l} \hat{\beta} < \hat{\alpha} \le \hat{\gamma} \\ \hat{\beta} \le \hat{\alpha} < \hat{\gamma} \end{array}\right\} \implies \overset{\triangle}{\beta} < \overset{\triangle}{\alpha}$$

Figure 9

Troisième exemple (problème à trouver) (*MRP, chapitre 3*). On considère 5 plans en position générale dans l'espace. En combien de régions l'espace est-il découpé ?

Comme il est difficile de se représenter la figure, commençons par des problèmes analogues, en considérant moins de 5 plans. Un plan divise l'espace en deux régions, 2 plans en 4, 3 plans en 8 ; en fait, à chaque étape, le nouveau plan que nous introduisons coupe les régions précédentes en deux. Ainsi, par induction, nous voyons que n plans divisent l'espace en 2^n régions. La réponse au problème est donc 32.

Passons maintenant à la dernière recommandation de Pólya: vérifier et critiquer. Pouvons-nous nous servir de la méthode pour un autre problème ?

Certainement. Au lieu de n plans dans l'espace à 3   dimensions, nous pouvons considérer n hyperplans dans l'espace à p dimensions. De nouveau et pour la même raison, le nombre de régions de subdivision de l'espace est 2^n. Cela vaut si grand que soit p.

A propos, cela doit valoir aussi pour $p = 2$ et pour $p = 1$. Or, n points ne divisent pas la droite en 2^n régions (sauf pour $n = 0$ et 1), mais en $n + 1$. Où est l'erreur ? C'est que le n-ième point ne divise pas toutes les régions qui précèdent, mais une seule.

De même, pour $p = 2$, la n-ième droite ne rencontre que n régions précédentes — puisqu'elle rencontre les $n - 1$ droites qui la précèdent en $n - 1$ points, donc qu'elle même est divisée en n régions. Pour $p = 2$, au lieu de 2^n, on obtient donc $1 + n + \dfrac{n(n - 1)}{2}$.

De façon générale, au lieu de 2^n, on obtient, par une induction correcte cette fois, $\binom{n}{0} + \binom{n}{1} + \binom{n}{2} + \ldots + \binom{n}{p}$, c'est-à-dire l'unique polynôme de degré p en n qui vaut 2^n pour $n = 0, 1, 2, \ldots, p$.

Les relations entre observations, hypothèses et vérifications forment le coeur de l'heuristique de Pólya, et de sa théorie du raisonnement plausible. La plausibilité est une sorte de probabilité qualitative. Pólya s'interdit de lui assigner une valeur, mais il étudie sa variation en fonction de différents types de tests. Sa théorie traite donc du plus plausible et du moins plausible.

Un exemple remarquable d'heuristique en action est son article sur les couples de nombres premiers p, $p+d$, dont la différence d est un entier pair donné (American Math. Monthly 1959). Il montre, de façon convaincante, comment l'usage des tables mène à la conjecture

$$\pi_d(x) \cong \pi_2(x) \prod_{p(\text{premier} \geq 3)/d} \frac{p-1}{p-2}$$

où $\pi_d(x)$ est le nombre de nombres premiers $p \leq x$ tels que $p+d$ soit premier. Puis, de façon heuristique, apparaît la conjecture

$$\pi_2(x) \cong C \frac{x}{\log^2 x}$$

$$C = \prod_{p \text{ premier} \geq 3} \left(1 - \frac{1}{(p-1)^2}\right).$$

Au cours de cette recherche, Pólya montre que les démarches du mathématicien et du physicien sont très voisines. *"Mathematicians and physicists think alike ; they are led, and sometimes misled, by the same problems of plausible reasoning"*. Il est très remarquable qu'indépendamment de Pólya et au même moment, un physicien et un mathématicien anglais, Lord Cherwell et E. M. Wright, aient donné deux *"preuves"* heuristiques d'une conjecture un peu plus générale, concernant la distribution asymptotique des systèmes de nombres premiers de la forme p, $p + a_1, \ldots, p + a_k$, où $a_1, \ldots, a_k$ sont des nombres pairs donnés (Quarterly J. Math. Oxford 1960).

Cela dit, il existe dans ce domaine une énorme distance entre la conjecture et la preuve. On ne sait pas encore si $\pi_2(x) \to \infty$, et on a tout lieu de penser que c'est un problème très difficile.

L'heuristique en théorie des nombres est à la fois une nécessité et une source de déboires. Nécessité pour avancer — Euler en est le magnifique modèle. Source de déboires comme le montrent deux fameux exemples.

Le premier est donné par Hardy dans son livre sur Ramanujan. C'est la comparaison entre $\pi(x)$, le nombre de nombres premiers $\leq x$, et le logarithme intégral de x, ℓix (*primitive de* $(\log x)^{-1}$). Le grand théorème des nombres premiers, prouvé par Hadamard en 1896, est l'équivalence de $\pi(x)$ et de ℓix. Mais l'observation —

déjà notée par Gauss — est qu'en vérité $\pi(x) \leq \ell i x$ si loin qu'on puisse consulter les tables. C'est vrai pour $x \leq 10^7$, et tous les sondages pour $x \leq 10^9$ confirment la conjecture. Cependant Littlewood a démenti la conjecture en 1912, sans pouvoir d'ailleurs donner une estimation du temps à attendre avant que l'inégalité s'inverse. Depuis 1933, on sait que l'inégalité s'inverse avant $x = 10^{10^{10^{34}}}$, un nombre tout à fait inaccessible. Rien d'étonnant à ce que tous les calculs directs aient semblé vérifier la conjecture !

Le second exemple est un problème de Pólya, qu'on a appelé conjecture de Pólya (1919): pour tout entier $n = p_1^{\alpha_1}...p_\nu^{\alpha_\nu}$ on pose $\lambda(n) = 1$ si $\alpha_1 + ... + \alpha_n$ est pair, et $\lambda(n) = -1$ si $\alpha_1 + ... + \alpha_\nu$ est impair. On considère $L(x) = \sum_{n \leq x} \lambda(n)$. A-t-on toujours $L(x) \leq 0$? C'est vrai, encore une fois, pour toutes les valeurs de x pour lesquelles $L(x)$ a été calculé. Cependant, en 1958, C. B. Haselgrove a montré que l'inégalité se renverse avant $x = 1,845 \times 10^{361}$.

Ces exemples doivent nous servir de garde-fous. Les conjectures en théorie des nombres ne peuvent se satisfaire de vérifications banales. C'est à la validité de conséquences très lointaines qu'on doit les tester. Ces tests à grande distance distinguent, dans ce domaine, l'attitude du mathématicien et celle du physicien.

Dans l'enseignement, la place de l'heuristique a été discutée. Certains se sont inquiétés qu'on superpose à l'enseignement des mathématiques un enseignement d'heuristique des mathématiques. Il est bon de poser la question et de s'en remettre à la préoccupation première de Pólya: aider l'élève, aider l'élève à penser par lui-même. Un enseignement approfondi de l'heuristique, surtout si l'heuristique se développe en discipline autonome, ne paraît pas s'imposer. Par contre, les moyens modernes — calculettes et ordinateurs — donnent aux élèves des possibilités d'observation, de vérifications, d'expérimentation qui étaient inaccessibles à leurs aînés. L'étude de la CIEM sur l'influence de l'informatique, de même que mon rapport au Congrès International des Mathématiciens à Berkeley, ont été consacrés pour une grande part à cette question, et je n'y reviens pas.

L'heuristique de Pólya a joué un rôle dans la philosophie des mathématiques. L'étude sur la formule d'Euler $F + S = A + 2$ qui se trouve dans *les Mathématiques et le raisonnement plausible* a son prolongement dans celle d'Imre Lakatos (Proofs and refutations, CUP 1976). Lakatos prend la question où Pólya la laisse, quand la formule, vérifiée dans des cas cruciaux, apparaît fondée pour une vaste classe de polyèdres. Comment alors l'énoncer, comment la prouver ? Des réfutations apparaissent. Le rôle du maître est de transformer chaque réfutation en condition, sinon en concept. Paradoxalement, réfuter une à une les étapes de la preuve n'anéantit pas l'énoncé. Réfuter l'énoncé n'anéantit pas l'articulation de la preuve. Peu à peu, par un va et vient entre erreurs et corrections, les méthodes et les définitions de la topologie apparaissent. Lakatos va plus loin que Pólya, mais poursuit la même démarche. D'ailleurs, quand Pólya, dans un de ses derniers écrits, *Guessing and proving*, revient sur l'exemple de la formule d'Euler, c'est en l'enrichissant de remarques de Lakatos, formulés *à la Pólya* (je pense au commentaire de la figure 4 de Guessing and proving).

L'heuristique de Pólya jouera peut-être un rôle en intelligence artificielle. C'est le pari de A. Schoenfeld dans son analyse de l'oeuvre de Pólya (1987). Elle donne en effet l'idée de stratégies générales, moins restreintes à leur objet que la plupart des systèmes experts.

Les réflexions de Pólya sur l'heuristique sont très liées à une idée majeure concernant l'enseignement. C'est la priorité à donner à la résolution des problèmes. Les *Aufgaben und Lehrsätze* de 1924, *How to solve it ?* de 1945, et de nombreux articles depuis, portent témoignage de la constance de Pólya à cet égard.

Au cours des années 1970, il s'est vivement opposé à la formalisation qu'on a associée aux mathématiques modernes, justement parce qu'elle rendait les problèmes pauvres et secs. Au cours des années 1980, il a pu se réjouir du regain de faveur pour la recherche et la résolution des problèmes. Il y a là un acquis irréversible.

Si maintenant nous nous interrogeons sur les années 1990, deux problèmes majeurs apparaissent. L'enseignement peut-il et doit-il faire face à la croissance exponentielle de la production mathématique ? L'enseignement peut-il et doit-il faire face à la croissance et à la diversification des besoins sociaux de connaissances mathématiques ? A mon avis, ces questions ne se résolvent pas par oui ou non, mais par quoi et comment. L'exemple de Pólya peut nous aider à y voir clair.

En 1920, les mathématiques contemporaines se trouvaient très loin de l'enseignement, même de l'enseignement universitaire. *Pólya–Szegő* n'est pas seulement un bon livre de problèmes qui s'enchaînent, c'est d'abord un ouvrage en prise sur la science contemporaine, qui vise et qui réussit à la rendre accessible. C'est parce qu'en son temps il a été très moderne qu'il est devenu et qu'il restera un classique. Nous aurions besoin, aujourd'hui, de centaines de livres de problèmes en prise sur les mathématiques de notre époque, et plus généralement de documents (imprimés, audiovisuels, informatisés) qui puissent nous familiariser avec les progrès contemporains. Quand je dis *nous*, il s'agit des mathématiciens, des professeurs de mathématiques, des étudiants, des élèves, de tous les publics concernés. Il y aurait là un vaste programme d'action concertée, au plan de chaque nation et au plan international.

Entre 1914 et 1940, Pólya a enseigné à des étudiants très divers. Il n'a pas enseigné les mêmes choses, ni par les mêmes méthodes. Quelles sont les mathématiques indispensables aux différentes activités humaines ? Quelle est la part qui doit être enseignée aux jeunes gens, et la part qui doit faire l'objet d'un apprentissage continu ? Qui doit enseigner, et comment ? Ces questions ont fait l'objet d'une étude récente de la CIEM, mais elles sont loin d'être épuisées.

Les mathématiques offrent bien des ressources pour l'enseignement, parce qu'elles sont en même temps un langage, un exercice intellectuel, et une science.

Des années 1970 nous devons retenir qu'elles sont un langage et que nous devons maintenir ce langage en bon état et l'enseigner.

Des années 1980 nous devons retenir qu'elles sont un exercice intellectuel, irremplaçable, parce que, sous la forme de problèmes, elles mettent les élèves, plus que toute autre discipline, en état de recherche scientifique.

Dans les années 1990 il faudra bien que les mathématiques dans l'enseignement apparaissent aussi pour ce qu'elles sont, c'est-à-dire une science en mouvement, avec des progrès rapides dans certaines directions, des retours en arrière, une réactualisation de son histoire, une prise directe sur presque toutes les autres sciences et les technologies modernes.

Je ne pense pas être infidèle à l'héritage de Pólya en projetant ainsi sur les années à venir ce qui me semble résulter de son oeuvre et de son exemple.

References:

Les références de base sont *G. Pólya Collected Papers (I, II, 1974 ; III, IV 1984).* *Obituary George Pólya. Bulletin of the London Mathematical Society 19 (1987),* *559-608.*

La plupart des articles ou livres cités s'y trouvent ; en particulier, on y trouvera les travaux de Pólya dont j'ai seulement signalés le sujet et la date. Voici des références complémentaires.

Matematikusok felhivása a fegyverkezési verseny megfékezésére, Matematikai Lapok 32 (1981–85), 215–216.

G. Pólya: On my cooperation with Gábor Szegő, in Gábor Szegő, Collected papers, volume I (1982), 11.

E. Cesàro: Remarques sur la courbe de von Koch (1905) in Opere Scelte, vol. II, 464–479.

W. Hurewicz: Über dimensionserhöhende stetige Abbildungen, Journal für die reine und angewandte Mathematik 169 (1933), 71–78.

P. Lax: The differentiability of Pólya's function, Advances in Mathematics 10 (1973), 456–464.

O. Frostman: Potentiel d'équilibre et capacité des ensembles, thèse, Lund 1935.

Lord Cherwell et E. M. Wright: The frequency of prime-patterns, Quarterly J. Math. Oxford 11 (1960), 60–63.

G. H. Hardy: Ramanujan, Cambridge University Press 1940.

I. Lakatos: Proofs and refutation, the logic of mathematical discovery, Cambridge University Press 1976.

I.C.M.I. study series (editors A. G. Howson & J.-P. Kahane):
   I. *The influence of computers and informatics on mathematics and its teaching,* Cambridge Univ. Press 1986.
   II. *School mathematics in the 1990's,* Cambridge Univ. Press 1986.
   III. *Mathematics as a service subject,* Cambridge Univ. Press 1988.

J.-P. KAHANE Enseignement mathématique, ordinateurs et calculettes, Proc. Intern. Congress of Mathematicians, Berkeley 1986 (1987), 1682–1696.

ACTION GROUPS

ACTION GROUP 1: EARLY CHILDHOOD YEARS

Chief organizer: Leslie P. Steffe (Italy)
Panel: Jarkko Leino (Finland), Francis D. Lowenthal (Belgium),
R. Ogbonna Ohuche (Nigeria), Bob Perry (Australia)

Preamble

The general goal of *Action Group A1: Early Childhood Years* was to identify the issues, problems, and opportunities presented by constructivism for mathematics education in early childhood and to make recommendations for the work in this area over the next four years — until ICME 7. In what follows, I have drawn selectively from three primary sources. First, the international panel who planned the work of the action group prepared a conceptual framework that was approved by the International Program Committee and appeared in the Program Statement issued to all participants of ICME 6. Second, eleven papers were prepared and distributed to the registered members of the action group prior to ICME 6. Third, abstracted accounts of the discussions that transpired during the work of the action group were prepared by the abstracters after ICME 6.

The revolutionary aspect of constructivism lies in the assertion that "knowledge cannot and need not be 'true' in the sense that it *matches* ontological reality, it only has to be 'viable' in the sense that it *fits* within the experiential constraints that limit the [human's] possibilities of acting and thinking" (von Glasersfeld, 1987). Accepting mathematical knowledge as being viable rather than true has far reaching consequences for mathematics education in early childhood. For example, the belief that mathematics is the way *it* is rather than the way human beings make it has permeated mathematics education at all levels and has served as the basis for the assumption that there is a standard school mathematics that students must come to know in the way intended. Not making this assumption opens up new possibilities for reform in mathematics education. Some of the possibilities were presented in a framework prepared by the international panel, *knowledge and learning, communication*, and *environment* were adopted as key organizing concepts. *Mathematical experience* was embedded in the discussion of these three organizing concepts and served as a basis for the deliberation of how children and teachers come to know mathematics. Although progress was made at ICME 6, almost all of the work to isolate the possibilities and carry them out remains to be done and it is in this spirit that I offer this overview.

Knowledge and Learning

In constructivism, children's mathematical knowledge is viewed as coordinated schemes of action and operation that are functioning reliably and effectively — what Sinclair (1987) has referred to as instruments of problem solving. The general question concerning knowledge that was posed for the work of the action group was: *What schemes (or coordinated schemes) of action and operation can be inferred from the language and actions of children?* The scope of this general question is indicated by the following specific questions.

1. How does the counting scheme develop and how does it relate to the development of numerical concepts and operations, including addition, subtraction, multiplication, and division?
2. What mathematical procedures do children use, including the ability to recognize and check counting and other arithmetical errors?
3. How can we characterize the thinking strategies used by children?
4. What constitutes logical thinking of children and how does it differ from mathematical thinking?
5. What constitutes spatial and geometrical knowledge of children?
6. What constitutes measurement knowledge of children and how does it relate to 1–5?
7. What is the role of technology in the mathematical knowledge of children?

Knowledge, learning and social interaction

The mathematical knowledge children can construct (learn) in early childhood and how learning might be fostered was investigated in the context of the above general and specific questions. Although mathematical learning can be viewed as consisting in the *adaptations* children make in their functioning schemes to neutralize perturbations that can arise in one of several ways, H. Sinclair (1988) emphasized that learning occurs as a result of the interaction between societal presentation and endogenous processes of abstraction. "The difficulty of studying learning — and teaching — lies, in my view, in the fact that it demands the study of the processes by which children come to know in a short time basic principles (in mathematics, but also in other scientific disciplines) which it took humanity thousands of years to construct" (p. 1). R. Lawler (1988) presented an example of learning not as socially directed but as, "a compensating adaptation to the deprivation of social interaction" (p. 4). The example consisted of a child who imitated her older sister by saying "knock-knock" and expecting an adult (her father) to complete the joke by saying "who's there?". The adult then deliberately changed his role to saying "knock-knock" and the child had no response, being unable to change her role in the social interaction. This apparently created a perturbation that the child resolved by re-enacting by herself, i.e., without the social situation, the two complementary roles of the past social interactions. Using this and other examples involving numerical and geometrical contexts, R. Lawler elucidated possible relations between social experience and personal construction.

F. Marton and D. Neuman (1988) emphasized the importance of mathematical experience that occurs as a result of children's interaction with their mathematical reality. "The child does not consciously reflect on how to solve the problem. They just explore the quantitative relations inherent in it, and suddenly something stands out as a figure against a background giving the child an intuition of how it could be solved." (p. 13). This description of the construction of quantitative relations by children is congenial with the notion of *bricolage* as explained by R. Lawler (1988): "The core idea of bricolage is the looseness of commitment to specific goals, the idea that materials and competencies developed for one purpose are transferable, can very easily be used to advantage in the satisfaction of alternative objectives" (pp. 18–19).

J. Leino (1988) also sees mathematical learning in part as occurring in, or as a result of, social interaction. "Learning process in school is not just a knower–known relationship, a pupil with his math problem, but ... a mediated learning experience, a pupil studying with his teacher and peers" (p. 7). However, mathematical learning cannot be said to be caused by social interaction and we should not expect it to be predictable in the sense that what a child does learn in a given situation is what an adult expects the child to learn: "Even though the concrete mathematical situation is simple there are always many logical ways to comprehend it. That is why pupils many times form a construct which is not the one the teacher desired or expected and why textbooks attempt to present extremely reduced and simplified models ... Their value is, to my mind, highly exaggerated" (Leino, 1988, p. 5).

Discussion of knowledge and learning

Two discussion groups were devoted to identifying the issues, problems, and opportunities presented by constructivism in the area of mathematical knowledge and its learning. Not surprisingly, the first issue that emerged was "Is there a body of mathematical knowledge all students should learn?" This issue has deep roots in the practice of mathematics education world-wide and is intimately connected with the assumption that there is a standard school mathematics that students must come to know in the way intended.

Although stressing the role of social interaction in mathematical learning is a fundamental opportunity presented by constructivism, there was agreement that social interaction and its relation to mathematical learning are not well under-stood. Children are not lonely voyagers in their construction of mathematical knowledge and their ideas are elaborated and authenticated through negotiation of individual meaning. But R. Lawler pointed out in a personal communication that our conventions — shared knowledge — are uncertain and our enculturation complex. Nevertheless, stressing the roles social interaction might play in mathematical learning brings with it possible transformations in mathematics teaching in early childhood.

One possible role of a teacher in a constructivist oriented classroom is to pose situations that might facilitate construction of mathematical ideas. Some of the participants felt that the way a situation is presented as well as the particular constraints are crucial because within each culture, there are situations which might be particularly helpful to children in their meaning making. These "cultural amplifiers" should be identified and used to establish culturally significant situations. For example, in some cultures children's money competence and currency value can be useful in the establishment and elaboration of basic numerical schemes in socially meaningful settings. This was identified as an area of research over the next four years — until ICME 7. "Cultural amplifiers" were viewed as having promise for producing educationally significant results because they provide a way to connect social interaction, anthropological contexts, and numerical development.

Another possible role of a constructivist teacher is to encourage children to reflect on their own thinking to encourage the construction and coordination of schemes. This turned out to be an issue that was hotly debated. Some argued that through reflective abstraction and conflict resolution, children do reflect on their own thinking — time to facilitate reflection should be an essential "built-in" component of constructivist environments. In this line of thinking, children are viewed as modifying current schemes in their construction of more effective, powerful, and general schemes. It was agreed that the conditions and products of reflective abstraction is an area where intensive research needs to be done. For example, social interaction among children in mathematical contexts might produce more reflection by children than social interactions among adults and children. But, in all cases, the need to coherently understand might be the most compelling force that drives reflective abstraction.

A third possible role of a constructivist teacher is to stress children's mathematical experiences. Bridging the gap between children's mathematical experiences in and out of school and how these experiences can be harnessed to promote learning was a fundamental problem identified by the discussion groups.

Several specific issues were isolated during the course of the discussion including how to identify when learning has occurred, what tasks with what constraints and form of presentation might encourage learning, what schemes children bring that they might use in mathematical learning, and what conceptual tools are available to make sense of children's informal knowledge. Some participants felt that learning cannot be readily measured using paper and pencil tasks. Alternative strategies for observing and assessing children's learning was identified as a continuing need.

Finally, it was recognized that while constructivists may view learning as consisting of adaptation and mathematical knowledge as a result of organizing experience, school mathematics is all too often viewed as a set of algorithms and rules. The constructivist approach to mathematical knowledge and its learning is a growing source of perturbation in the field of mathematics education and the modifications and reorganizations engendered by this perturbation is a fertile area of investigation.

Communication

In any communication between two human beings, signals can be transmitted between the communicators but not the intended or received meanings. This is why social interaction cannot *cause* learning to occur as intended by a transmitter of signals. On the receiver's side, the transmitted signals have to be interpreted, establishing received meanings. Although the received meanings might be *viable* in so far as there is no necessity to modify them in further experiential encounters, this would not mean they *match* the intended meanings. There may be future experiential encounters where the received meanings are modified. From the constructivist point of view, meanings are conceptual structures (Steffe, 1988; von Glasersfeld, 1987) that serve in the interpretation and organization of experience that are subject to being changed and modified by the very experience they organize. Within this general understanding of communication, the following two questions were posed for the work of the action group: *"What does mathematics teaching consist of in early childhood and what forms could it take?"* and *"How can children's informal, spontaneous thinking be combined (or linked) with teaching practice?"* Because interpretation depends on the particular context in which a communication occurs — the social setting, the problem situation, the goals and intentions of the communicators, the language used by the communicators, etc. — the following focal points were suggested.

1. The mathematical knowledge and beliefs of teachers, both actual and possible.
   a) How might teachers of mathematics in early childhood view mathematics and its meaning?
   b) How might teachers of mathematics in early childhood view *children's mathematics* — their mathematical knowledge and its elaborations as a result of the social interactions of the children with each other and with their teacher?
   c) How might teachers of mathematics in early childhood foster the independence of children when doing mathematics, their generative power, their ability to reflect on their own problem solving attempts, their abstractions from their own mathematical activity, and the role of play and emotions in their mathematical activity?
2. The nature and forms of mathematics teaching.
   a) How might teachers use their knowledge to foster the mathematical knowledge of children in mathematical communication?
   b) What features of children's language and actions are indicative of certain elements of knowledge the teacher might focus on?

c) What decisions might teachers make about what children could or should do?

d) What problematic or interactive situations might teachers use to encourage children's mathematical learning?

Conceptual and contextual diversity

E. von Glasersfeld developed the idea "that language is not a means of transporting conceptual structures from teacher to student, but rather a means of interacting that allows the teacher here and there to constrain and thus to guide the cognitive construction of the student. This guidance, as good teachers have known all along, necessarily remains tentative and cannot even approach absolute determination" (p. 12). J. Easley and H. Taylor (1988), working as teachers of first grade mathematics classes in the United States, demonstrated the conceptual diversity among children pointed to by von Glasersfeld and the "unreasonableness of the common assumption that a teacher, aiming each lesson for a particular conceptual goal for all pupils, should expect nearly all of the pupils to succeed and to understand the same concepts" (p. 1). Through their teaching experiences, they observed "the relative ease with which children appear to communicate novel mathematical ideas with their peers compared to communicating with their teacher" (p. 1). For them, the interesting question is "to explain how young children are able to form and apply so many mathematical ideas to these problems which they have not been explicitly taught, ideas which one might think have a history of development in scholarly traditions that go back hundreds if not thousands of years" (p. 12). Their examples sharply illustrate the necessity of interpreting the myriad of verbal and nonverbal signals children send to establish possible meanings of the children's mathematical language and actions. As P. Cobb (1988) put it, "The trick is to imagine a world in which the child's activity does make sense" (p. 4).

As essential as these interpretative activities are for the teacher, they offer only a partial account of mathematical communication. Cobb forcefully argued for multiple perspectives including experiential, psychological, sociological, and anthropological. The advantage of assuming multiple perspectives is illustrated by F. Lowenthal's (1988) nonverbal communication devices. Emphasizing the experiential and psychological contexts, he reports working successfully with children who, in verbally oriented "normal" school contexts, could be said to be at risk. These children were called "problem children" and were all at least two years behind in their studies relative to normative standards. Lowenthal described how, after 18 months of using NVCD's, the children were no longer behind in mathematics. These considerations highlight the necessity for reform in mathematics education that emphasizes contextual as well as conceptual diversity.

Discussion of communication

In the three sessions devoted to discussing mathematical communication, one issue that emerged was the contrast between traditional pedagogic discourse in schools and discourse outside of schools. Some participants felt that improvements in the former might emerge if the latter was analyzed from the point of view of its participants, especially from the point of view of children. There seemed to be agreement among the participants that children are not given much of a chance to express their ways and means of operating in school and that teachers need to involve children in interactive (nonverbal as well as verbal) mathematical communication. How such communication might be carried out was identified as a major problem on which work needs to be done. Interactive mathematical communication is quite different from simply showing or telling children what to do. How teachers might establish and sustain such communication might be enhanced by examples as well as by models. The participants felt that the difficulty of learning to engage in interactive mathematical communication should not be underestimated even in the case of mathematics teacher educators.

Investigating how mathematical communication is like, or different from communication in other domains of human experience could reveal why mathematical communication seems to be so difficult. Like language and music, mathematics could be a powerful medium of communication among human beings. But, unlike language and music, experiential bases for mathematics beyond the most basic conceptual operations seems to be lacking. Even though mathematics is universally taught, it is regarded for whatever reason as not being useful in the organization of experience or even relevant in experiential contexts. How this lack of experiential bases for mathematics might be changed through interactive communication is a primary problem of education in mathematics.

Several specific issues also emerged. How the social norms of the school influence the fabric of interactive mathematical communication in classrooms became a major issue because the participants in a mathematics classroom must engage in open and honest mathematical exchanges. Teachers' beliefs in their role, the student's role and the nature of mathematics in traditional classrooms are such that discussions about mathematics are almost non-existent. In settings where negotiation of meaning occurs the teachers view their role as that of a facilitator in helping children to understand mathematics. As a facilitator, the teacher coordinates her knowledge of mathematics with her knowledge of the processes children use to solve problems. In so doing, she draws on her understanding of children's processes to probe their solutions and to offer suggestions which serve in creating learning environments in which problems are debated and resolved. Crucial to these discussions is the teacher's concern for listening to the solutions children offer and to consider them from the perspective that children are trying to make sense of their experiences. As the teacher and students engage in interactions about mathematics, they mutually construct the social norms necessary for the smooth functioning of the discourse. The obligations and expectations for collaborative dialogue are established as the teacher intervenes with the students in cooperative learning activities. Cooperative activity is characterized by temporary working interims that are terminated by the occurrence of problematic situations that arise in the course of the interactions. In the process of resolving these situations and achieving a new working interim, the individuals renegotiate both classroom social norms and institutionalized mathematical practices.

Children with special needs also provide constraints in developing alternative forms of communication. The discussion of these forms led to a broader concern that teachers need to be aware of attitudes and beliefs which might negatively influence their interpretations of children's mathematical understanding. Focusing on children and their sense-making constructs was considered to be essential in developing robust teacher–child relationships that mediate day-to-day mathematical communication.

Environment

An individual establishes an environment by using available conceptual operations. It is nothing more than the result of an assimilation, "the integration of new objects or situations or events into previous schemes" (Piaget, 1980, p. 164). The result of an assimilation of a particular situation is an experience of the situation and this experience constitutes an environment in the here and now. So, a mathematical environment is relative to a particular situation or event and to the individual's available mathematical schemes. In constructivism, situations do not come ready-made. "The problem situations themselves ... are seen, articulated, and approached differently by different cognizing subjects" (von Glasersfeld, 1988, p. 12).

Our understanding of mathematical environments includes those cases where the acting individual uses his or her mathematical schemes in assimilation of representations of sensory experience as well as sensory experience itself (visual, auditory, or tactual, or some combination of these three). One person cannot create a mathematical environment for another person; each person creates his or her own mathematical environment. A teacher can, however, conceive of possible mathematical environments for children based on their observations of children's actual environments. An observer's interpretation of an observed individual's mathematical environment is itself a mathematical environment. Because these two mathematical environments are relative to the assimilatory schemes of the individuals, the two individuals might "see" different things in what to a third person is the "same" mathematical situation.

Given these notions concerning environment, the following two questions were posed for the work of the action group: *"How can we conceive of mathematical environments of children?"* and *"How can teachers influence mathematical environments of children to facilitate their construction of mathematical knowledge?"* Although the first question might seem to be already answered, H. Sinclair (1988) shattered any such illusion: "Unfortunately, it appears to be extremely difficult to define 'mathematical contexts', especially with reference to young children. Given the very general basis for the construction of logico-mathematical operations ... almost any situation that can be commented upon, asked about, indicated as desirable etc. can lead to actions, utterances, gestures or other communicative acts that have something to do with logic or mathematics" (p. 11). Sinclair's observation has to serve as a counter-point for those mathematics educators who would use only a particular collection of tasks as constituting what children should encounter in school mathematics.

The two questions, especially the second one, do imply reciprocal relations of a child with his or her external reality. When a child uses his or her mathematical schemes successfully in a given situation, the schemes are viable with respect to that situation. It might be possible that certain elements of the situation that are apparent to an observer are unintentionally disregarded by the child and could be seen only if the child modified his or her schemes. These elements are candidates for mathematics learning if certain constraints can be established in the child's environment that the child wants to overcome and that might lead the child to modify his or her ways and means of operating. If successful, the modifications would permit the child to see the situation from a new perspective — the child would reorganize a given environment. Given that an environment as we perceive it is our invention (von Foerster, 1984), two issues were identified for the work of the action group.

1. The nature of mathematical curriculum in early childhood.
   a) What might "mathematics curriculum" mean?
      I. How does the knowledge of children and teachers relate to mathematics curriculum?
      II. How does mathematics learning and teaching relate to mathematics curriculum?
   b) What do we mean by "possible mathematical environments" and how do they relate to curriculum?
   c) How can a teacher establish a possible mathematical environment to facilitate the children's construction of some part of their mathematical knowledge?
2. The nature of curriculum in early childhood teacher education.
   a) What is the nature and object of communication in the teacher education classroom concerning:
      I. Mathematics?
      II. Teaching mathematics in early childhood classes?
      III. Children's mathematical knowledge and learning?
      IV. Available technology adapted to early childhood?
   b) What are the products of such communication from the point of view of:
      I. The preservice teacher?
      II. The mathematics educator?
   c) What methods and technology could we use to help preservice teachers of early childhood learn with respect to (1) above?

Curriculum as a post hoc activity

If the conceptual schemes that constitute mathematical knowledge cannot be conveyed or transported from one person to another by words of the language, then according to von Glasersfeld (1988), the task of education becomes one of "inferring ... models of the students' conceptual constructs and then generating hypotheses as to how the students could be given the opportunity to modify their structures so that they lead to mathematical actions that might be considered compatible with the instructor's expectations and goals" (p. 6). Although von Glasersfeld's comment identifies what has to be done, it leaves mathematics educators with the job of learning how to do it. Toward this end, T. Kieren (1988) identified a view of mathematics that seems to be essential: "One such view of mathematics ... is to see mathematics as the personal building of patterns involving distinguishing, modifying, and using patterns and forming patterns on patterns and patterns of patterns. Such a view of mathematics sees it as interactive with the rest of a knower's life ... and is clearly historical in a personal sense ... Finally, it is a personally active view of mathematics" (pp. 4–5). Within this view of mathematics, Kieren isolated several features of a facilitating mathematical environment in which a young child is to be an active abstracter and language user in building personal mathematics. He calls the features "categories of awareness" — multifaceted knowledge (Lawler's bricolage); complementarities of understanding (form and substance); mathematical variations and complementarities (e.g., fraction as quantity and ratio); development (Easley and Taylor's conceptual diversity); and potential for pattern and pattern growth.

Another exciting possibility for reform in mathematics education resides in the transformation of what is taken as mathematics curricula in teacher education. L. Burton (1988) challenged several beliefs about early childhood mathematics teacher education. She identified these beliefs as the simplicity belief (*because mathematics becomes more complex throughout schooling, its learning 'must develop from simple to difficult*), the belief in enjoyment (*e.g. focusing on making books more brightly colored rather than focusing on meeting mathematical challenges*), the belief in 'reality' (*objective mathematical or physical reality*), and the empty vessel belief (*X knows something that Y must learn. It is X's job to fill Y up with the knowledge and skills*). After developing several consequences of holding these beliefs, she concludes that "Whereas a syllabus is a list of desirable knowledge/skills, the achieved curriculum is demonstrated by the pupil's engagement and resultant understandings. A teacher is in no position to dictate that output in advance. So, defining a curriculum for each individual is a *post hoc* task for the teacher. He or she does this by closely observing and listening to children as they engage with the challenges provided and discuss their actions. Central to this style of teaching/learning is social interaction, control and personal interpretation within a group. Equally important is for the teacher to interact with children intermittently and *not* to dominate the exchanges or the interpretation" (p. 14). Burton is especially optimistic that such an approach in the education of teachers holds the promise of recognizing "the ceiling which mathematics schemes can place on children's mathematical behaviour, constraining them to perform within prescribed limits" (p. 15). But she interprets these "limits" as going far beyond the limits imposed by computationally oriented curricula.

R. O. Ohuche (1988) provided a valuable insight into the nature of social interaction in mathematical learning and development in Third World situations. "Open-ended inquiry is a non-starter in societies where children are not usually encouraged to ask questions of adults or of other children. There are indeed investigations in Third World situations which have tended to point out that individualized instruction may not be as prize-worthy in the Third World as it is in the Industrialized countries of the Western World" (p. 4). Based on Ohuche's observation, it seems critical to investigate the nature and role of social interaction across different cultures to more fully understand its relation to mathematics learning and development. Ohuche also provides an interpretation of "intended curriculum", "implemented curriculum", and "attained curriculum", all of which have interpretations in curriculum theory based on traditional theories of knowledge. His interpretations provide serious challenges to more traditional interpretations and point to the necessity for a reformulation of curriculum theory within constructivism. For example, in the case of intended curriculum, Ohuche comments that "This is usually a national, regional, or local prescription of content, objectives and suggested activities and materials. Yet, in the constructionist view the child is expected to learn through his personal constructions. Thus, some have considered it healthier to view the intended curriculum from the perspective of setting the criteria to be used in selecting the concepts and processes that may form the basis of the activities of the learner" (pp. 3–4). This is a much more difficult task and cannot be accomplished in the absence of actual teaching and learning. As pointed out by both Cobb and Ohuche, it is contextual and we should not expect to be able to prescribe a universal curriculum.

Discussion of curriculum

The choice of mathematical content and its nature that might be appropriate for early childhood was debated at length in the discussion group devoted to mathematics curriculum in early childhood. Taking the environments of children as one basis for their construction of mathematical concepts and operations was advocated because the belief that children *do construct* their mathematical concepts and operations outside as well as inside of schooling makes everything about early childhood mathematics curricula problematic. The notion of created curricula — those mathematical situations created in the context of ongoing living and learning — is essential in a constructivist framework. Created curricula are not *a priori* and can be only anticipated — not prescribed. They are compatible but not identical with a child-centered approach to curriculum that was advocated by M. Sumio in which mathematics, science, and language would be drawn out of everyday situations. They are compatible also with the relatively free forms of early childhood mathematics curriculum in most English and some Australian schools. However, they were not compatible with the mathematics curriculum in countries where there is a "top–down" approach or where mathematics textbooks functionally determine the curriculum. Representatives of these countries or traditions were concerned with the traditional scope and sequence of mathematics curriculum and felt that something would be needed to replace it. The notion of abstracted curricula was suggested.

Abstracted curricula are models created as a result of participation in ongoing mathematics teaching and learning that contain a specification of (1) mathematical concepts and operations of the involved students and itineraries of their construction including time estimates, (2) possible learning periods of the concepts and operations along with the goals and intentions of the teacher within these learning periods, (3) problem situations that can serve in the establishment and modification of mathematical learning environments by students in the learning periods, (4) critical decisions that can be made in posing the problem situations to students, and (5) sample interactive communication that can be encouraged among the participants within the learning periods.

Specific features of mathematics curricula in early childhood were identified. Some group members favored eliminating standard computational algorithms to allow children to construct numerical concepts and operations in various experiential contexts. Another specific recommendation that emerged was that much more time be given over to the development of mathematical concepts and operations in spatial and measurement contexts, both of which are currently neglected in favor of computational algorithms.

Several recommendations emerged in the case of mathematics teacher educators in early childhood. The argument was made that since teacher education students bring their own experience of the learning and teaching of mathematics to higher education, these personal experiences should be a starting point for their further explorations in mathematics education. This was taken as a crucial feature of the learning environments they establish in their teacher education courses.

Conceiving of possible learning environments that take into account the knowledge of their students and situations which might lead students to create such new environments should be expected of teacher education students. Such situations should "fit" certain elements of the mathematical knowledge of their students but should contain constraints the students might want to overcome but cannot unless they modify those elements of their current knowledge. This implies that teacher education students should work first-hand with children. These experiential encounters with children might be directed toward determining the children's current levels of development of certain elements of their mathematical knowledge, isolating zones of potential development, establishing actual learning environments, and encouraging mathematical communication among and with the children. It is through these types of experiences that teacher education students can learn what it means for children to learn mathematics as well as what it means for teachers to teach mathematics.

An issue that arose concerned the nature, quality, and extent of the mathematical knowledge teachers need in order to function independently in the context of ongoing teaching and learning. This was identified as a critical issue that is currently receiving little or no attention in the mathematics education community world-wide and it is recommended that work be undertaken in attempts to provide insight into the issue. The issue might be recast as isolating networks of mathematical concepts and operations that could serve in the organization of experience and deepen, unify, and extend the teacher education students' conceptions of mathematics curricula. This network of mathematical relations would transcend the mathematical knowledge of children but include it as a special case.

It was suggested that teacher educators make explicit the actions they find themselves carrying out as they conceive of possible learning environments for their students and as they encourage their students to establish and maintain mathematical learning environments. There are classroom management issues, styles of questioning, and expectations of the students implicit in the actions of mathematics teacher educators, that, if made explicit, would go a long way toward describing the nature and object of communication in mathematics classrooms. It was also recommended that teacher educators specify the special problems associated with their students making a transition from a "transmission-of-knowledge" view of teaching to a constructivist view. The majority of the participants pointed to the need for mathematics and mathematics education classes not to work in opposition to one another. As one participant put it, "Teacher lust (the desire to have the learner see things the teacher's way) stops pupils and teachers alike from exploring their own problems of learning and teaching mathematics".

The processes by which students construct their knowledge in mathematical learning environments is a topic of central concern in teacher education. Case studies which focus on processes of construction would help to specify them more fully. Finally, because constructing knowledge about mathematics and how it is learned and taught is a process which continues throughout one's professional career, it was recommended that forums be provided where practicing teachers of early childhood and early childhood teacher education can share their experiences.

In many of the countries where is a "top–down" approach to curriculum or where textbooks functionally determine the curriculum, the teacher education programs for early childhood require only one or two courses in each of mathematics and mathematics education. These educational programs are founded on the beliefs L. Burton challenged and are compatible with the assumption that specialized knowledge is not required for teaching mathematics in early childhood. The substantive results of the work of the action group challenges this assumption in the most fundamental way possible.

Final Comments

The issues, problems, and opportunities presented by constructivism for mathematics education in early childhood offer hope that the mathematical education of children can be continually improved in the future. Although there might be some sense of dejavu when trying to improve children's mathematical experiences using constructivist' principles, the principles encourage us to decenter and try to see the children's experiences from the children's point of view. The deceptively simple insight that the mathematical experiences of children are inaccessible to us and that we can only establish a consensual domain of experience is enough to initiate reforms in how we adults practice mathematics education. When coupled with the revolutionary aspect of constructivism, we begin to appreciate more fully how limited our educational practices are when children's mathematical realities are essentially ignored.

If children's mathematical knowledge is accepted as being legitimate mathematics, not mathematical misconceptions, then we accepting adults have the exciting challenge to learn it. This challenge is accentuated if we realize that children's mathematics can be an active, dynamically changing mathematics that is used by children in adaptations in their environments. When viewed in this way, understanding how the adults who are responsible for children's mathematics education might facilitate those adaptations becomes equally challenging. Reform in at least the constitutive characteristics of mathematics curricula and mathematics teaching is called for. But the most important reforms are perhaps those that occur in our own knowledge and beliefs concerning mathematics, children, and education.

I found writing this report to be a learning experience. Although it is no substitute for the contributions and comments made in Budapest nor for the experience of "being there", I hope that it will be useful to other participants of the action group as a retrospective document. It does provide a "thumbnail " sketch of possible reforms concerning early childhood mathematics education. Perhaps by engaging in the work to carry out some of those possible reforms, we can improve the education of children in mathematics and come to a better understanding of what it might mean.

Acknowledgement: Appreciation is extended to George Stanic, Ernst von Glasersfeld, and Robert Lawler for their comments on earlier drafts of this report. I especially thank Paul Cobb, Terry Wood, and the abstracters (Brenda Denvir, Nick James, Niel Pateman, George Stanic, Max Stephens, and Grayson Wheatley) for their written documentation of the discussions that transpired during ICME 6.

References

(Papers denoted by **(A1)** were specially prepared for this Action Group)

Burton, L. (1988): What could teacher education be like for prospective teachers of mathematics in early childhood — with particular reference to the environment. **(A1)**

Cobb, P. (1988): Multiple perspectives. **(A1)**

Easley, J. – Taylor, H. (1988): Conceptual splatter in peer dialogue in selected Japanese and US first grade mathematics classrooms. **(A1)**

Kieren, T. (1988): Children's mathematics/mathematics for children: Notes on mathematical environments for young children. **(A1)**

Lawler, R. W. (1988): Constructing knowledge from interactions. **(A1)**

Leino, J.: Knowledge and learning in mathematics. **(A1)**

Lowenthal, F. (1988): Communication in early childhood. **(A1)**

Marton, F. - Neuman, D. (1988): Constructivism versus constitutialism. Some implications for the first mathematics education. **(A1)**

Ohuche, R. O. (1988): Early childhood mathematics and the environment. **(A1)**

Piaget, J. (1980): Schemes of action and language learning. In Piatelli–Palmarine, M. (Ed.): *Language and learning: The debate between Jean Piaget and Noam Chomsky*. Cambridge: Harvard University Press.

Sinclair, H. (1987): Constructivism and the psychology of mathematics. In Bergeron, J. - Herscovics, N. - Kieran, C. (Eds.): *Proceedings of the Eleventh Meeting of the International Group on the Psychology of Mathematics Education*. Montreal.

Sinclair, H. (1988): Learning: The interactive re-creation of knowledge. **(A1)**

Steffe, L. P. (1988): Lexical and syntactical meanings: Brenda, Tarus, and James. In Steffe, L.P. - Cobb, P. (1988): *Construction of arithmetical meanings and strategies*. New York: Springer Verlag.

von Foerster, H. (1984): On constructing a reality. In Watylawick, P. (Ed.): *The invented reality*. New York: Norton.

von Glasersfeld, E. (1987): Constructivism in education. For *International encyclopedia of education*, Pergamon Press.

von Glasersfeld, E. (1988): Environment and communication. **(A1)**

Supporting survey presentation:

Bob Perry (Australia): Cultural Prospectives on Success in Early Childhood Mathematics.

ACTION GROUP 2: ELEMENTARY SCHOOL

Chief organizer: Jacques Colomb (France)
Hungarian coordinator: Zoltán Kovács
Panel: Emilio Lluis (Mexico), Edward Rathmell (USA), Hilary Shuard (UK), Leen Streefland (Netherlands)

The organizers of Action Group 2, which had 350 participants, had first to identify topics of importance in elementary school mathematical education for the topic group organization.

According to recent developments of research in education, they chose topics centered on the learning process, is presently a major direction of work.

It is now generally admitted that mathematical concepts must be *constructed* by the pupils themselves in various situations. This construction requires a very long time. So for the teachers the main problem is to manage the "long term learning process", which was the first group's topic.

In other respects, this construction process implies that the teachers give full attention to "problem solving and errors in the learning process", which was the second group's topic.

The third group's topic dealt with "the place of new technologies in curricula", a most important problem in all countries.

In the following pages you will find the reports of the three topic groups. These reports were compiled by Leen Streefland (Netherlands), Walter Szetela (Canada) and Michel Merigot (France) respectively.

1. Long Term Learning Processes

In this topic a number of different aspects of the curriculum have been considered:

Ed de Moor (Netherlands) spoke about mental computation and estimation — "an everlasting care". He stated that mental computation and estimation are the most important subjects in primary and secondary mathematics education. The reasons for this are:

- wide applicability, as well in mathematics and applied situations (physics etc) as in everyday situations (horizontal mathematization);
- they are a foundation of necessary conditions for progression in mathematics itself (vertical mathematization);
- development of good number sense and insight into the number system;
- development of a critical and flexible attitude towards mathematics, especially for numbers and computation with numbers.

Mental computation was characterized by using number properties, relations between numbers and operations, knowledge of basic facts and practicing estimation. Mixtures of mental computation, paper and pencil procedures and using a (simple) handheld calculator can be used in estimation activities. Mental computation and estimation should not be considered as separate subjects. In most applications both skills are required. Measuring numbers are frequently used in applied problems. The essence of solving such problems concerns having the courage and ability to be able to choose an appropriate measuring unit (standard or non standard) and doing a rough calculation with these measuring numbers. In order to achieve such an attitude the children should be confronted with problems which lack number information ('About how many cars are there in a traffic jam 3 kilometer?'). In this way the students are forced to create their own personal system of measures as a set of points of reference. Most of this measuring, working with large numbers included, should also be done mentally. Mental computation and estimation need to be subjects integrated in a vertically planned mathematics curriculum. Activities should be done every day during short, oral and interactive sessions. Basic facts should be taught repeatedly in order to become memorized facts. In the meantime the problems which are offered may increase in difficulty and complexity. All that has been learned earlier can be deepened and understood better and better. Facts and concepts should be connected with other knowledge in order to produce relations between all the things learned before. From this point of view mental computation and estimation can be regarded as a long term learning process.

Hans ter Heege (Netherlands) reported on a new course for teaching elementary multiplication in which children's own constructions are stressed. Three stages can be indicated in the instruction process. In the first stage orientation on the concept of multiplication is stressed. The formal notation of multiplication is introduced meanwhile.

Secondly the emphasis is on the construction of thinking strategies for the solution of multiplication problems. The basic multiplication facts become more and more factual knowledge. During the third stage the emphasis is on the consolidation of the declarative and procedural knowledge. The knowledge of strategies and multiplication facts can be applied in problem situations.

The essence of this new approach is that children are offered opportunities to construct the knowledge of basic multiplication facts in their own way, starting from their informal knowledge. The flexibility to apply the acquired knowledge in new situations is important.

A number of contributions were concerned with the algorithmization of the operations of subtraction, multiplication and division with natural numbers.

Eigenheer Bertoni (Brazil) discussed subtraction. The present project aimed at the improvement of the teaching of mathematics. Like the following contribution about multiplication the search was for a natural transition from intuitive understanding and the manipulation of actual data and/or concrete material to a mathematical formalization. In addition involving tens and ones a successful construction of the standard algorithm was obtained. Concerning subtraction, however, in situations referring to taking a number away from another and seeing how much will remain, the children constructed alternative algorithms and rejected the standard one. Moreover they exploited several other situations of comparison, constructing their own structures for each one of the situations and began to understand analogies between these situations and the previous one. They obtained a convergence to some algorithms, which they used indistinctly in any of the situations. It has been concluded that the children's own ways of thinking must be learned and respected. Thus, the teacher will be able to argue in accordance with their logic and to help them to find out ways of systematization and more concise structures.

Mousinho Guidi (Brazil) spoke about multiplication, a passage between the concrete and the formal. It often happens that children who are able to do mathematical operations in daily life cannot do these operations as learned in school. This occurs because they cannot construct the relation between their way of thinking in real life and the processes they learn at school.

In a special project the child and its thinking were brought together and ways were found to obtain the desired transition. Concrete situations served as a starting point and were connected to the students' reality through the use of didactical materials on which they operate informally. After the students have grasped the process involved, symbols were introduced gradually as a representation of what they were doing in the concrete situations.

When the students are familiar with the symbols, activities will be suggested on the symbolic level, although the concrete material can still be used. During this phase, in which the students use both the symbols and the materials, they gradually evolve efficient algorithms, which can vary from child to child and which (can) converge to one or more standardized ones. After this stage we try to arrive at a standard procedure which will be normally applied later.

When the student has successfully passed through this phase, he can apply in real life what has been learned in school, passing smoothly from the concrete to the formal.

Els Feijs (Netherlands) considered long division according to progressive schematization . The traditional algorithm for long division causes great difficulty for many children, as shown by international research. In new standards for mathematics education in some countries the importance of column arithmetic and especially of long division has been reduced substantially. In an alternative approach to the teaching–learning process however it is not necessary nor preferable to do so. A course has been developed in which long division starts with repeated subtractions: piecewise distribution produces a long tail, which by taking ever bigger groups of tens or hundreds, and so on, is shortened further. This so called column arithmetic according to progressive schematization can be characterized as follows:

1. The course of column arithmetic is based on clever calculating by using relevant properties and rules. Estimation permanently plays an important part.

2. Context problems serve the process of algorithmization. They evoke schematizing and shortening at every important stage of the learning process. Applicability is increased by connecting the formal arithmetic procedures and the pupils' informal methods of solving context problems.

3. In the process of algorithmization the informal methods pupils apply to solve context problems are being used. This approach offers children the opportunity to develop their own standard algorithms. As a consequence they discover rules for calculation, which in the course of the historical learning process only achieved their final shape after centuries of development.

4. From the beginning problems with rather large numbers are offered which will be solved at an adapted level of schematization and shortening.

5. Calculations and notations are being schematized and shortened more and more as the course proceeds. Pupils may run through the course at their own pace.

6. The levels of competence can vary according to the objectives.

Tom Kieven (Canada) talked on fractional number knowledge: a growing whole? The purpose of this contribution was to contrast an example of a traditional, computational language oriented hierarchy of fractional knowledge with a holistic growth oriented scheme. To be sure that the traditional sequence represents change and even a hierarchy, knowledge at a higher level should imply knowledge at a lower level. But because the knowledge is not well connected, and because such purely computational language-based knowledge can easily be only nominal, this sequence is not likely to promote long term growth for most people. On the other hand the growth scheme sees fractional knowledge as a personal integrated growing whole with later knowledge elaborating previous knowledge in a connected manner. In general, the first sequence sees as legitimate only the mathematics of symbols. The second sequence legitimates and uses mathematical language patterns and properties with the *mathematics of physical action*: the mathematics which integrates thinking tools such as dividing up equally, imagery and informal use of mathematical language: *intuitive mathematical knowledge*. It also legitimates and ties to the other mathematics of fractions, knowledge *built* through the study of manipulation of *symbolic patterns*. If one intends to see a person's knowledge as a growing whole in action, intuitive and symbolic fractional knowledge are all close together at some level. Thus one can proceed in instruction from the active to the symbolic, but also in the other direction as long as the connections are there. The hope for long term retention of fractional knowledge, rather then easily forgotten nominal knowledge, is through development of curricula which foster such growth.

Alan Bell (UK) dealt with decimal numbers. Studies of normal conceptual development and common conceptual obstacles in this field were presented. As a consequence the design of teaching material was considered, including games and conflict situations. Stress was put on the use of calculators in this concept-based teaching of decimals. Several teaching situations were presented. In order to understand the decimal notation one should start with situations of mensuration. The use of calculators demands great insight in number structure and operations. That is why among other things attention was paid to the production of decimal number situations by the pupils themselves, based on advertisement material. Pupils' errors and successes and the role of the calculator to help to develop understanding of decimals and their operations were emphasized in particular.

Kath Hart (UK) discussed ratio and proportion. Proportional reasoning is usually thought to be indicative of high level cognition and in Piaget's theory it was indicative of the stage of formal operations. Young children can judge qualitatively whether an object is similar to another but they have considerable difficulty in calculating exact relationships. The research projects "Concepts in Secondary Mathematics" (CSMS) and "Strategies and Errors in Secondary Mathematics" (SESM) have investigated the performance of children aged 11–16 years. During this secondary school period there is an emphasis on the calculations involved in ratio and proportion questions.

There is however, within the performance of this age range, evidence of naive methods being used rather than more generalizable methods of solution. These are limiting in a way very similar to those methods used by children who can only reason qualitatively.

Guy Brousseau (France) discussed the problem of linking up school levels (problems of apprenticeship or problems of didactics). He argued that more often than not secondary teachers do not have possibilities to recall, to apply and to check their pupils' knowledge, which is tied to contexts, personal experiences and so on, unless this knowledge has been institutionalized and connected with cultural examples, which then can be recognized as such. There exist epistemological and didactical discontinuities between school levels which are unavoidable. This is shown by what pupils learned and understood in the past, but what they are unable to recall and apply at the next school level or in the next class. As a consequence the teachers will tend to infer unjustly from this, that the goals of the previous class have not been attained. Because of attempts to reach the goals of the previous level the teachers will get into a vicious circle. The knowledge of the pupils, which is tied to contexts and personal experiences will disappear because of striving for the lower goals, which means algorithms and rules. In order to master this problem it has been suggested that the phenomenon as considered seems to demand specific invented solutions for each type of transmitted knowledge, provided they are discussed and negotiated publicly with everyone involved in instruction.

Following the presentations, the discussions fell into three main categories, detailed below:

Content

Using problem situations in contexts as starting points for mathematical activity was considered as very important. The processes of algorithmization, subtraction, multiplication and division for instance should start that way with rather large numbers. Mental computations and estimation will shape the course of clever calculation, which eventually can and will lead into procedures serving as (personal) algorithms. So this is a matter of child generated algorithms instead of (or vs.) the conventional ones. Within this process children's strategies (informal methods, intuitive procedures, natural approaches) play an important part. In other words: children's own constructions (which can be characterized by their gradual abbreviation during the course, growing in effectiveness and efficiency) are at the core of the teaching/learning process. This also applies to the other topic areas as mentioned. In general one might say a conflict exists between the way textbooks have been organized topicwise and the demands long term learning asks for. It was felt that the didactics of long term teaching (and learning) have been neglected for too long.

Children's own strategies

The discussions were led by the question: how to deal with children's own strategies in the teaching/learning process? Among others the following questions were raised in relation to the previous one:

- What teaching approaches are compatible with the development of children's own strategies?
- Children's informal strategies and procedures, can they be changed into institutionalized strategies and knowledge; or, in other words, will it be possible to adapt *long term* teaching to the children's strategies?
- Which qualities do children's strategies have?

Of course children's own strategies reflect different levels of mathematization and moreover their strategies are not always correct or they have other drawbacks. But on the other hand they can long persist against teaching methods, if teaching does not take them into account, as was shown by research (Bell, Hart). Hart's research on ratio and proportion problems for instance, showed the persistence of naive methods over a long period of time, faulty additive strategies included. These results plead strongly in favour of using the observed naive solution methods as starting points for teaching and to try to reshape these methods in more generalized, multiplicative ones. (It is unnecessary to say that the pupils will need mathematical tools, for instance the ratio table, which fit the observed naive methods).

This will also favour recognizing the pedagogical argument of giving the children confidence by taking into account their methods and strategies (mistakes included) as worthwhile and as building blocks for teaching. With respect to this the following dilemma will have to be faced:

- on the one hand attaining the goals of instruction, which at present often means going too fast through the curriculum, and
- on the other hand meeting the need of the development of children's own strategies.

Several different solutions were suggested in the group discussion, like:

- taking into account children's approaches unconditionally (many children have success with their methods, although teachers may not always be aware of this fact);
- putting the emphasis on the teacher and the teaching process in order to reach the goals (cf. Brousseau).

In general one might state that the work in the group dealt with bridging the gap between teaching and learning, or, to express it in a different way: How can teaching and learning be brought into line in such a way, that proceeding in teaching will mean meeting children's progression in learning mathematics in an unhindered manner, even when subsequent school levels are involved? (With respect to this last point Brousseau argued that teachers in secondary education more often than not do not deal with the personal frame of reference within which pupils' knowledge has grown in primary school. Their starting point almost always concerns the formal system of mathematics and as a consequence an unbridgeable gap is raised between the knowledge of the students and the teacher. As a solution for this problem he suggested the invention of specific approaches for each type of knowledge to be transmitted.)

As provocative statements for the discussion the group as a whole suggested:

- Most of the actual teaching methods (approaches) are not compatible with the development of children's own strategies. (This especially concerns one's point of view with respect to mathematics (human activity vs. acquired body of deductively structured knowledge)).
- Teachers need to have confidence in their own possibilities in teaching in order to be able to have confidence in the effectiveness and quality of children's own strategies.
- The goals for primary mathematics education can be attained by *reconstructing* children's own strategies into the more institutionalized ones in the teaching/learning process.

In the plenary discussion it was suggested that these statements should have been introductory ones for this group instead of having the character of concluding remarks. One might object that no mathematics programs have been developed thus far, which take into account children's informal knowledge and procedures, for instance with respect to the visual perceptual reality of the child as a source for geometrical activities which aim at spatial orientation or as a source for geometrical ratios, to start with at the age of 5–7 year old, or which meet children's informal knowledge of fractions. This means that the goals to be attained at primary level will also have to be reconsidered, if children's own strategies are token be taken seriously.

Proposed theoretical framework for future development and discussion: Reconstructive Learning.

In order to make reconstructive learning possible, which means dealing with children's own strategies in a long term perspective, education developmental research will have to be carried out. This is research in action. Such research aims at developing prototype courses, especially for refractory subject areas like fractions and ratio and proportion or algebra in secondary education, and theory building for teaching and learning these subjects. Experiments in instruction start with provisional material based on a thorough mathematical/didactical analysis. The teaching/learning process needs to be both closely as well as permanently observed. Continual observation and recording of individual learning processes is at the heart of the research. What matters is that the strategies of the pupils (which means their constructions and productions) *are used for both building and shaping the teaching course.*

In the variety of children's possible responses, that is responses to open, construction and production evoking problems, one will find a rich choice to find what is best, enable one to determine and in the long run the most effective. Blocking and diverting material will be eliminated. In this kind of design children, by their learning processes, decisively influence course development, which nourishes the source for creating reconstructible instruction. The prototype as developed can serve as a model for establishing and developing derived courses. Such potential instruction is predesigned in textbooks and manuals. This demands realistic mathematics education based on the following tenets:

- reality serves both as a source for the mathematics to be constructed (concepts, operations, ideas, structures) and as a domain of application;
- children's own constructions and productions are the backbone of the teaching/learning process;
- the working apparatus, that is mathematical tools like terminology, symbols, (visual) models and schemes, needs to be developed in the teaching/learning process;
- the teaching/learning process develops a full measure of interactivity, in order to negotiate, cooperate and to discuss the wealth of constructions and productions to be expected; and finally
- related learning strands like those for mental computation, estimation and algorithmization of the operations with natural numbers or for fractions and ratios or other relative numbers, need to be interwoven from the beginning.

2. Errors in mathematics and the learning process; problem solving and the learning process

Errors and misconceptions in mathematics are of great importance in helping teachers to provide remediation and to plan subsequent instruction. In recent years the traditional negative attitudes toward errors and misconceptions have begun to give way toward much more positive views. These positive views are reflected in several of the papers which were presented in this subgroup. In facing today's challenge of improved instruction in problem solving, errors and misconceptions must be addressed including errors of reasoning and inadequate understanding of problem situations. In this subgroup, eight papers were presented on errors or on problem solving. Following the paper presentations, the participants were divided into three smaller groups to discuss questions raised in the paper presentations. We shall summarize here the paper presentations and then report the substance of the small group discussions.

Roland Charnay (France) presented four types of errors of which the two most important were errors due to lack of knowledge and errors caused by faulty reasoning or misunderstanding. He found that more than two thirds of the errors among French students aged 10–12 were simply due to lack of knowledge of mathematics. The results were based upon a sample of 42 teachers and a total of 110 class sessions. Charnay discussed how teachers tend to intervene when students make errors. Rather than encourage the student to explain his thinking and discover his error, teachers tend to obtain the correct answer without investigating the cause of the error. Thus, the answer is more important than the reasoning. This practice tends to discourage students from answering questions. That is, error is considered an abnormal phenomenon to be quickly erased.

Dianne Siemon's (Australia) paper on error patterns among students performing calculations took a constructivist view. She suggested that students make errors by constructing interpretations that often have some logical basis, but which may not be apparent to the teacher. When teachers do observe that the knowledge constructed by the student is different from the knowledge they have intended to impart, they can analyze the error, get to know what kind of thinking led to the student's error, and provide a better basis for future instruction to prevent such errors.

Tadato Kotagiri (Japan) displayed a variety of elementary computation errors made by Japanese children with whole numbers and decimals. Children learn calculation procedures by rote and their failure to attain understanding of numeration concepts and place value makes it impossible for them to recall the procedures once they are forgotten. Kotagiri's suggestions to improve pupils' understanding by "curative teaching" relies heavily on such manipulative materials as place value boards.

Raffaella Borasi (USA) presented errors as "springboards for inquiry". She views errors as analogous to "getting lost" whereby the student regards the error as a stimulus to work his way toward a goal rather than looking upon the error as a dead end to be corrected by the teacher. She used an example of student definitions of a circle to show how different interpretations may produce different misconceptions. Yet, the different interpretations may help students to attain a correct and deeper understanding of the concept of circle and of related figures when they are discussed. By using errors and misconceptions as springboards for inquiry she suggests that students' motivation will be increased, that they will be more creative, and that they will be more reflective.

Tom Brissenden (UK) focused upon the teacher's instructional model. He proposes a teaching model in which the teacher begins with a situation that promotes the learning of a concept using the whole class, small groups in the class and individuals. The instruction demands a great deal of student–teacher interaction with questions and suggestions. As the lesson continues, the teacher introduces a problem related to the concept introduced earlier. Again the discussion with the problem may involve any unit from the individual to the whole class. In this model it is essential that student participation is maximized. Thus discussion and inquiry play a much more dominant role in this model than in the more traditional expository model. With this approach relational understanding is more likely to occur than rote or instrumental understanding.

Michel Blanc (France) presented the notion of the "situation problem" as a means of expanding conceptual learning to problem solving. He used the example of the concept of an angle in a limited sense (something you measure with a protractor), as opposed to developing the angle conception problem solving situation. Here the students are active learners, not passive receivers of teacher knowledge. The situation problem may lead to other discoveries in addition to the particular goal planned by the teacher.

George Lenchner (USA) provided a variety of examples of interesting problems which would help to enhance problem-solving ability. Problems necessitating the use of the strategies such as guessing and testing, looking for a pattern, solving a simpler problem, etc. were suggested. Lenchner emphasized the importance of the teacher and that the teacher must first become a problem solver before success in teaching problem solving could occur.

Walter Szetela (Canada) analyzed errors in solving problems made by grade 5 students. He showed that the majority of errors were not mechanical but representational. Thus students failed to solve problems not so much because they lacked the mathematical tools but because they failed to obtain a mental picture or understanding of the problem situation. For example in a problem with four numerals, forty percent of the students simply added the four numerals although the problem required two multiplications. His conclusions were in contrast to those of Charnay who had indicated that most errors were due to lack of knowledge rather than lack of reasoning or understanding.

Conclusions and suggestions from small group discussions

In the discussions that followed the paper presentations the following conclusions and suggestions were made:

1. It is necessary to define more clearly what is meant by mistakes as compared to errors and misconceptions.
2. Errors may be dependent on what the teacher expects. An error may be the result of a different yet logical interpretation.
3. Language is a source of many misconceptions.
4. The idea of having children explain their work may be promoted by choosing problems in which the answer is not unique.
5. The role of the teacher is critical. Willingness to proceed slowly with thinking activities may produce better long term knowledge, retention and initiative.
6. The idea of thinking of an error as "getting lost" is useful. Engage the students in "debugging" and challenging common conceptions.
7. Get students to develop as much as possible in developing concepts and principles. Provide opportunities for discovery, for thinking, and for sharing.
8. The classroom should be an enriched environment. Problem solving can provide that kind of environment.
9. We may have limited error analysis to numerical situations. Efforts toward examining errors related to spatial concepts and reasoning may be fruitful.
10. It is important to provide the experience of showing how errors may lead to valuable discoveries.
11. In new situations errors are expected. They help to refine a rule or a theory.
12. We need to provide tools or techniques that allow teachers to examine errors in a more time-efficient way, not like methods used by researchers.
13. It is valuable to highlight certain errors to promote better understanding or clarify misunderstanding.
14. Consider what research tells us about error correction in learning a foreign language. Error correction is sometimes counterproductive.
15. Teachers must acquire skill in analyzing each situation to be able to interpret an error and decide whether or not to intervene. Teacher trainers must take account of this.
16. Problem solving strategies might degenerate into rules.
17. Children should construct their own problems.
18. Small groups may help students to collaborate and communicate as they try to solve problems.
19. Errors are opportunities to investigate misconception of students. They are not really errors but attempts which have failed to construct shared or common meanings. We should invite students to explain what they have done and to reflect upon their actions.

The set of suggestions made in the small groups clearly shows agreement that errors can be used in a positive and productive way but that teachers need to have more positive attitudes toward errors themselves and that they must be able to identify those errors which can benefit students through discussion. The potential for problem solving to enrich the classroom environment and provide a better basis for learning is also clearly recognized.

Questions raised in small group discussions

Following are important questions raised in the same small group discussions.

1. How can we change the negative attitudes toward errors in mathematics as held by teachers and students?

2. It is well-meaning to suggest that errors should be explored rather than immediately corrected. How is this possible in the real world of the classroom where more "time-efficient" procedures are required?

3. How can we promote the use of materials for exploration than simply for demonstration?

4. How long should we allow children to explore before the teacher intervenes, especially when mistakes are being made?

5. How do examination systems, e.g. in the UK, hinder or help instruction for understanding in mathematics?

6. Errors are part of the learning process. How can we choose teaching situations which exploit errors effectively?

7. How can we distinguish between productive errors and trivial or unproductive errors?

8. Are more errors those of logical reasoning or due mainly to lack of knowledge?

9. What types of errors are most significant? How much more might we focus upon reasoning errors rather than arithmetic errors?

10. How and when should strategies for problem solving be taught?

Conclusions and recommendations

The suggestions made by the small groups as given above appear to be sound and promising for improving mathematics instruction. However, the questions raised by the small groups suggest that although there is agreement that teachers should take positive actions in utilizing errors and in providing good problems, the practical implementation of techniques for exploiting errors and problems productively is difficult. Such questions may be answered at least in part by research or may be solved through improved teacher education programs.

Many of the concerns and suggestions that arose in the small group discussions focused upon the ability of the teacher to adopt a teaching style which would provide more opportunity for students to engage in explanations of their own work so that students could discover their own errors and arrive at a fuller understanding of a skill, concept, or principle. A teaching model like that suggested by Brissenden may be effective, but only if the teacher is confident enough to engage in the considerable variety of teacher–pupil interaction which characterizes the model. In order to help teachers develop confidence with such a model, teacher education institutions should concentrate more on the development of better techniques for teaching problem solving than on isolated algorithmic skills. Perhaps greater use of calculators in schools can reduce the time needed to teach calculation skills and provide more time for teaching of problem solving techniques. The amount of in-service assistance normally available is too little to have a significant effect on changing traditional expository teaching techniques to more open and interactive teaching styles.

Teacher educators are more likely to emphasize interactive teaching methods if there is sufficient research evidence that such methods are superior to the traditional expository methods. For example, more research to investigate the effectiveness of teaching with small group discussions might provide teacher educators and their students with enough evidence to motivate instruction to include small group investigations.

In order to develop more positive attitudes toward errors among students, teachers themselves need to develop such attitudes. Again teacher education institutions can be most effective. Every teacher should take a course in problem solving, creative thinking, or some equivalent which can provide situations where errors are naturally made and naturally analyzed and discussed. Inappropriate representations of problem situations, unfruitful strategies, incomplete development of good strategies, insufficient monitoring of progress toward solution, incomplete or unreasonable answers, etc., can provide a wide range of errors that may be fruitful not only in solving a problem but in helping students to acquire a deeper understanding of a principle.

3. The place of new technologies in Curricula

In the 3rd topic, seven speakers presented 6 different subjects about new technologies but only one is a TV project.

H. Shuard and A. Walsh (UK) discussed The Calculator–Aware Number Curriculum. The PIME (Primary Initiatives in Mathematics Education) project in England and Wales, is conducting in major experiment to develop the "Calculator–Aware Number Curriculum". This curriculum takes it for granted that calculators are available to children whenever they are needed. Calculations are normally done in the mind or with a calculator. Work started with about 400 children and their teachers (Sept. 1986); these children were 6 year old. It is intended that the work will continue until the children enter secondary school. Further groups are joining the project each year. Emphasis is placed on the style of working undertaken by teachers as well as on the contents of the curriculum.

The evidence from the teachers and the evaluator suggests that almost all children are making greater progress in number than they did under the normal diet. Many of them have already encountered several mathematical ideas and concepts which would not normally be met until much later. They appear to be able to successfully operate with these more advanced ideas and to have developed intuitive approaches.

The "I.P.T." plan was introduced by L. Corrieu (France). In France computer science was introduced in schools about 20 years ago. In January 1985, the Prime Minister, Laurent Fabius, launched the so called "Computers for everyone" plan ("Informatique Pour Tous") based on three ideas.

1) Initiation of pupils and students of all levels to the computer

2) Training teams of teachers

3) Intention of giving computing facilities to all citizens.

The objective was to place 120000 computers in schools. The software equipment of schools is given in the form of a base of standard software and a right to buy products listed in a special catalogue.

The "I.P.T." plan also includes another base, a technological one. The 1985 primary school syllabus defined a large teaching field called "Sciences and Technolgy" in order to promote scientific ability from primary school onwards and allow the robotic field to develop.

The "I.P.T." plan has enabled a rapid development of computers in the education system and has posed a challenge in teaching methods, but it is still early to assess the progress in teaching methods related to new technologies.

The well known programming language NOLOG was presented by M. Thorne (UK). It allows manipulation of all the elementary properties of numbers including factors, divisibility, remainders and the four fundamental arithmetic operations. Its focus is on creating patterns on the familiar 10×10 number square and relating these to the underlying numbers. Some interesting examples are provided.

NOLOG is very similar to LOGO in its provision of a problem solving environment in which the creative aspects of a child's mathematical thinking are also allowed to flourish. Pupils can have access to a more concrete representation of many numerical relationships.

The next topic discussed was the role of pencil and paper algorithms in a calculator integrated curriculum. Through its influence, modern technology is causing changes in society, in educational objectives and in pedagogical possibilities, especially regarding the role of the traditional pencil-and-paper-computational algorithms of arithmetic.

Effective change will only be possible if we delineate the changed objectives for teaching and learning written computations and teachers change their beliefs and perceptions accordingly. Now, however, the calculator is simply a more efficient tool for certain computations. The change of objectives means that the approach to the teaching of algorithms should be to make the construction of algorithms a problem solving activity for pupils.

The objectives for teaching written-computational algorithms should be to facilitate understanding of number and to develop algorithmic thinking as a mathematical process and not because the algorithms are useful tools. The standard written-computational algorithms should be abandoned in favour of alternative informal and ad hoc algorithms developed by pupils themselves.

Some thoughts about Treasure–Keeping was the title of D. O'Brien's (USA) presentations of some interesting software. The interactive challenge such software provides to young (and old) minds amounts to no less than an educational revolution, allowing educators to do old things better and to do new things never before dreamed of. Then he presented some thoughts about two sorts of knowledge:

– often education consists of teachers' transmission of associations (or chains of associations) to children;

– rather, the knowledge more closely resembles a fabric.

It is a network of information, images, relationships, errors .... They differ very much in their origins, in their accessibility. One of the most important characteristics of mental fabrics in that they begin and end; they generate new knowledge.

There are at least four very important actions that the mental fabric conducts in its encounters with the world: construction, coordination, elaboration and differentiation.

The next question is how to care for, nourish and enrich the treasure entrusted to us for so short a time.

J. Schneider (USA) introduced "Square One TV", an interactive videodisc project. Square one is a TV program about mathematics of 75 half hour TV shows and a total of thirty minutes to be used in the interactive system.

The evaluation consists of three phases:

– in the first phase video selection and disc production tasks were completed;

– in the second phase, the development of the interface, data base and interactive activities will be completed;

– the third phase of the project addresses the pedagogical concerns.

Conclusion

Changing the curriculum in mathematics in order to meet the demands of a rapidly changing world is becoming the concern of all those involved in mathematics education. It seems quite viable to suggest that the general availability of calculators and computers could have a wide ranging effect on the mathematics curriculum (in particular the learning of arithmetic and the new procedures in teaching problem solving).

The re-thinking of the curriculum of these powerful machines would of necessity involve:

– The introduction of new topics

– The other topics would be given less or more emphasis than at present

– The role of professional development of teachers in the use of new technology

– The place of programming in primary education.

ACTION GROUP 3: JUNIOR SECONDARY SCHOOL (ages 11–16)

Chief organizer: **Ichiei Hirabayashi (Japan)**
Hungarian coordinator: **Mara Kovács**
Panel: **Patricia M. Hess (USA), Vadim Monakhov (Soviet Union), Elisabeth E. Oldham (Ireland), Lucia Grugnetti (Italy), Walther L. Fischer (FRG), Hans-Joachim Vollrath (FRG)**

1. The Aims of the Activities and its Evaluation

The aim of this action group was to discuss problems of mathematics education in junior secondary schools, focussing on the characteristic features of this school level and to exchange ideas from many countries on the resolution of these problems.

At ICME 5 Alan Osborne, the chief organizer of the corresponding action group, characterized this school level as *transitional*. We accepted this and sought to develop it further in considering new ideas on mathematics education at this level. Osborne's article in the "Pre-congress Reading" for ICME 5 enumerates the chief problems of mathematics education at this level, and the nine items from that article were very helpful in orienting our discussions. They provide a standard against which to measure the success of work in this field, and heightened our awareness of the many problems with which we were concerned.

Among other things we noted the transition in the nature of mathematics itself, from *arithmetic* to *mathematics*. To mention only one of Osborne's examples, the concept of model changes drastically: in arithmetic in primary school it is a physical reality used to learn mathematical ideas, while in junior secondary and higher schools mathematical ideas themselves gradually become useful models with which to investigate many physical variables.

In my introductory lecture I drew attention to these differences between arithmetic and mathematics and suggested that this change could be symbolized by two concepts: *variable* and *proof*. In trying to understand these concepts many pupils fail in their learning of mathematics. I myself do not believe that it is necessary that every pupil should learn these two concepts in mathematically strict form: rather I think that they should learn mathematics of another kind. However, we should not make any pupil fail to learn if that pupil has enough ability and wishes and needs to learn.

What alternative ways might be available to us for teaching these concepts, and what is the mathematics of another kind that is suited to those pupils who will not need academical mathematics? These problems were discussed in various forms throughout our deliberations, and the aims which I proposed in the first session were, I believe, substantially fulfilled.

2. The Organization of the Action Group

The number of provisional registrations for the group was almost 500, and so the original plans had to be modified. Several sessions were arranged, each chaired by a member of the panel as indicated in the following. To summarize, the first (plenary) session (Hirabayashi) was devoted to a *review of problems;* the second and third sessions ran in parallel and were concerned with *contents* — with subsessions on algebra (Monakhov), geometry (Grugnetti) and probability and statistics (Hirabayashi) — and *methodology* — with subsessions on teachings methods (Fischer) and attitude and apparatus (Oldham). The final (plenary) session (Hess) was used to formulate *conclusions.*

3. Reports from the Sessions

Session 1

The aim of this session was to review the problems mentioned by Osborne at ICME 5 and to recognize new problems that remain unresolved. I hoped that we would be able to report on the question of the progress that had been made in the intervening four years, but as Osborne wrote to me "such a question is ephemeral and difficult to assess". Among other things I wanted to concentrate discussion on the following two problems:

- What are the characteristic features of mathematics education in junior secondary school as the transition from primary school to senior secondary school or to society?
- How can we cope with the wide differences in pupil's mathematical abilities, in designing the curriculum and teaching method?

A. Osborne (USA) spoke on Transition Mathematics in the Junior Secondary School. He recognized firstly that the transition from the study of arithmetic to the study of variable and proof is a most critical stage for a student. Because of this, curriculum development has always been his major concern. He reported his experiences with the project AAN concerning the weaknesses that exist in the teaching of mathematics and some of the strategies to cope with them.

The AAN project uses problems and technology to develop mathematical concepts. The materials for grades 7 and 8 establish the concept of variable before students first encounter algebra. The project has produced significantly greater growth in numerical and algebraic concepts and skills than traditional approaches.

G. Malaty (Finland) considered How Mathematics Education Can Promote the Mind. In this presentation he discussed the question of what is appropriate in teaching mathematics in relation to Pólya's famous words: "Mathematics promotes the mind provided that it is taught and learned appropriately."

As a result of examining the teaching of geometry, it was found that the guided discovery of the idea of proof can make the students *use heuristics* and thus *promote the mind*. For example, if the idea of "bisecting the base" occurs in a proof, the question would be "WHY?"

The general finding is that the guidance in *causal thinking* is one of the main attributes of teaching, which can promote the mind, when the question "WHY?" becomes the core in teaching a term, a symbol, a principle, in building structures, solving a problem, etc.. The normal language, etymology and the history of mathematics can help in answering the *WHYs*, but the main tool which has to be used is the *students' guided causal thinking*.

S. O. Ale (Nigeria) talked about Junior Secondary School Mathematics in Nigeria — Issues and Problems. He distributed a paper and explained the present state of mathematics education of Nigeria and indicated serious problems that they have. In his summary we read:

"It is only about six years ago that the educational system changed, to become 6–3–3–4. In part, this implies that a student spends three years each in Junior and Secondary Schools. Unfortunately this system did not start uniformly in every state of the country. In particular, the Junior Secondary Mathematics is on trial. It contents and philosophy are mathematically sound but there are problems in implementation and methodology."

Session 2

Subsession 1 — Algebra

The main themes of this session were:
– What kind of algebra should be taught?
– How do we establish the idea of variable?

P-T. Chang (USA) discussed Strategies for Teaching Algebra in Junior High School. He reported his experiences with making algebraic concepts more concrete by relating applications to the everyday life of the student, by using a matrix to illustrate multiplication of polynomials, and by providing investigations in Pascal's Triangle.

J. D. Austin (USA) talked about Algebra Teaching in the United States — Practice, Problems and Selected Research. He also reported the poor problem solving ability of American students as evidenced by the results of several national studies, and on the efforts to improve this situation under the leadership of National Council of Teachers of Mathematics by developing standards in teaching algebra that will emphasize applications and the use of technology.

M. Koyama (Japan) spoke on Intuition in Learning Algebra. He argued that there is a significant "jump" in mathematical thinking between arithmetic and algebra, which many students fail to make. To achieve this jump we should call into question not only the students' formal skills but also their recognition in learning algebra. He tried to approach the problem of recognition by focussing on intuition. He defined *intuition* as being like a function of recognition which grasps directly the entire picture, essence, significance, meaning and structure of a perceived or concrete object. Human thinking could be developed productively through the cooperation of intuition and logic. In teaching and learning algebra, not only numerical–literal language which has rigid syntactical rules but also diagrammatic language which is reach in intuitions plays an important role.

M. Avital (Israel) spoke about Educating Children to Read Mathematics. He edits a periodical in Canada — *Fun With Mathematics* — for grades 7-10, and one in Israel — *Gilyonoth Leheshbon* — for grades 9-11. A survey of subscribers in Israel has shown that even slightly above average students in Middle School can be educated to read and comprehend mathematics written at his/her level. The major factors involved are: attractive material — not only problems but also of an explanatory nature; an appropriate classroom climate; keeping subscription rates down so that the students themselves buy the journal; good explanatory material to be spread out over a sequence of issues (examples: strip ornaments, tiling the plane with regular polygons, searching for Pythagorean triples, etc.)

Subsession 2 — Geometry

In this subsession the following two problems were addressed;
- How do we help pupils in understanding the nature of *proof* through teaching of geometry?
- Euclidean geometry, is it really needed? For what pupils and how far?

P. Lobry (France) raised some Geometrical Problems Connected with the Representation of Space. In the spirit of R. Bkouchke's remark, "the idea of space is born when it becomes necessary to make the mutual relations between bodies explicit or by the planar representation of spatial situation". She discussed the problems through some examples: to draw the contour of cubic, the intersection of a sphere with a plane, etc.

F. Verrier (France), in discussing How to Draw Objects in Space, argued that the teaching of geometry depends on political opinion, and that once the teaching of geometry has been decided upon, the starting point for thinking about its teaching cannot depend on didactical observation of pupils but on the subject matter of geometry itself. Two major problems arise, "measurement" and "plane representation of spatial objects". He gave some thoughts about the nature of a few representations, the type of geometry connected with them and the determination of the didactic steps to be taken with pupils through their curriculum. He also suggested that the problem of the true representation of the objects from a given perspective returns us to the problem of interpreting a drawing, which is too often dependent upon a code.

B. Denys (France) reported on Procedures used by Japanese and French Junior High School Students to solve the same given Geometrical Task. The students' work on orthogonal symmetry was analysed with three purposes in mind:

- to relate the characteristics of the given figure and the procedures used by the students,
- to identify, after teaching of the notion, the properties and knowledge available to the students,
- to determine the similarities and the differences of strategies and conceptions of both population.

The similarities confirmed the results of the French research: availability of "conservation of distance", etc.. The differences related to what they were taught and their strategies appeared to be influenced by learning and culture. This presentation was related to the following one.

D. Grenier (France) spoke about An Activity of Construction and Formation after Teaching (12 year old pupils). In undertaking a didactical analysis, she argued that two aspects of orthogonal transformation will interact, one related to the familiar notion of bilateral symmetry, and the other to the mathematical concept. Some properties of the cultural notion correspond to the mathematical concept but others are not compatible and may lead pupils to locally erroneous conceptions.

"Research undertaken in several countries revealed that, even after a teaching period, some conceptions related to the cultural notion remain very influential and may mislead pupils..."

In the discussion the role of models and real problems was emphasized. Some remarks was made by L. Grugnetti (Chair) about the role of geometric transformations and the possibility of introducing some "Euclidean" proofs by geometric transformations. Some examples were given.

Finally it was emphasized that a complex network of inter-relationships exits between three key elements:

- space, the intuitive aspects, real problems and models,
- the plane, with consideration of transformations and measures,
- space, concerning drawing and also transformations and measures.

Subsession 3 — Probability and Statistics

Two main problems were posed for discussion in this subsession:

- Are not these branches of mathematics the most useful for all pupils?
- How can we use the calculator or personal computer to achieve a good command of these branches of mathematics?

The presentations and discussion ranged widely, and the second question was not tackled to any significant extent.

A. Pesci (Italy) gave some details of a Proposal of Curriculum for Teaching Probability and Statistics. She reported on the activities of the Didactic Research Group on this matter, in particular concerning 11–14 year old pupils. They have designed a series of work-sheets for the pupils. Their main thesis is that statistics and probability are separate themes. Teachers can examine their connection through the "law of great number" if they wish. As far as the didactic methodology is concerned, they tried to construct a theoretical framework on the basis of problems arising in concrete situations.

N. A. Malara (Italy) gave An Example of Research for and with Teachers: Analysis of Textbooks on "Probability and Statistics". She reported on aspects of the Italian school situation as regards teaching and learning for 11–16 year old pupils. She described some features of her group's work (objectives, cultural and methodological choice, etc.) Finally she gave some details of the actual lines of research.

M. Rouncefeld (UK) spoke about Teaching Statistics through Practical Work. She suggested some steps involved in practical projects: 1) formulation of the problem, 2) data collection, 3) to decide appropriate means of approaching the problem, 4) analysis of data, 5) interpretation of results, 6) communication of meaning. She stressed that more concentration should be on step 4 and that the question: "*who* collected the data?" and "*how* was the data collected?" are basic to analysis. Positive aspects are that pupils are active, gain more concrete experience, are in touch with reality, work at their own pace, gain confidence to tackle more challenging projects, *enjoy* their works and gain *understanding* of the work they do. She concluded by saying that the methodology of teaching should shift from "Theory–Example–Exercises" to "Problem–Real Data–Discussion–Model and Theory".

A. Olecka (USA) talked about a Probability Abacus as an Intermediate Stage between Simulation and Theoretical Probabilities. Using the probabilistic abacus that was introduced by Arthur Engel, she demonstrated that this opens up new exciting avenues that can be introduced and explored at the school level. Flow graphs for different games can be constructed and isomorphisms discovered. This flow graphs can be transformed into tree diagrams, which can be used to find probabilities of some "local" outcomes, using the Multiplication and Addition Principle. The Multiplication and Addition Principles themselves can be introduced through probabilistic abacus.

M. Halmos (Hungary) reported on Probability Teaching in Hungarian Schools. The Hungarian government instructs schools to teach probability from age 6 to 14, but it seems that not many teachers adhere to that. The two important highlights were:

- Probability should be taught not from the point of view of combinations, but we should establish the *concept* of probability before moving on to combinatorial aspects.

- The approach to probability should concern the whole of a distribution rather than only parts. For example, the following question would be presented: If we throw a die 360 times, how many times will be "2" , and "3" be observed altogether?

She also pointed out that in lessons on probability extensive discussion of decimals and percentages can be undertaken.

In conclusion the chairs (I. Hirabayashi and P. Hess) commented that the concepts in probability are not skills or techniques like computation. They are conceptual results to be understood. They stated as a problem for the future that of determining how we can establish a sound concept of probability and its usefulness in pupil's mind. We should not set out to develop theoretical presentations but to create interest in the subject and recognition that it permeates every aspect of our diverse societies.

Session 3

Subsession 1 — Teaching Methods

A short introductory lecture by the chair (W. L. Fischer) pointed out that in many discussions the term "method" and "teaching method" are used rather unspecified manner. He proposed to use the characterization of this school level as "transitional" as the guideline, and the scheme of the semiotic triangle to channel the discussions.

E. Pehkonen (Finland) reported on a project for Low Attainers "Open Problem Solving in Mathematics" (with B. Zimmermann (FRG)). He discussed the *development of prerequisites for teaching problem solving* and the development of *problem solving persistence*, exemplified by a variety of concrete tasks. Recreational mathematics and open problem solving situations lead to the constitution of a problem field, in which several problems are interrelated. By this means not only is the amount of knowledge increased (quantitative changes), but qualitative changes occur as well. In addition skills are developed for mastering difficulties related to problem solving at a variety of levels including low attainers. Even routine exercises should be changed into problem solving situations involving reflective thinking by means of these activities.

S. Krulik and *J. A. Rudnick* (USA) spoke on Problem Solving — The Focus of the Curriculum. Following Pólya's scheme in terms of sequence "Think–Plan–Solve–Look Back", they demonstrated *how problem solving is and has to be the focus of all mathematical curricula*. They pointed out the distinction between the process and method aspects of problem solving. They used various instructive examples to clarify how open problem situations give access to subject matter and attitudes within the mathematical classroom. The necessity to reconstruct and to revise our traditional curricula was explicitly demanded.

K. Schultz (USA) discussed The Teacher as a Model of Problem Solving. She reported on the result of a project "Problem Solving and Thinking", an experienced-based approach for *modelling reflective learning through problem solving*. This project was motivated by the concern about those teachers who do not see themselves as mathematics thinkers, but rather as the dispensers of abstract symbols and rules so prevalent in curricula and textbooks. Both students and teachers are viewed as learners in a problem solving environment. The teacher must not tell the student how to solve a problem, but must clearly understand the solution – paths of the students and the reasoning behind their work, thus motivating them to monitor their own thinking. *Constructivism Through Metacognition in Problem Solving Environments* is the shorted 'chiffre' for the aim of the project, passing through the sequence of modelling, experiencing and reflection, through a hierarchy of learners (the teacher educator, the teacher, the student each as a learner) and observing how the teacher educator models mathematical thinking to teachers and how the teacher models mathematical thinking to students.

Finally the chair (W. L. Fischer) summarized the main features of the presentations. They have covered between them a variety of aspects of problem solving as a teaching method for improving our classroom work in mathematics. It has been stressed that teaching — indeed, all instruction — should be as problem oriented as possible. Attempts had been made to characterize the concepts involved and to classify cases. Theoretical ideas were illustrated with examples of problems and problem fields and their treatment. All contributors emphasized that the "transitional" characterization of the junior secondary school level implies that in using heuristics we must move from isolated problems to problem fields, achieving in the end, both with respect to subject matter and to the students, a favourable classroom atmosphere, readiness and flexibility for creative activity, and fostering positive attitudes and a development of cognitive skills.

Subsession 2 — Pupil's attitudes: Teaching apparatus

This group gave its attention to the following problems:

– How can we reform pupils' attitudes, where necessary, towards mathematics?
– What are the characteristic features that teaching material should have at this level?

Three contrasting but complementary presentations were given.

P. Shannon (UK) reported on Investigative Methods of Learning Mathematics in Secondary School Using Three Mathematics Teaching Aids — Trigonometer, Negator, Geotiles. These were designed for teachers wishing to introduce investigative methods and to encourage discovery learning. The "negator" and "trigonometer" provide simple concrete ways of developing basic concepts of integer arithmetic and the meaning of trigonometric functions, respectively. Geotiles are brightly coloured triangular tiles which can be moved around, on a magnetic board for example, to build up various geometric figures. They can be used, not only for investigating transformations, but also in the context of proving theorems.

M. Harris (UK) spoke about Mathematics and Textiles, taking up the aspect of aesthetic appeal. She described the project "Maths in Work", which has been running for ten years. It aims to make closer links between the mathematics taught in school and the mathematics actually used in the outside world. It does so not strictly for vocational reasons, but rather to broaden the pupils' view of both mathematics and work. In her talk she showed that the ordinary activities of textile work — for instance, knitting a sweater — are rich in mathematics. Such activities are non-threatening, and provide motivation. Boys as well as girls are responding positively.

E. A. *Silver* (USA) talked about Stimulating Mathematical Inquiry through the Use of Open–ended Problems and Post-solution Conjecturing. He did not use concrete teaching aids, but asked participants to use their minds. He posed some open–ended problems for which children could easily offer a wide range of conjectures. In particular the problem of making as many conjectures as possible about the product of four consecutive integers was considered. This has been well received by junior secondary classes in the USA. Children are stimulated to justify their own conjectures to their friends, and hence are motivated to provide some form of proof.

E. *Oldham* (chair) commented that the presentations offered ways of developing positive attitude in three different contexts: teaching *standard context,* introducing *non-standard context,* using *open–ended problems.* Although it had not been possible to reach definitive answers to the initial questions, certain important common aspects had emerged offering tentative solutions for further discussion.

As regards *attitudes,* the mathematics should in some sense be *the children's own,* made and used by them. Also if children are to enjoy doing investigative and thought-provoking mathematics, as opposed to long routine computations, say, then the situations in which this mathematics arises must be both *open and non-threatening.* Finally *aesthetic appeal* may be a more important than has been realized.

As regards *apparatus,* again *aesthetic appeal* may be helpful, and apparatus should be easy for teachers and pupils to use.

Although these conclusions may seem obvious and not particularly new, it is relevant to note their diversity and complementarity. Good pedagogical principles spanning three such different contexts are perhaps rather robust, and therefore widely applicable.

Session 4: Reports, New Proposals and Conclusion

The aim of this (plenary) session was to summarize the group's activities. In addition to comments from the chair (P. Hess) there was discussion on the structure of future action groups, including a wish for more discussion time, and a suggestion that one overall theme be fixed. There were two presentations at this session.

M. *Hallez* (France) spoke about a Historical Perspective in the Teaching of Mathematics. She reported on classroom experience in making mathematics pleasurable as well as an efficient technical tool. The research group "Mathematics: an Approach by Historical Texts" has chosen to study such texts to provide a temporal, cultural, social and economical context. As an example she quoted her experience in telling her pupils about Christian Huygens and the scientific turmoil of his days. (The school is in Rue Huygens.) The objectives were to introduce a new mathematical tool, to show that building machines is part of mathematical activity, to show how scientists lived and worked in the past.

K. Shigematsu and *H. Iwasaki* (Japan) reported on Recent Reformation of Mathematics Education in Lower Secondary School in Japan, from the viewpoint of "transition" of objectives and of content. They considered the former historically. Despite a realistic necessity, the traditional academic character of secondary school mathematics was accelerated in the postwar school system through two causes: a surprisingly high advancement rate for upper secondary school and modernization movement in mathematics education.

As to the transition of content, they gave the results of two surveys concerning attainment in the current curriculum. Both indicate that achievement on the academic aspect is very low and it is necessary to build a new framework for the curriculum at this level.

Final Remarks

This report was compiled by the chief organizer on the basis of many sources. Responsibility for any imperfections remains mine however.

I would like to express my heartfelt thanks to all participants and contributors, and members of the organizing panel, particularly to Mara Kovács, the Hungarian coordinator, who worked devotedly on our behalf.

Related activity: ATM Children's Workshop

During the working days of the congress the Association of Teachers of Mathematics (UK) had continuous workshop sessions for children from a Hungarian school, and for the children of the delegates at ICME. Generally they worked with 10–15 children.

The sessions offered activities which were very active involving materials, drawing etc. The visitors took part, interacting with the children and with each other. There were opportunities for discussion of the pedagogical implications of this mode of working.

This was an especially good opportunity for informal meeting of teachers and educators, and was one of the most popular presentations for the Hungarian participants.

The members of the Workshop Team were: Anthony Brown, David Cain, Geoffrey Faux, Gillian Hatch, Christine Hopkins, Jan Stanfield-Potworowski, Robert Vertes.

Hungarian coordinator: Julianna Szendrei.

ACTION GROUP 4: SENIOR SECONDARY SCHOOL (ages 15–19)

Against formalism, for more students, using new technology

Chief organizer: *Jan de Lange (Netherlands)*
Hungarian coordinator: *Norbert Hegyvári*
Panel: **Bernard Cornu (France), Terence Heard (UK), Anna Sierpinska (Poland), Osame Takenouchi (Japan), Zalman Usiskin (USA),
Coopted members: Michiel Doorman (Netherlands), Daniel Reisz (France), Diana Rosenberg (Argentina), Heleen Verhage (Netherlands)**

Major trends

The final plenary panel-session of Action group 4 (15–19) made clear that we might distinguish three major trends in mathematics education.

In the first place there is a widespread *rebellion* against the structuralist approach that had its starting point in 1959 at a Seminar initiated by the O.E.E.C. for the purpose of improving mathematics education.

According to the structuralist conception, mathematics as a cognitive achievement is an organized, closed deductive system. For mathematics *education* this means that the structure of mathematics is of fundamental importance for this systematically directed education.

Not surprisingly, one of the most outspoken advocates of this approach was Dieudonné. In his address to the conference, entitled: "New Thinking in School Mathematics" he proposed to offer students a completely deductive theory, starting right from basic axioms. [1]

Critical remarks appeared only a couple of years later. In 1962 a memorandum was published in the 'Mathematics Teacher', and the 'American Mathematical Monthly':

"Mathematicians may unconsciously assume that all young people should like what present day mathematicians like or that the only students worth cultivating are those who might become professional mathematicians" [2]

This memorandum 'On the mathematics curriculum of the High School' — which was signed by Bers, Birkhoff, Courant, Coxeter, Kline, Morse, Pollak, Pólya and others — is still worth reading, as we will see later in this paper.

Lakatos launched an attack against formalized mathematics and Dieudonné; in particular, in his book 'Proofs and Refutations' [3]. He points out that informal mathematics is *also* mathematics. Informal mathematics is a science in the sense of Popper; it grows by a process of a successive criticism and refinement of theories and the advancement of new and competing theories.

In many countries there is now a clear trend away from formalism toward informal mathematics. Also two other trends are noticeable: mathematics education becomes more *process* oriented instead of *product* oriented, this in relation to the problem solving and applications and modelling trends.

There are strong indications that we should not only teach mathematics, but also mathematical thinking.

We follow Burton [4] when she states:

"I draw a clear distinction between mathematical thinking and the body of knowledge (content and techniques) described as mathematics. Mathematical thinking is used when tackling appropriate problems in any context area".

It has been argued that mathematical *thinking* is the means by which infants first organize information they gather through their senses in order to learn from their environment [5]. This comes close to *mathematizing* which is an organizing and structuring activity according to which acquired knowledge and skills are used to discover regularities, relations and structures [6].

During the last 80 years there has been a discussion about the desirability of including applications in mathematics education. In 1976 Pollak noticed a worldwide drive to make mathematics in school more applied. In some cases this was part of a reaction against the 'New Math', as we can see in Kline's 'Why Johnny Can't Add' [7]. In others it is a recognition of the increased mathematization of many other fields than physics. Many experiences and some research seem to indicate that the problem of motivating students becomes easier when applications are taught [8].

There are suggestions that a greater element of problem-solving through *mathematical modelling* should be introduced into 16–19 mathematics. As Burghes stated: "We surely want our students to be able to put their mathematical skills into practice, and it is only through active problem solving that they be able to do this — the problems can be real or purely mathematical — what unites them is that they will give students the chance to

- apply their mathematical skills
- show creativity, imagination, innovation, critical judgement
- motivate further mathematical study" [9].

Here we are back to the need for more process-oriented mathematical education.

Where viewpoints seem to be wider apart is on *how* to teach mathematics in the real world. In the classical view applications can only be taught *after* learning has occurred.

Lesh [10], de Lange [11] and others believe that applications and problem solving should not be reserved for consideration after learning has occurred; they can and should be used as a context within which the learning of mathematical ideas takes place. A similar view was expressed in the already mentioned memorandum [2]: "We wish that the introduction of new terms and concepts (to be extracted from a concrete situation) should be preceded by sufficient concrete preparation and followed by genuine, challenging applications and not by thin and pointless material."

The second major trend that emerged from the discussions in our Action group was that there is an increasing need for students to take a significant amount of mathematics. This was also noted at ICME-5 in Adelaide in 1984 by the organizers of A4, Egsgard and Fletcher. They noted that in former times courses catered for a group of moderately elite students; now the group is much wider, and in some countries the 'average' students, for whom a fresh style of course is needed, constitute a majority.

There is general agreement that it is necessary to develop fresh styles of courses for students participating in senior secondary education, but the approaches adopted in different countries vary greatly. Some countries are attempting to accommodate variations in attainment by designing differentiated courses — that is, students are grouped by attainment and follow courses at different levels of difficulty. In other countries this is unacceptable and it is policy to develop modules providing a core and options [12].

Another solution is to make different mathematics programs: one for students heading for exact sciences, and one for students heading for economic, social and life sciences.

Indeed, one of the reasons for the increasing numbers of students in upper secondary education is the fact that mathematics is used in almost all scientific disciplines. Linguistics, Anthropology, Archaeology, Medicine, Geography, Biology — to name but a few — use or sometimes misuse mathematics.

This brings us to the point that there is a second explanation for the increasing need for mathematics education.

It becomes clear that to fully understand society, politics and economics one needs mathematical skills, thinking, and in particular, critical judgment.

Mathematics cannot be regarded as an isolated subject — the fifth day of this ICME-6 on Mathematics, Education and Society made this perfectly clear. Mathematics Education has to do with political, ethical and cultural issues as well.

Emotions and debates may get very heated as was the case in the recent Koblitz–Lang–Huntington debate.

Professor Samuel Huntington lectures on the problems of the developing countries. In his book *Political Order in Changing Societies* (1968) he suggests various relationships between certain political and sociological concepts:

(a) "social mobilization"
(b) "economic development"
(c) "social frustration"
(d) "political participation"
(e) "political institutionalization"
(f) "political stability".

He then expresses these relationships in a series of equations:

$$a/b = c; \qquad c/d = e; \qquad e/f = g.$$

Does Huntington mean that: "Social mobilization is equal to economic development times mobility opportunities times political institutionalization times political instability!?"

According to Koblitz mathematics is used as a means for mystification, intimidation, an impression of precision and profundity. He then comes to the following implications for teaching: "Whether mathematical devices in arguments are used for fair ends or foul, a well-educated person should be able to identify misuses of mathematics as well if he is to know the correct uses" [13].

The third major trend is the use of new technology in mathematics education. In his opening address for our Action Group Tall made this very clear to his audience [14].

In his opinion — shared by many — the arrival of the new technology brings us to the threshold of an exciting new phase of development in the history of mankind, which will affect every aspect of life, not least in mathematical education, where it heralds the second major revolution in thirty years. Whereas the introduction of 'Modern Math' in the 1960's was largely a product of internal forces, caused by mathematicians, the new technology will change our cultural environment in a way which will compel mathematics education to respond.

To respond in a proper way may be difficult. As the work in our Group on Computers showed, there is a pool of suggestions, ideas and possibilities. But at this moment we can have no reliable idea to what extent any of the suggestions that are put forward will prove feasible.

As Burkhardt pointed out in the ICMI study on Computers in Mathematics Education the translation from an idea to a small-scale pilot experiment with exceptional teachers and facilities, and then to large-scale reality will involve critical distortions of the aims of the exercise which may call into question its value.

Another unusual factor makes curriculum development involving advanced technology more difficult than usual. It is the mismatch of time scales between technical change (one year) and curriculum change (ten years). The curriculum designer cannot assume a specific level of technological provision and sophistication in schools — it will vary widely both in time and from place to place [15].

It is clear that we should be very careful in talking about computers in *reality*. According to Winkelmann [16] this means: observations on the percentage of the teachers of a certain discipline who actually use the new technologies in their teaching, on contents and themes which to a certain extent are really taught and/or changed by computers; considerations on necessary equipment, organization, teacher training, curriculum development.

There are many problems in any of the above mentioned fields: on computers, they were in some cases introduced into schools at an early stage — and had to be replaced when MS-DOS more or less became the standard. But these machines are still considered by many as very user-unfriendly and not well suited for mathematics education. At this moment there seems to be a trend towards Macintosh Computers for education. For certain schools this means the third generation of computers at school. But waiting for the ultimate computer seems also not a very wise choice. It may be that the introduction of the pocket graphics calculator will give rise to more changes in a proper direction. It was discussed widely in our Group on computers [17] [18].

The problems concerning software are well-known: in many cases hardware was installed at schools while there was no software available. At this moment there is a wide choice of reasonable and good software. Software designed to help to develop basic skills, to handle real applications, to develop mathematical concepts (the work of Tall and others), to generate graphics (differential equations, solid geometry) and many more. All of these have one factor in common — in order to be used in classrooms in *reality* those programs should be very user-friendly and very flexible. Otherwise we are asking too much from teacher organization. As indicated in the discussion in our Group one of the key questions is how to generate the necessary pre-service or in-service training programs for the teachers.

We conclude our discussion on the three major trends with the following observation: The computer can and should be used for *conceptual learning*; for using the computer to *apply* mathematics requires the students to understand the underlying concepts even better than before.

We will now turn to the discussion and presentations in the five sub-groups of our Action group. Of course we have no intention to be complete — we will just highlight some of the questions raised — especially those that are related to the three major trends.

Curriculum Contents and their Evolution

Curricula may change for a wide range of reasons. As mentioned previously, the introduction of modern mathematics was initiated by dissatisfaction among *mathematicians*. The structure of mathematics was imposed on the mathematics teacher — many of them sharing the enthusiasm of the mathematicians: "The whole edifice must be rebuilt from the foundations and erected in accordance with modern ideas." [19]

The changes that are now taking place all over the world are in part a rebellion against Modern Math, but there are many other factors — that are not always distinct — that play a role in causing curriculum changes.

There are *movements* for more accessibility to mathematics education; many students terminate their study too soon, not realizing the importance of mathematics in later schooling, on the job and in society. Mathematics for All, or Mathematics for the Majority are familiar slogans in this respect.

Sometimes there is a need to adjust the content of the curriculum because of recent *developments in mathematics*. Today discrete mathematics, statistics, exploratory data analysis are subjects that are already found in new curricula and considered for others.

A fourth point worth mentioning when looking for probable causes for curriculum change is the role of the different *tools* that are used in mathematics education. Paper and pencil, cubes and compass are still used, but graphic calculators, computers, video and laser discs are replacing them quickly. This will have effects on the curriculum too. Some subjects will disappear, others will get more attention, and a few will be introduced for the first time.

The fact that mathematics is increasingly used in *other disciplines* than physical sciences is also a factor that has affected the curriculum recently. The Netherlands (1985) and Denmark (1988) have recently introduced new curricula for upper secondary education in which applications and modelling in other disciplines play a vital role, while the U.K. has had a long tradition in this direction.

The usefulness of mathematics is not restricted to applications in other disciplines but there is also a growing awareness that mathematics is useful in *daily life* and *on the job*. Mathematics education is necessary for intelligent citizenship, for a critical attitude, to judge political and economic decisions. This point of view also has its impacts on the curriculum.

Finally *didactical, psychological* and *epistemological* points of view may have influences on curricula. This is for instance the case in the new curriculum in the countries mentioned above, where a *realistic* approach to mathematics is advocated by exploring real world problems intuitively before mathematizing them, in sharp contrast to the structuralistic approach [20].

It is interesting to see that some countries have another factor that influences the curriculum. Curriculum development in Taiwan for instance has been much affected by America [21]. English speaking countries from Africa use Cambridge examinations. Indonesia has adapted S.M.P. materials. We see them not as real factors but as derived ones. Increasing internationalization will make the curriculum market more open. But there is also noticeable a growing awareness that countries should conserve their own cultural–mathematical traditions. Curricula worldwide are dominated by Western thinking. This is not always a desirable situation.

The innovation of the curriculum in Taiwan was influenced by the School Mathematics Study Group (SMSG) in the United States. Another influential project from recent years is the University of Chicago's School Mathematics Project (UCSMP) which was discussed in our Group [22].

Recommendations by various national commissions in the United States have called for updating of the content of the secondary curriculum by including or increasing emphasis on (1) applications and modelling, (2) statistics and probability, (3) discrete mathematics, (4) mathematics now accessible or prominent because of calculator and computer technology, (5) estimation and approximation, (6) transformation and coordinate approaches to geometry. To provide time for these newer topics, there are calls for (7) decreased emphasis on algebraic manipulation and (8) less work with formal geometric proofs.

Usiskin, co-director of the program, stressed the balanced approach: recent history in the USA teaches us that a narrowly-focused curriculum will not last. The new mathematics curricula of the 60's, in retrospect, were too exclusively based on the pure mathematician's view of mathematics. The back-to-basics curricula of the 70's were too exclusively based on the typical citizen's view of mathematics as a dissociated collection of paper-and-pencil skills.

Some in the United States currently are recommending that all topics be approached concretely; this idea seems — according to Usiskin — to be dependent on the psychologist's point of view. The UCSMP also does not wish to make the mistake of organizing too narrowly on the applied mathematician's view.

The last two points — starting in the real world to extract via mathematization and reflection the necessary mathematical concepts and applying them later to reinforce the concepts — form exactly the heart of the new Mathematics curriculum for students aiming at social, and life sciences in the Netherlands [20].

Although there seems to be considerable agreement on the *contents* of the U.S. and Dutch curricula (applications and modelling, statistics and probability, discrete mathematics, use of computer or graphic calculator) there are also differences. The coordinate approach and transformations in geometry are very distinctive in the UCSMP materials, while in the Netherlands three dimensional geometry gets most attention ("grasping space" [23]).

But it hardly came as a surprise that the discussion in the Group focussed on the methodology. Usiskin feared that in the Netherlands the pendulum was swinging too far to the other extreme. This remark provoked a very good discussion on what mathematics education ought to be, for whom it is designed, and what the influence of mathematicians should be. Of course, there was only the beginning of a discussion — as usual on conferences like this one — time was running out. But on one important subject everyone seemed to agree: in process oriented mathematics education one of the crucial factors is assessment. If we cannot change testing, we will not succeed in implementing process oriented curricula.

Didactical, Cognitive and Epistemological Points of View

At some time during the conference the question was posed as to what the reason is that there seems to exist only a limited amount of research on upper secondary education. There exists an abundance of research reports on the 'additive conceptual field', on fractions, on multiplication — on primary education in general. Looking at the Research Journals one cannot but conclude that indeed relatively little research exists for upper secondary level.

At the ICME 3 in Karlsruhe in 1976 a movement grew to form a working group to study the psychology of mathematics education, and every year since that time regular conferences have been held under the auspices of the International Group for P.M.E. Also here most of the attention was devoted to primary education.

However, recently a subgroup was formed that addresses itself to the problems of "advanced mathematical thinking", which look at mathematics from age 16 through to university and research level.

A major activity is the investigation of student's misconceptions and how these are often related quite naturally to their previous experiences. We follow Cornu [24] when he states that going to a higher level of knowledge usually consists in passing an 'epistemological' obstacle, that is an obstacle constituting the knowledge to be acquired. This is done by de-stabilizing incomplete and insufficient knowledge. Such knowledge, up till now sufficient within a given range of problems, becomes insufficient. The pupil has to solve a problem where the new knowledge is a necessary tool.

Cornu shared his ideas and results with those of Sierpinska who recently carried out research on the concepts of limits.

An interesting way to deepen students' understanding and appreciation of modern mathematical concepts and methods was presented in this group [25]. The French group "Mathematics: Approach through Historical Texts" (M.A.T.H) has been working since 1982 on a study that aims at the introduction of mathematics from a historical perspective by students reading fragments of original texts. Up to now the authors of the French report have been making an effort to develop a technique of integrating a historical perspective into their mathematics teaching. In the discussion it was suggested that their research should be extended or accompanied by a reflection on their own thinking when choosing historical texts and preparing the activities in order to make explicit criteria of choice and a methodology of preparation of such activities.

In the United States research has been carried out on the interpretation of modelling as a Local Conceptual Development [26]. According to the researchers involved the idea of interpreting the multiple-modelling-cycle 'Applied Problem Solving' as "local conceptual development" has many practical and theoretical implications. The mechanisms which contribute to general conceptual development can be used to help clarify the kinds of problem solving processes which should facilitate students' abilities to use mathematical ideas in everyday situations. Besides this these mechanisms, which are important in "*local* conceptual development sessions", can be used to help explain *general* conceptual development. The calculator or computer is used in these sessions.

In the discussion in the Group several questions linked with the use of computers in problem solving were raised. It was mentioned that the presence and use of a computer during a problem solving activity may substantially change the problem itself. And, as indicated before, there is the problem of organization for the teacher. It has been argued that a computer releases the teacher from the less human part of his work and allows him to be counsellor, mentor, and tutor and not just a conveyor of knowledge. In the plenary discussion of the Action group there seemed to be wide disagreement on the previous statement. It is clear that the discussion on the role of the computer in the mathematics classroom has not reached its zenith yet. With the arrival of the graphic calculators this discussion has reached a new chapter as will be shown in the discussion of our next subgroup.

Influence of the Computer

The computer introduces many changes. Changes in mathematics itself [27]. Changes in the mathematics to be taught, and in how to teach mathematics [28]. This results in a different task for the teacher — we mentioned this point already in the discussion of the previous subgroup.

But the most important change may be the effects or impact of any of these computer programs or ideas on students' learning of mathematics. This seems to be the key question.
There is indeed an abundance of software to compute, calculate, graph, draw solid bodies, and much more. But the software itself does not suffice to improve the learning as many teachers-of-good-will have experienced. The pedagogical and didactical principles and strategies play an essential role. During the discussion in our Group it became clear that we need more *substantial evidence* than storytelling about experiences. One of the participants in the Group stated his feelings on the discussion and presentations in the following way:

- My computer can do more than yours
- If we can't teach it to our kids, we teach in to our computer
- If I can do something with my students, every teacher must be able to do the same
- If anybody questions any of the above he just is not good at computers [29].

One of the problems at this stage is that only isolated pieces of work are carried out in often more or less ideal situations. There are many ideas on how to use the computer

- in problem solving
- in solving equations
- as a superboard
- in drawing and rotating solid bodies
- to draw graphs
- as an instructional tool (in pre-calculus)

– to teach quadratic polynomial functions.

Examples of all of the above were presented in this group — all with promising results. But more research is needed. The rule that one needs to make first a didactical and cognitive study of the concepts to be taught when designing suitable and efficient didactical products, holds also for software. Regrettably — because of the pace of technological development — this kind of developmental research is not always possible. It seems — as was noted in the discussion — that we are not able to keep up with the pace of the hardware changes. It is clear that we cannot make the necessary changes in our teaching and examining quickly enough. And finally there is little doubt that there remains a lot to be done for the training of teachers. But also here we ran into problems. Designing a training course for teachers is a very difficult and time consuming task. By the time the course is ready there is a fair chance that the hardware and software will be quite different.

It might even be the case that those who still do not use a micro at this moment will never have to use it because of the presence of the new graphic calculators.

At this moment there are at least four graphics calculators (Casio fx-7000G, Casio fx-8000G, Hewlett-Packard 28C, Sharp EL-5200) available in many countries. These calculators are so powerful and offer so many possibilities that one wonders why we should use micros any longer. In one of the first articles on the HP-28C Nievergelt not only describes the power of the calculator (on calculus, statistics, numerical analysis and linear algebra and computer science) but also gives some first academic reactions [30]. He describes how four freshmen familiarized themselves with the HP-28C in just over two hours, with the help of a well-written Starting Manual. However, they felt that they needed a better understanding of the mathematics involved in order to use the calculator more intelligently. "Anyway", sighed someone, "teachers won't allow it on tests, or will they?".

At this stage nobody seems to know. Should we change the curricula? Should we change our problems and exercises? Should we be happy that all computations and graphing will be taken care of by the calculator?

Shumway presented some ideas on the use of the graphic calculators with a positive view [31]. He warns that we should not propose to ban the graphic calculators because of a confusion between *skill* learning and *concept* learning. It is interesting to note that we always try to introduce micros in school, but at the same time note the proposals for banning (graphic) calculators at certain schools or classes.

Because skill learning is the primary goal of many textbooks, few textbooks provide divergent examples or matched non-examples. Just as drill and practice strategies are not good strategies for concept learning, divergent examples and matched non-examples do not provide quick, repeated practice of skills. The calculator allows one to add emphasis on concept learning because the skills are performed by the machine. Conceptual understanding is required to determine which skills are appropriate and consequently divergent examples and matched non-examples are appropriate. Following Shumway changes might be made in:

- exercises given to the students
- test items used to measure understanding
- the teaching strategies employed
- the learning strategies employed by students.

Some of the answers will possibly be given in Quebec in 1992. Maybe they will be presented in our next Subgroup: Teaching, Assessment and Evaluation.

Teaching, Assessment and Evaluation

One of the issues that seems to command increasing interest in mathematics education today is assessment. We just mentioned that the introduction of the graphics calculator might change test items used to measure understanding. In general there seems to be a growing awareness that

- we should look at forms of assessment that contribute to learning; instead of a test at the end of a course that tends to influence the teaching
- we should look at forms of assessment that test mathematical thinking, critical attitude, concept learning or, in general, higher order goals [20]
- we should look for what students know rather than what they do not know [32].

Reasons why there is an increasing interest are varied. We mentioned the trend towards more applications and modelling — assessment here is a very complex task. Another trend is towards more process-oriented mathematics education. There is a feeling that tests are often a sorting device with bad influences on teaching. This feeling has been strengthened in some cases by computer-testing related to computer-teaching.

Finally we note that it has been recognized that we cannot only test basic skills.

Evaluation of students should include the *full range* of any program's goals, including skills, problem solving, and problem solving processes, according to the recommendations of the National Council of Teachers of Mathematics in the 'Agenda for Action' [33]. Other recommendations of this 'Agenda' deserve attention as well: Teachers should become knowledgeable about, and proficient in, the use of a wide variety of evaluative techniques. And the evaluation of mathematics programs should be based on the program goals, using evaluation strategies consistent with these goals.

Mathematics teachers often state their goals of instruction to include all cognitive levels. They want their students to be able to solve problems creatively. But too much of their testing consists only of recall of definition, facts and symbolism [34].

The English "Mathematics Counts" report discussed the restriction of timed written papers: examinations in mathematics which consists only of timed written papers cannot, by their nature, assess ability to undertake practical and investigational work or ability to carry out work of an extended nature. They cannot assess skills of mental computation or ability to discuss mathematics nor, other than in very limited ways, qualities of perseverance and inventiveness. Not only do written examinations fail to assess work of the kind described above, but they lead teachers to emphasize in the classroom work of a kind which is directly related to the type of question which is set in the examination [35].

An added problem is the role of the computer. It is very unclear how the computer should be used in assessment. By this we do not mean the standardized tests — often multiple choice — that are given to the students, but we mean the fact that we seem to expect from our students some computer proficiency in mathematics education.

Apart from assessment, major questions to be discussed in this group were
- how can we make and understand our teaching better
- how can we improve the performance of our students
- how can we break the cycle: shortage of teachers → poor teaching → poor image of mathematics → shortage of teachers → ... ?

Social interaction can and should play a major role when teaching and learning mathematics. When students are forced to explain to and question students, contradictions coming from the partner are more likely to be perceived than contradictions confronting the solitary learner.

The contradictions are also harder to refute than in a conflict resulting from individual and temporary hesitations between two opposing points of view [36]. We may speak of a socio-cognitive conflict [37]. Research has shown that inter-individual encounters lead to cognitive progress when socio-cognitive conflict occurs during the interaction. The social and cognitive roles are inseparable, because it must clearly be a matter of conflict between social partners about the way to resolve the task.

Interaction (also) plays a major role in research carried out in order to investigate and analyse the different strategies of explaining mathematics in the classroom [38]. Also the word "explaining" by itself does not indicate much of interaction. The theoretical framework — the speech act theory — emphasizes that 'understanding mathematics' is not considered as a psychological–educational concept that is connected to some hidden mechanism of understanding in the brains of students, but that it gets its meaning in the interaction between teacher and students. As a consequence they can and should negotiate about it.

In certain countries — New Zealand, the Netherlands — central authorities have taken actions to improve the participation and performance of girls.

In New Zealand by the action "Girls Can do Anything" and in the Netherlands by "Choose Exact". In New Zealand it is clear now that such actions, however successful in the beginning, cause major problems later on. In the first place there is the problem for girls of remaining in mathematics, and in the second place the more serious problem is the shortage of teachers [39].

This shortage of teachers seems a major problem in many countries now. Contributing factors that were mentioned in the plenary discussion were: low pay compared with other professions, low prestige of the profession of teaching, poor image of mathematics itself, discipline problems in schools, and an increasing work load.

Another major reason for the shortage of mathematics teachers is the loss of personnel to the rapidly expanding field of computer technology.

Not only do many mathematics students find these jobs more exciting, but computer firms are also hiring teachers with years of experience. This becomes an even more serious problem when we know that salaries for mathematics teachers have gone down in many countries.

We can wholeheartedly follow the conclusions from ICME 5, where it was stated that it seems clear that improving working conditions for teachers, increasing salaries and raising the prestige of the profession would do much to improve the prospects for an adequate supply of teachers.

The question remains as to whether ICMI, or its parent IMU can do anything to make politicians aware of the problems in this field.

Teachers and Teacher Training

It is interesting to note that while there is a continuous discussion on Teacher Education — a key variable without any doubt — that at present there are hardly any teachers to train. In the USA for instance, a special mathematics program had to be designed for teachers with very little background who — suddenly a decade ago — found themselves about to teach at secondary school level [40]. And the USA is not unique in this respect.

The basic question when thinking of teacher education is: what do teachers have to know and how must they be trained to be able to guide their students? The answer to this question is certainly not obvious. One could think of: what is unique in mathematics? What mathematics should a teacher know? What is unique about the audience? What tools does the teacher need? What should the teacher know about pedagogical and didactical principles? What should be said about assessment?

Some of these questions were discussed in this Group. The participants agreed on the fact that every teacher should know enough about computers that he feels confident at least. So it is necessary to train, within a few years, almost all the teachers in the possibilities and restrictions of the computer. Not so clear however is how much the teacher should know about computer science. Some state that mathematics teachers need a wide overview of Computer Science, to avoid reducing the use of the computer to programming or using software [41]. Others make the point that using (excellent) software in a proper way should be all that we should expect in the mathematics classroom.

During the discussions in the Group different points of view also emerged on the role of computers in developing countries. It was argued that developing countries should stick to pure mathematics without using the computer. Most participants didn't quite agree with this point of view.

Another point that was stressed was the necessity for teachers to reflect. To reflect on their own education, on their views of education and on their own history. Reflection is research into one's own presuppositions and subjective theories, and not an ability one possesses by nature. A teacher-training program has to take into account the fact that students have already had years of instruction before they enter a teacher-training program.

During that instruction they have acquired not only specific knowledge and skills. They have also been continually confronted with specific ways of teaching. When students enter a teacher-training program, they already possess common knowledge held about education in society. This influences *what* they learn and *how* they learn. This means that there is a meta-contract of a teacher–pupil interaction. We can recognize this in remarks like "as a beginning teacher you must at the start keep the students well under your thumb" [42].

Teacher education is a key variable. This fact is not always reflected in the attention it is given during conferences on mathematics education. This is a situation that leaves much to be desired. We can design beautiful curricula, develop excellent software and do research about how students learn, but if we fail to recognize the restrictions that are imposed on the teacher, the pressure of the examinations, large classes, poor working conditions, low prestige, problems with discipline of students, low motivation of students, new curricula imposed on teachers without consultation, bringing into the classroom micros without software or without telling the teacher how to organize the activities, we will never succeed in successful mathematics education.

Contributors: Z. Usiskin, Y.T. Lue, J. Fishman, J. de Lange, M. Buhler, F. Jozeau, B. Cornu, R. Lesh, S.J. Bezuszka, J. van Hamme, H. Kakol, I. Nishitani, H. Kanaya, R. Shumway, O. Takenouchi, A. Velez-Rodriguez, M.M. Sharma, B. Waits, K. Yamashita, G. Barr, J.S. Engel, S.D. Forbes, N. Hikita, S.L. Kemme, B.K. Lichtenberg, M.J. Long, H.N. Schuring, J.J. Vas, H. Broekman, M. Kazadi, D. Rosenberg, P. Laridon, D. Tall.

The Group had twelve meetings. The plenary opening meeting (address by D. Tall) was attended by more than 500 persons. The five subgroups each met twice with an attendance of some 40–100 people at each meeting. The final plenary report and discussion session attracted some 400 people and was presided over by the Chief Organizer.

I would like to thank all people mentioned above and of course all participants of Action Group A4 for making this part of ICME 6 a success — if I may rely on the many positive reactions that reached me in Budapest and soon thereafter. It was a pleasure organizing A4 for you.

References

[0] Lange J. de and M. Doorman: *Senior Secondary Mathematics Education, Reader prepared for Action Group A4 of ICME 6*, OW&OC, Utrecht, 1988.

[1] Fehr, H.F.: *New Thinking in School Mathematics* O.E.E.C, Paris, 1961.

[2a] Ahlfors, L.V. et al: On the mathematics curriculum of the High School, *Mathematics Teacher* 55(3), 1962, 191–195.

[2b] Ahlfors, L.V. et al: On the mathematics curriculum of the High School, *American Mathematical Monthly* (3), 1962.

[3] Lakatos, I.: *Proofs and Refutations*, The Cambridge University Press, Cambridge, 1976.

[4] Burton, L.: Mathematical Thinking: The Struggle for Meaning, *J.R.M.E.* (15,1), 1984, 34–49.

[5] Gattegno, C.: *The universe of babies*, Educational Solutions, New York, 1973.

[6] Treffers, A. and Goffree, F.: Rational Analysis of Realistic Mathematics Education, in: Streffland, L. (ed): *Proceedings of the Ninth International Conference for the PME*, OW&OC, Utrecht, 1985, 79–122.

[7] Kline, M.: *Why Johnny Can't Add: The Failure of the New Math*, Vintage Books, New York, 1978.

[8] Kaiser-Messmer, G.: *Literaturbericht zu Empirischen Untersuchungen über Anwendungen im Mathematikunterricht*, Gesamthochschule, Kassel, 1986.

[9] Burghes, D.S.: *Modelling and Case Studies at A-level. The Reform of A-level Mathematics*, Everton, 1987, 65–67.

[10] Lesh, R. et al: Conceptual models and applied mathematical problem solving research, in: Lesh, R. and Landau, M.: *Acquisition of mathematics concepts and process*, Academic Press, New York, 264–345.

[11] Lange, J. de: Real World Mathematics for Real Understanding, in: [0] 14–24.

[12] Egsgard, J., Fletcher T.: Action group 4, Senior Secondary School, in: [0] 229–240.

[13] Koblitz, N.: Mathematics as Propaganda, in: Steen, L.A.: *Mathematics Tomorrow*, Springer, New York, 1981, 111–120.

[14] Tall, D.: Mathematics 15–19 in a Changing Technological Age, in: [0] 2–12.

[15] Burkhardt, H.: Computer Aware Curricula: Ideas and Realisation, in: [0] 105–113.

[16] Winkelmann, B.: Informatic Technology across the Curriculum, *I.F.I.P*, 1987, 81–88.

[17] Nievergelt, Y.: The Chip with the College Education, in: [0] 129–135.

[18] Shumway, R.J.: Graphic Calculators: Skill versus Concepts, in: [0] 136–140.

[19] Servais, W.: in: [1]

[20] Lange, J. de: *Mathematics insight and meaning*, OW&OC, Utrecht, 1987.

[21] Lue, Y.T.: High School Mathematics Curriculum Development in Taiwan, in: [0] 25–28.

[22] Usiskin, Z.: The UCSMP: Translating Grades 7–12 Mathematics Recommendations into Reality, in: [0] 29–35.

[23] Freudenthal, H.: *Mathematics as an Educational Task*, Reidel, Dordrecht, 1973.

[24] Cornu, B.: Learning Limits, in: [0] 50–54.

[25] Buhler, M. and Jozeau, M.F.: Towards an Historical Perspective in the Teaching of Mathematics, in [0] 47–49.

[26] Lesh, R. and Kaput, J.: Interpreting Modelling as Local Conceptual Development, in: [0] 60–68.

[27] See. [15]

[28] Cornu, B.: The Computer: Some Changes in Mathematics Teaching and Learning, in: [0] 114–121.

[29] Goddÿn, A.; OW&OC, Utrecht.

[30] See [17]

[31] See [18]

[32] Cockcroft, Sir W.: *Mathematics Counts*, H.S.M.O. London, 1982.

[33] N.C.T.M.: *An Agenda for Action*, N.C.T.M. Reston, 1980.

[34] Wilson, J.: Evaluation of Learning Secondary School Mathematics, in: Bloom, B.S., Hastings, J.T., Madaus, G.F.: *Handbook on Formative and Summative Evaluation of Student Learning*, McGraw-Hill, New York, 1971, 643–696.

[35] See [32]

[36] Balacheff, N. et al: *Social Interactions for experimental studies of pupils conceptions: its relevance for research in didactics of mathematics*, I.D.M.–T.M.E. Bielefeld, 1985, 1–5.

[37] Doise, W. et al: *The social development of the intellect*, Pergamon Press, Oxford, 1984.

[38] Kemme, S.L.: Explaining Mathematics, in: [0] 167–173.

[39] Forbes, S.D.: Ethnic and Gender Differences in Performance in Mathematics, in: [0] 156–161.

[40] Long, M.J.: The College Mathematics Major Reconsidered: Teacher vs. Researcher, in: [0] 174–178.

[41] Cornu, B.: Some Remarks and Questions about Teacher Training, in [0] 220–223.

[42] Broekman, H.: The prehistory of teacher trainees and the consequences for teacher education, in: [0] 204–219.

ACTION GROUP 5: TERTIARY (POST-SECONDARY) ACADEMIC INSTITUTIONS

Chief organizer: John Mack (Australia)
Hungarian coordinator: Gábor Székely
Panel: Michele Artigue (France), Marjorie Carss (Australia), Jack Gray (Australia), Glyn James (UK), Miguel Jimenez Pozo (Cuba), Lynn Steen (USA), Vinicio Villani (Italy)

A brief plenary session, in which Vinicio Villani spoke about and distributed a summary sheet on the structure of post-secondary education in seven different countries, illustrating differences, was followed immediately by separation into the six announced area groups, who then commenced work, which is reported upon here. The Chief Organizer thanks his panel, all area group organizers, leaders and reporters, and especially the Hungarian hosts, for their contributions to the successful work of this group, embracing some 250 participants. Especially, thanks go to Lynn Steen for his comprehensive and lucid survey lecture and to László Babai for a stimulating demonstration of some exciting undergraduate mathematics.

Transition Secondary–Postsecondary

Leader: John Mack (Australia)

Two subgroups operated throughout the four sessions. Group A, led by John Mack, considered examples of activities and programs that reach across the boundary in order to improve transition. Group B, led by Milton Fuller (Australia), looked at the provision of support structures, within postsecondary institutions, designed to assist students with weaknesses and/or deficiencies in their mathematics.

Group A.

Three sessions involved discussions based on presentations describing a variety of recent experiences with transition programs. The presenters were Joan Leitzel (USA), Marie-Francoise Coste-Roy (France), Eileen Poiani (USA), Terry Pearson (Canada), Dan Lorenz (Israel) and Elias Toubassi (USA). The final session was a round-table discussion, led by Hugh Burkhardt, on the possible roles technology might play in improving (or worsening!) transition problems.

In general, transition is formally determined and controlled by procedures which may vary significantly between countries, within countries and from institution to institution. Formal avenues of contact between those responsible for secondary education and for higher education, and the extent to which cooperative planning of curriculum matters across the transition years is possible, also differ significantly, although there seems to be little evidence of the latter in most education systems. The activities, described in our sessions, are generally examples of initiatives taken by particular institutions in order to establish and foster better communication and information flow outside "formal channels". The purposes of these activities are to provide prospective students and their teachers with more precise information on mathematical requirements, to identify and provide help for areas of the secondary mathematics curriculum which seem to produce problems, to help students appreciate the significance of mathematics as part of education in many fields and (not least) to demonstrate a willingness to create a useful and friendly working relationship across the secondary-tertiary interface. All such activities require a commitment by staff, concerned with these problems, to devote time and effort towards them. Recognition of the need for such programs, on an institutional (or even a national) level, is becoming more common as governments address matters such as completion rates in higher education, the growing importance of mathematics in higher education, the need to attract students into mathematical studies (especially so for students from under-represented groups) and the problems of providing an adequate supply of mathematics teachers at all educational levels.

Some examples, drawn from the presentations, that illustrate different types of activities, are now briefly described. Several speakers indicated the powerful effects of providing high school students with "early placement" or "readiness" tests which give them direct, individualized information on their mathematical competencies and the possible courses open to them. Such tests, given in a supportive inter-institutional arrangement, have proven effective in encouraging students to improve their mathematical preparedness. Arising from the information obtained from such tests or from a frank appraisal of first year performance, areas of weakness or of mismatch can be identified and addressed. (For example, four high-school curriculum projects in Ohio (Leitzel), changes to the structure of and instructional environments for entry level courses at Arizona (Toubassi) and various universities in France (Coste-Roy), better matching (most places, including Canada (Pearson)). At the Technion, Haifa, Lorenz indicated the positive effects, obtained within a fairly rigid competitive entry system, by adding extra weight for results on higher-level school mathematics courses. Poiani stressed the tremendous importance of "humanizing" mathematics (in content, teaching methods, institution-student-community interaction) if more women and students from minority groups are to be successfully attracted into mathematics-based courses.

In the final session on technology, the major issues discussed were the stress that will be placed on transition if there is a technological mismatch across it, the need to plan a transition curriculum incorporating content and skills changes that have arisen or will arise from the general use of successive generations of hand-held calculators, the even larger problems posed by graphics/symbolic/numerical devices (e.g., HP–28 series) and the paramount problem of learning how to use technology to enhance concept formation and mathematical thinking. In many cases, it was felt that the challenges at tertiary level will prove to be considerable and there is an urgent need for those using or experimenting with mass use of new technologies to communicate freely their experiences.

Group B

Presentations were made by Lyn Taylor (USA), Deann Christianson (USA), Mary Barnes (Australia), Harm Jan Smid (Netherlands), Osamu Kota (Japan) and Milton Fuller (Australia). The topics addressed in them and in the ensuing discussions were as follows.

1. Descriptions of existing remedial and support units — examples are Mathematics Confidence Workshops, Mathematics Resource Centres and Mathematics Learning Centres.
2. Testing and Placement Programs, which ranged from quite sophisticated placement tests, used to assign students to courses, to tests intended to assist the building of study skills and confidence in mathematics.
3. The attitudes of new students towards mathematics were considered to play a major role in students' success in overcoming deficiencies. In general, these attitudes are negative.
4. Teaching methods in entry level courses and the overloading of the curriculum are contributing factors to poor performance.
5. Liaison by tertiary teachers with high school teachers is usually on an individual basis and needs expansion via formal institutional support.
6. Ongoing support throughout a student's first year was considered an essential component of any remedial program.

Observations and recommendations.

(a) Formally organized remedial and support units have existed for many years, mainly in the USA and in Australia. Several participants indicated they would be making submissions to establish such a unit.

(b) Countries with traditional school examination systems appear to be able to rely more on results as a measure of student mathematical competence than those with school-based assessment.

(c) There is a need for professional mathematical associations to be supportive of remedial programs within tertiary institutions.

(d) Any "recipe" approach to school mathematics teaching contributes positively towards student failure in first year.

(e) Mathematics faculty need to rethink first-year (including service) programs, taking (d) into account and attending to teaching methods, motivational and attitudinal problems.

(f) Remedial and support services require staff who perceive this work as valuable and crucial in maintaining standards at tertiary level.

(g) Sophisticated calculators can play a useful role in support programs and in confidence building.

(h) Some established remedial units are already overloaded.

In summary, it was agreed that tertiary institutions should adopt the mathematics support model, accepting that there is often a real gap between actual and expected mathematical knowledge at entry. The group resolved to form a Panel of Interested People and seek ways of establishing a formal association of providers of mathematics support programs.

Mathematics for Non-Specialists

Leader: Glyn James (UK)

Two splinter groups were formed, with themes as shown:

A: (i) Impact of the computer on the mathematics curriculum in engineering;
   (ii) A mathematics core curriculum for engineering programs.
B: (i) Mathematics for Actuarial and Business courses;
   (ii) Mathematics for the Liberal Arts.

Group A. The obvious areas of overlap between themes (i) and (ii) were drawn out during the interesting and lively discussions that occurred in all four sessions.

A (i) Impact of Computers.

The two sessions were led by Mike Beilby (UK). Aspects relevant to mathematics for Physical Scientists were also addressed. The following issues, having implications for the theme, were particularly identified:

(a) increasing pressures on the total engineering curriculum (e.g., inclusion of managerial and marketing skills) threaten the time allocated to mathematics. Class sizes are already large (50–150–300) and in some countries diminishing resources are increasing student-staff ratios.

(b) in some countries (e.g., Australia and UK), the desire to increase intakes at a time of falling school-leaver populations leads to a greater spread of students' mathematical backgrounds, requiring bridging work in the first year. This is not a problem elsewhere (e.g., Peru, China) where there is intense competition for entry and less of a problem in Hungary and the USSR.

(c) in order to exploit computer technology fully, there is a need for more mathematically skilled engineers and scientists.

The use of CAL to help overcome the heterogeneous intake described in (b) was illustrated by Ed Robson (UK), who reported on an individualized remedial instruction program for first-year students.

Generally, engineering students are highly motivated and look for justification and development of mathematical ideas within their own subjects. Thus tedious manipulation is best delegated to the computer as and when this becomes possible. Pam Surman (Australia) described diagnostic tests in calculus, based on CAL modules and suitable also for external students. Tom Scott and Diana Mackie (UK) showed examples of realistic applications exercises using worksheets to guide exploration. Feedback from an evaluation programme is being used to adapt the material. Robert Harding (UK) described a series of Computer Illustrated Texts, on mathematical topics, useful as coursework. Matt Farrell (Australia) reported success in using shapes generated by a graph plotter — the character of functions is discussed, thus stimulating interest in their mathematical properties.

The use of computers to provide more complex and realistic, but less laborious, applications was stressed. Collaboration on realistic problems, involving mathematics and engineering staff, is practicable with positive attitudes to cross-disciplinary approaches, joint or team teaching and agreed notation.

A (ii) Core curriculum.

These sessions, led by Glyn James, used working papers of the SEFI (European Society for Engineering Education) Working Group on Mathematics as a discussion framework. There was general agreement that the subject areas identified in these papers formed a suitable core curriculum outline; these are calculus and analysis, linear algebra, statistics and probability and discrete mathematics. It was also generally agreed that numerical methods should be integrated into the curriculum, for example by use of case studies/modelling exercises. The matters of depth of treatment and of a balance between skills training and concept development were discussed extensively, as were the effects due to the change of emphasis from a field theory to a systems approach. Michel Helfgott (Peru) stressed the importance of applications and outlined a Chemical Engineering program successfully integrating mathematics exercises with specific presentations devoted to practice. The session concluded with a discussion on the future impact of symbolic algebra, simulation and expert system packages.

B (i) Actuarial and Business courses.

Wolfgang Ettl (Austria) led the discussion on actuarial courses by contrasting the two systems operating in the western world. In continental Europe, most studies are full-time at university, there are no correspondence courses, examinations are university-controlled, work experience is obtained after completing study and is essential before being appointed an actuary by the appropriate government authority. In all English-speaking countries, most courses are part-time and by correspondence, work experience is gained concurrently and examination and certification is provided by the Society of Actuaries. The main subjects studied are common to both systems. It is important that teachers and actuaries at the universities maintain active practical experience and often they join with company actuaries in research groups working on large scale problems.

Martha Siegel (USA) introduced the Business courses session by presenting a paper discussing the initial mathematical needs (content and mathematical thinking skills) and appropriate higher-level courses. She noted that, in many cases, courses are prescribed by other bodies and mathematics teachers provide a strict service role only. The uses of research in teaching and learning, and of calculators and computers, in efforts to improve the mathematical skills of these students, were also canvassed. Discussion of these issues extended into the B (ii) sessions, as follows.

Jim Tattersall (USA) described experiences running individual and small-group help sessions for business mathematics students. A lot of psychology was needed and going over homework and tests was valuable. In ensuing debate, many argued that calculus was "overkill" for these students and perhaps more statistics and probability would be more useful. Later, Christina Broomfield (UK) related the problems of teaching an overloaded business mathematics course that also had to be integrated with information technology. The problems of such courses were further discussed by Martha Siegel and Robert Bruhschwein (USA).

B (ii) Liberal Arts.

These sessions, led by Jim Tattersall, looked at ways of introducing mathematics to students (undergraduate, graduate and adult) with little affinity for the subject. Ruth Hubbard (Australia) reported on work devoted to teaching mathematical reading and teaching skills via a remedial maths facility. Prior to this, students relied entirely on memorization. She developed a course based on slow and careful (multiple) readings of mathematical texts with particular attention to all theorems, and discussed several examples. In the discussion, the use of newspaper articles to reinforce the utility of mathematical understandings was mentioned.

John Iacono (Australia) described using SPSSX to teach statistics to large numbers of adults, in 4-hour sessions once weekly for a year. He emphasized the importance of affective techniques to achieve a good rapport with students and thus help them to accept the mathematics.

Bob Hayes (Australia) showed how a historical approach in teaching is successful with trades people who need mathematics in their work. He mentioned use of a newspaper article on marathons to try and predict when a two-hour marathon will be run. This involved not only curve fitting and prediction but also time and history.

Ed Barbeau (Canada) talked about two courses — *Mathematics in Perspective* (little assumed knowledge, a problem-solving focus, assessment based on problem solutions, films, book reviews, essay and exam) and a new course, *Integrated Study,* stressing human experience and knowledge. Yoshio Kimura (Japan) discussed the Japanese school system and how students there hate mathematics — he has introduced comic books and allows 8 years to complete a 4-year program.

The closing discussion led to agreement on the need for a more humanistic teaching approach and a concern for teaching thinking and operational knowledge, rather than content, in all courses for liberal arts students.

Computers in Postsecondary Mathematics Education

Leaders: Paul Zorn and Warren Page (USA)

This group studied curricular, pedagogical and technological issues related to computing in mathematics at this level. The group split into two subgroups (A and B), enabling more topics to be covered and to facilitate discussion on them. Work began with a short general presentation, by Paul Zorn, on issues and open questions surrounding instructional computing.

Later, brief individual presentations were combined with open discussion. Papers and discussion fell into three broad categories: *educational technology* (computer graphics, symbolic manipulation, calculators, etc.); *course-specific issues* (calculus, linear algebra, discrete mathematics, number theory, etc.); and *pedagogy* and *philosophy* (cognitive issues, the role of programming, general curriculum issues, etc.). Issues in all categories are related to each other, and cut across artificial boundaries, so discussions were free-ranging.

Subgroup A sessions

1. Educational technologies for teaching and learning mathematics.

Warren Page surveyed the types of software of importance to mathematics instruction, comparing and contrasting the criteria and salient features for TUTORIAL programs, TOOLS software, SIMULATION packages and INTEGRATED software combining two or more of the above (e.g., a simulation package with a built-in tutorial). He also described the three mathematics software packages considered "distinguished" in EDUCOM's 1988 Higher Education Software Awards Competition: MAX — the MAtriX Algebra calculator, Phase Portraits, and Graphical Aids for Stochastic Processes (which received the top award).

Frank Demana described the use of hand-held graphics calculators at precalculus and calculus levels. He illustrated how syntax errors revealed conceptual errors, showed the benefits of a graphics ZOOM feature and gave examples of the kinds of pitfalls inherent in a too cavalier use of graphics calculators. David Clegg also stressed the motivational benefits of graphics displays, illustrating their use in studying approximations and Gibbs' phenomenon. He stressed the convenience of allowing students to work in an unstructured, "own time" environment.

David Smith demonstrated the utility and diversity of MathCAD (a scratchpad with facility for text and formula manipulation) by way of a problem in population dynamics.

John Kenelly surveyed the evolution of hand-held calculators, up to and including the present HP 28-S, which has integrated numeric, graphic and symbolic processing capabilities. He also discussed the specifications for a simple, user-friendly, hand-held Computer Algebra System, formulated by the MAA's Committee on the First Two Years of College Mathematics.

2. Computer Algebra Systems (CAS's)

The purpose of this session was to initiate dialogue in and raise awareness of the many critical issues related to the introduction of CAS's into the mathematics curriculum. It began with a demonstration of the HP 28–S by John Kenelly. Warren Page then listed the critical issues needing attention now, including *curriculum concerns* (the changing nature of, and rationale for, present and future course offerings), *impact on students* (which student populations will be affected, and in what ways), *impact on teachers* (changing epistemological roles, teachers at risk in dealing with bright students using sophisticated CAS's, teacher training), and *sociopolitical implications* (institutional support of mathematics departments to enable them to meet responsibilities and obligations to students and staff in this technological age, and issues of equity — providing equal access to CAS's).

Page's presentation was followed by a panel of respondents. Ian McGee (Canada) considered the critical mass of faculty needed to maintain the use of CAS's throughout students' mathematics courses. Phoebe Judsen (USA) responded to some of the cognitive and affective issues — in teaching a Business Calculus course, she found that use of the CAS MAPLE increased students' motivation and enthusiasm while alleviating anxiety over the need to memorize formulae and do lengthy calculations. Eric Muller (Canada) reported on a survey of a CAS calculus course, taught by him, where CAS laboratories replaced tutorials for 100 students. Robert Kuhn (USA) expressed some concerns about the use of CAS's in education as distinct from research, especially with regard to poor user interfaces and the temptation to use CAS's as "oracles".

The panel presentations aroused some lively and open discussion. Key points made included:

- CAS's can make courses more relevant to the world students live in.
- There may be increasing pressure, particularly in service courses, to reduce mathematics requirements since CAS's can replace routine processes.
- Finding some standard common language or syntax for use by CAS designers is highly desirable.

Finally, Page reminded participants that a report on the July, 1988 Colby College Conference on Computer Algebra Systems is expected to appear soon.

3. Pedagogy and philosophy of computing.

Brief presentations were made by Daniel Poisson (France), Donal O'Shea (USA), Qi-Xiao Ye (China), and R.M. Kuhn (USA). Their topics included: various ideas regarding the use of computers in mathematical experiments; the role of computers in mathematical proof; problems of implementing computing in the classroom; and decisions that must be made as computers find a more extensive role in mathematics education.

In general discussion, the following points were made:

- Computer use in the classroom takes time. Trade-offs are inevitable, but time for teaching mathematical *concepts* should not be sacrificed.
- There is no convincing evidence, positive or negative, of the efficacy of computer laboratories in mathematics instruction.
- In some countries, students avoid mathematics courses with a *programming* component, but appear willing to take courses in which software *packages* are used.
- There is no consensus among teaching faculty that computers really improve mathematics instruction. Several participants recommended *calculator* (rather than computer) use, for reasons of portability, cost, and acceptability to other faculty.
- Computers are only tools; they will not replace teachers in explaining mathematical ideas and developing mathematical understanding.

Subgroup B sessions: 1 and 2. Course-specific issues — calculus

This focussed on how CAS's can be used in elementary calculus courses. Several participants reported their own experiences with such systems and Harley Flanders (USA) described the calculus software package MICROCALC.

Various issues emerged in subsequent discussion:

- How much knowledge of and teaching experience with CAS's exists among faculty? Several participants mentioned their own uses of CAS's, mainly for research, but for now, cost remains an obstacle, especially in developing countries. Still, we must anticipate the fact that students will soon use CAS's widely.
- Should students learn calculus techniques that machines can perform better? Lively discussion — but no consensus — occurred, except on the point that computing will force us to make and defend our own positions on this important question.
- What are the advantages — and costs — of using CAS's in calculus and elsewhere? Among benefits cited: the possibility of concentrating courses on ideas, not techniques; reducing drudgery of routine computation; allowing more realistic applications and modelling; studying algorithmic aspects of calculus. Among the risks, costs, and questions CAS's raise, the following were cited: the dangers of unthinking or too-early use of powerful technology; diversion from mathematical issues; unfriendly or non-mathematical user-interface problems of some systems; whether CAS's should be optional or compulsory; whether CAS use requires minor or fundamental revision of standard calculus curricula.

In the second session, elementary calculus continued as the principal focus. Louise Raphael (USA) described the background of the calculus reform movement there, particularly as regards computing issues. She mentioned criteria which funding institutions might use to assess grant proposals, especially those related to computing. James Leitzel (USA) described problems (cost, faculty training, co-

ordination with "client" disciplines) that arise in implementing calculus reforms
— especially those involving computing — at large public institutions. Norman
Biggs (UK) argued for incorporating structured programming into the mathematics curriculum. He recommended that neither computer science nor mathematics
be regarded as experimental disciplines, but rather as deductive ones, in which programs and proofs play analogous roles. James Hurley (USA) described a two-year
computer-enhanced sequence in calculus, linear algebra, and differential equations,
which uses true BASIC. Students work directly with program code, modifying it
and interpreting results, but not creating programs from scratch. David A. Smith
(USA) summarized a proposal to develop an entirely new calculus course, which
would use the computer as a laboratory instrument for discovery, and also as an
assistant in *writing* sound mathematical prose.

In open discussion, participants emphasized the need for care in introducing
packages at appropriate *times* and in ways that truly reinforce *mathematical*, rather
than *computing*, concepts. Numerical programs were cited as leading to useful
analytic insights (e.g. on differentiability and continuity), but care must be taken
to avoid misleading round-off errors.

4. Computers in courses other than calculus

Warren Page (USA) described a course, "Conjectural Geometry", based on
laboratory modules. High school teachers used these modules to explore concepts
and to acquire new ways of thinking in geometry, and also learned to create new
modules to enhance students' learning of geometry. Renfrey Potts (Australia)
reported on experiences with a linear algebra package (MATRIX), discussing the
results of a survey of students who had successfully completed the course in which
it was used and the disappointment of the faculty teaching the course with the
students' results. Peter Giblin (UK) described a number theory course in which
students wrote their own programs and modified existing programs. Reaction was
quite positive.

Carole LaCampagne (USA) briefly surveyed how computers were used in
post-secondary mathematics education, citing as examples: tutorial, classroom
management, programming, and computer based instruction. At the University of
Northern Illinois, students work in groups and, according to LaCampagne, this has
positive impact on their responses and social interactions.

In the ensuing discussion, concern was expressed about the ability of students
to write effective programs. Tools such as mathCAD eliminate the need for programming. Staff enthusiasm by itself is insufficient to guarantee the successful use
of computers in instruction and and the purpose and function of computers in a
mathematics course must be clearly thought through in planning and teaching such
courses.

Revitalizing Undergraduate Mathematics: Curriculum and Faculty

Leader: Jack Gray (Australia)

Introduction

This theme is based on a perceived world-wide dissatisfaction with the under-graduate curriculum, particularly with the first few tertiary-level offerings in calcu-lus and algebra. Too often our courses have become ossified , remaining unchanged in spirit and content. (In my own institution, a 2nd year course on mathematical methods for engineers has remained unchanged since 1954, as if engineering prac-tice had not changed and the computer had not been invented.) Too often, our courses lack vigour and vitality both in content and teaching. Too often, the focus is exclusively on technique at the expense of ideas. Too often, we leave students with the impression that mathematics is like latin — extremely dead and not very useful.

How can we revitalize our offerings? How can we present mathematics as the attractive, alive, growing discipline we know it to be? In particular, how can we show students in their early years the profound impact mathematics has on modern technology? Shouldn't students of linear algebra see the significant role played by error-correcting codes in enhancing the fidelity of their CD's? Our calculus courses in particular have become encyclopedic and often cluttered with uninteresting pseudo-applications. These issues are not new but have acquired some urgency due to falling enrollments in later years and the increasing gap between what we teach and what we research: between the old and the new.

For revitalization to proceed it must overcome the usual educational and institutional inertia. Pressure must come from teachers themselves who in turn need to articulate coherent programmes of change. They must, for instance, develop materials on modern aspects of mathematics suitable for presentation in the early years, something that necessitates committing the cardinal sin (in calculus at least) of omitting some established topics. (In my own university I managed to introduce into the first year courses two 3–hour enrichment topics (eg: coding theory and chaos) designed to stimulate and reveal the modern excitement of our discipline. Such a piece-meal approach is far from satisfactory, yet one spin-off, revealed by a student survey, showed that the vast majority of students had previously been unaware of the existence of research in mathematics!)

Faculty, too, need rejuvenation for reform to be successful. Teaching and scholarship in the broadest sense need to be institutionally recognized as legitimate paths to promotion in order to enhance teachers' self esteem.

All these issues, and more, were discussed in the subgroup sessions, which continued with a presentation by Bernie Madison, Project Director of the U.S. "MS 2000" project, the focus of which "is on talent — students and faculty — with special attention to increasing the participation of under-represented groups, women and most minorities. Activities and reporting are occurring and planned in availability and development of mathematical talent, curriculum reform, resource needs and availability, and public awareness. The first major public activity, held in October, 1987, was a national colloquium, *Calculus for a New Century* (MAA Notes #8 (1988); Lynn Steen, ed.) for discussing reform in the teaching and content of calculus. Extensive descriptive and discussion activities are under way, leading to a set of recommendations for action by those responsible for higher education in the United States."

In discussion, Professor Madison commented on the growth in the percentage of low-level students entering college and university and how this depresses faculty. For students, calculus is often the peak of their mathematical attainment; for faculty, the pits. This gap raises serious morale problems. The issue of morale was further focussed by his data that only some 30% of the teaching faculty are active in research — 70% are professionally stagnant.

Annie Michel–Pajus presented the French view. She concentrated on high drop-out rates, the formal nature of the mathematics taught and its lack of coherence with other subjects, and the inability of students to think in an "experimental" way. (An intriguing study at Grenoble, on "scientific debate in the classroom", aroused much discussion. There, in classes of up to 100, students are asked, under guidance, to formulate conjectures and offer supporting arguments.)

The French government has given an impetus for renewal, based upon decentralization. Through a variety of on-going classroom experiments, including syllabus variation, the history of mathematics and increased use of computers, revitalization is proceeding and is aimed at giving more meaning to mathematics and inciting students to show more initiative.

Reuben Hersh (USA) spoke briefly on how to teach the philosophy of mathematics. He would like to see philosophical material in all courses as such material can provoke thought in faculty and students alike.

Tony Gardiner (UK) led a workshop showing how, with appropriate open-ended materials (cf: A. Gardiner "Discovering Mathematics"), one could insist on mastery and meaning even with a moderate-sized class. Particularly emphasized was Freudenthal's remark that "teaching mathematics is fraught with relationships", the task being when and how to introduce them. It is these relationships that are memorable, not the linear chains of theorems and proofs.

Ted Byrt (Australia) similarly involved the group in solving elementary statistical problems, again revealing how, even with a class of 70, one can engage students in meaningful mathematical activity.

The last session was a "round-table" discussion on revitalizing faculty. Four main issues emerged:

(i) The lack of reward and encouragement for good teaching in a research-only ethos.

(ii) The difficulty of describing the essential but diverse characteristics of good mathematics teaching. Two notions were highlighted: (a) a list of 101 "do's" and "don'ts"; (b) an anthology of essays on good mathematics teaching, akin to Joseph Epstein's "Portraits of Great Teachers".

(iii) The lack of opportunity to see others teach and of criticism of one's own and others' teaching. Videos may help here.

(iv) Large, rigid and crowded syllabuses which successfully prevent innovation and a scholarly approach. Some suggested that the forces for rigidity, that seem to stem from users (particularly Engineering schools) are perhaps not as strong as has been believed.

Thanks go to all 40-odd participants from 16 countries, and especially to Joe Gilks for recording.

Research into teaching and learning

Leaders: Michèle Artigue (France) and Marjorie Carss (Australia)

The purpose of the first session was to introduce our theme and to launch a preliminary discussion on the following points:

- What are the specific elements of research into teaching and learning at this level?
- What are the main problems?
- What are the main directions of current research?
- What results have been obtained?

In her opening address, Aline Robert (France) surveyed the extant research on teaching and learning at this level of education. Principal characteristics of this level of mathematical education seem to be the complexity of the subject matter (including its abstraction) and the metacognitive abilities of the students. She classified published research as follows:

(i) content-oriented (e.g., some of the work of PME). Much of this concerns calculus. The research utilizes distinct, non-contradictory theoretical models, some developed specifically for analyzing student difficulties (for example, "epistemological obstacles" — Cornu, Sierpinska; or "conflict between concept images and concept definitions" — Tall and Vinner). Some researchers (Artigue, for example) assign a central role to analyzing the operation of the whole educational system relative to the "topic" considered.

(ii) research on teaching methodology whose aim is to produce and test organizational models (e.g., Artigue on differential equations, Robert and Robinet on linear algebra, Tall on calculus).

(iii) recent research on "new levels", to be discussed in subsequent sessions (work of Schoenfeld, Dubinsky, Alibert). Robert described her own work (with Tenaud) on the teaching of geometry, in which small-group work on "well-selected" problems are combined with practical instruction in methods, the two aspects allowing a simultaneous acquisition of understanding of the new material and an absorption of this into students' mathematical knowledge. On the third point, Robert emphasized that research allows us, in some areas, to identify the main difficulties encountered by students and thus to suggest suitable learning environments for overcoming them. She concluded by observing that research at this level is crucial for the overall advancement of education, because it is the university which predominantly shapes future teachers, and each one tends to imitate the teaching styles experienced as a student.

The second session involved three presentations and discussion. Alan Schoenfeld first described his course structure (based on small-group work on problems and extended discussion of the mathematical reasoning involved) to help students develop mathematical thinking embracing subject matter, problem solving strategies, efficient behaviour and a mathematical epistemology.

In his second presentation, Schoenfeld identified three directions of current research:

(a) social/espistemological work to create new learning environments that assist students to develop a mathematical epistemology,

(b) "applied" work, either using psychological approaches to design curricular materials, or software design, and

(c) detailed psychological studies of thinking and learning.

Each of these directions is important, but tertiary teachers regard (a) and (c) as irrelevant, leading to conflict between them and the researchers. Schoenfeld suggested that researchers need to provide "good" examples and "ready-to-use" applications while respecting the integrity of their own work.

Ed Dubinsky (USA) then described his work in developing a theoretical framework (the genetic decomposition of concepts) useful for design of teaching practice. He utilizes the programming language ISETL to develop appropriate teaching activities and described an instructional treatment for predicate calculus based upon his theory and ISETL. In discussion, the question arose as to the need to use the computer. Dubinsky replied that, for him, the programming of activities in ISETL helped the different cognitive processes and in particular that one (called "encapsulation" by him) by which students construct objects from processes. At present, he saw no way of avoiding the use of ISETL.

The third session began with Julian Smith reporting on examination results for first year students at a technical university in South Africa, showing they had least success with calculus and trigonometry problems. Video interviews with a group of six students, in which they identified misconceptions underlying incorrect responses, were then shown.

Daniel Alibert reported on research at Grenoble, based upon the introduction of "scientific debate" as the vehicle for developing mathematical thinking with a class of some 100 first year students. The idea is to introduce a problem whose solution requires a new mathematical tool and to encourage students to make conjectures about the properties of this tool. These conjectures are discussed, refined, proved or disproved via class debate, with the teacher sharing thoughts with the students and the class voting on the status of conjectures and arguments. So far, it has proved successful in that the students involved perform well in subsequent courses and exhibit greater mathematical maturity than other students with a "conventional" first year.

In the vigorous ensuing discussion, several participants reported using similar strategies and raised the issue of the attitude of colleagues to such experiments. Others asked about research on the role of the teacher in this method, to which Alibert replied that, at Grenoble, this critical issue is still considered on an empirical basis. He emphasized the positive effects of giving equal standing to all the students' conjectures, over the usual method of presenting only true assertions, because misconceptions emerge naturally and are discussed openly.

The final session, moderated by John Mason (UK), provided a wide-ranging discussion on the connections between teaching and research, using some of the analyses and examples from earlier sessions as starting points. The notion of a "shift in attention", to which John Mason assigned the term "framing", stimulated considerable debate. An example of framing is that of a teacher "seeing his/her teaching as an object of study". While this happens in the classroom, A. Robert indicated the difficulty of doing this in real time, in a class, where one tends to make snap decisions. Thus effective research, both before and after teaching, is needed for this kind of framing. Framing also happens frequently with student-teacher interaction, involving changes in metacognitive levels, and work needs to be done on the extent to which students (and teachers) notice these shifts in attention. Several participants felt that teaching situations should be designed to expose such framings.

Another discussion point concerned that of "didactic tension" between student and teacher, involving changing levels in question-answer exchanges. Further discussion examined the relation between research and the "scientific debate" idea. A major idea here is that of obtaining "commitment" from the students, and this can be achieved in other ways. Other points raised were those of teacher-control, the need to achieve useful learning, and other strategies involving framing (including ways of asking students to reflect on their learning, such as the technique of reconstruction used at the Open University). Finally, there was discussion on the "Yes, but" response — e.g. "I would like to use scientific debate, but in a class of 300... .". The need to make precise one's difficulties and to examine possible strategies for overcoming them was emphasized and illustrated.

In conclusion, Michèle Artigue noted that, despite a rapid growth in educational research at this level, much remains to be done. For example, much research is devoted to calculus, little to other subjects. She observed that more and more, even in research on specific aspects, consideration is given to the global educational environment and that more attention is devoted to the study of teaching strategies supporting the development of a mathematical epistemology. Questions of characterization of teaching and learning and of teaching methodology at this level are far from settled.

The situation in developing countries

Leader: Miguel Jimenez Pozo (Cuba)

The subgroup was devoted entirely to discussion seeking to identify common major issues, both internal to each country and generally in a global context. The spectrum of topics is so broad that it was difficult to decide whether to discuss all problems simultaneously or to focus on one at a time - exactly the position one faces in each underdeveloped country when deciding what to do.

I. The internal situation — identified problems.

It should be noted that many underdeveloped countries have extraordinary curricula "on paper", but that practical circumstances impose a more realistic implemented curriculum. The subgroup believes that a diversified mathematics curriculum, including courses in statistics and computer science, is essential in order that graduates have a wide scope for employment. To achieve and maintain such a curriculum requires a broad and strong mathematical culture, but as demand for mathematically competent graduates increases, one may expect mathematics enrollments to decrease in favour of enrollments in related fields such as computer science, statistics and engineering. Indeed, it may be expected that, more and more, the less able students will confine themselves to mathematics and mathematics teaching. Unless there are positive incentives (such as good teaching positions at elementary, secondary and post-secondary levels) for good students, the problem of maintaining good mathematics teaching will become worse. This will be particularly severe in the countries where a University degree is the royal road for movement from one social class to another.

Another serious issue is that of providing a suitable professional journal for mathematicians and mathematics educators. The criteria adopted in developed countries for such publications are usually inappropriate to the diverse needs of developing countries. Besides original research, national or regional journals should contain explanatory articles and articles on teaching, heuristics, history and philosophy of mathematics and other cognate areas.

Recommendations

1. Curricula in mathematics in developing countries should be broad and should include statistics, computer science, data analysis, operations research, numerical analysis.

2. UNESCO should be requested to compile and maintain an updated list of journals in mathematics and mathematics education in developing countries, together with the language/languages used, and send copies of the list to universities and mathematical societies, who may then explore the possibilities of exchange of journals or publishing new ones.

3. Universities (mathematical societies) from developing countries must increase efforts to obtain collections of existing international journals, perhaps via cooperative arrangements.

4. Suggested editorial guidelines for scientific journals in developing countries:

(i) Journals should not be obliged to imitate the characteristics of journals from developed countries, but should develop their own characteristics.

(ii) Recognizing limitations on reference materials available locally and the small numbers of specialist workers, papers should be written in a more self-contained form, with at least an account of earlier relevant results.

(iii) Articles are not obliged to contain only new work. In particular, a mathematician who publishes an abstract in an international journal may write an expanded version of it in his national journal, perhaps in a language common to his countrymen.

(iv) Stability of publication and quality of content must be emphasized. To help achieve this, national mathematical societies and university mathematics and education faculties must help in obtaining suitable contributions from their members.

II. Developing Countries in the world mathematical community — the situation.

The already poor development of mathematical education in the third world continues to be hampered as a result of economic and manpower deficiencies, as well as problems of culture and scale. Such difficulties will have to be dealt with on a world-wide basis if bodies concerned with mathematics education in the developing world should ever achieve a state of "critical mass".

Recommendations

1. The subgroup urges that ICMI, or the IMU Commission on Development and Exchange, immediately constitute a survey committee, comprised of appropriate active representation from developing countries of Asia, Africa and Latin America, to undertake a study of the situation in mathematics education in these countries. This study should report on the literacy rate, enrollments at first, second and third level education, enrollment at post secondary level mathematics, the rate of growth of enrollments, the curriculum and the languages of instruction.

Since there is wide variation in goals and aims, strategies and development in mathematics education in the tertiary level among the developing countries, this survey should be done with the cooperation and assistance of existing mathematical societies, the Third World Academy of Sciences and others, and make use of earlier results of ICMI and other groups.

The report of this study should provide a basis for future recommendations, whose implementations would depend on ICMI, governments and local groups.

2. The representation from developing countries in ICME–6 is insignificant, compared with that of the developed countries. IMU, particularly ICMI, is requested to develop ways to ensure significant participation from developing countries in the future. Such efforts should include:

(i) improved communications *within* regional groupings, particularly among countries sharing a common language. (ICMI's role could be to provide, if not travel funds, at least postage money, to a representative willing to distribute a newsletter or other information.)

(ii) improved communications *between* regional groupings. (ICMI's role could be to ask the regional groupings to share information. Especially ICMI should ensure that information about financial aid be disseminated, both travel aid for conferences, distributed in a timely way, and aid for further study.)

♣ ♣ ♣ ♣ ♣

Supporting survey presentation:

Lynn Steen (USA): Mathematics for a New Century.

Special lecture:

László Babai (Hungary): Linear Algebra Methods in Combinatorics.

ACTION GROUP 6: PRE-SERVICE TEACHER EDUCATION

Chief Organizer: Willibald Dörfler (Austria)
Hungarian Coordinator: Gábor Hetyei
Panels: Trygve Breiteig (Norway), Terezinha Carraher (Brazil), John A. Dossey (USA), Graham A. Jones (Australia), Geoff T. Wain (UK)

The first session of the group was a plenary session with three presentations. The second and third sessions involved working groups for short presentations and discussions of the main topics of the Action Group. The final session was organized as two parallel subplenaries (chaired by *Wain* and *Dossey*) during which the working groups reported on the main results from their topics. The following report was prepared by *Breiteig, Dörfler, Dossey* and *Jones* based on these reports and on the papers or abstracts of the oral contributions.

In the initial plenary talk *Hilary Shuard* (UK) examined the challenges for primary teachers in the 1990s and their implications for pre-service teacher education. She partitioned these challenges into two categories, internal challenges associated with mathematics and external challenges associated with society.

The major internal challenge related to the use of new technology, both computers and calculators. While there are some limitations associated with use of computers in primary mathematics, viz. availability of resources and reluctance of some teachers to enter a computer environment foreign to their own upbringing, but not to that of their pupils, Shuard argued that teachers will be able to find software which enhances their style of teaching, e.g., programs which support problems solving and languages such as Logo which facilitate both free exploration and programming. Calculators, on the other hand, have become thoroughly accepted. This acceptance will inevitably produce changes in primary classrooms. Shuard reported on the approach being used to address these changes in the PRIME project through the adoption of a Calculator-Aware Number Curriculum. In this project young children are free to use a calculator whenever they wish. They are not taught traditional paper and pencil algorithms. Greater emphasis is devoted to early introduction of large numbers, negative numbers and decimals, all of which are consistent with the present day experiences of young children.

In relation to external challenges, Shuard pointed to the imposition by the government of a National Curriculum. This curriculum was devised for the ages 5 to 16 in England and Wales and will operate for mathematics, science, and language. The National Curriculum requirements will only be concerned with content, and teachers will be free to present this content using any methods they wish. The curriculum will encompass elements of accountability for teachers as all the children will be tested at ages 7, 11, 14 and 16 and most test results will be published.

In summary, Shuard believed that the challenge for primary teacher education in the 1990s will be to produce teachers who are: *enthusiastic* about mathematics, its learning and teaching; *flexible* enough to accept new challenges and discard redundant activities; *willing* to encourage children in mathematical thinking and problem solving; *adaptable* in their teaching styles to accommodate new situations such as a National Curriculum.

The second speaker was *Günther Malle* (Austria) on the theme "The Question of Meaning in Teacher Education". He emphasized that the construction of meaning associated with mathematical concepts and methods should be the central goal of teaching and learning mathematics at all levels. This applies specifically to the mathematics taught to future teachers, since they will then be responsible for the transmission of adequate meaning to their students. Malle asserted that meaning has to be constructed consciously and deliberately and that the education of teachers has to address questions of meaning explicitly and on many occasions. He proposed a general model of teaching which employs extensive discussion on a meta-level to lead students to acquire insight and thereby their own meaning for the respective subject matter. They should then be able to justify for themselves and to other (pupils and their parents) the teaching and learning of specific topics and of mathematics in general. Students should learn that meaning is not unique, but open to debate, and they should become prepared to engage their future pupils in discussions of meaning, relevance, feasibility, and the like. Malle exemplified his general model with two examples: a seminar on algebra(the meaning of the concept of normal subgroups) and a seminar on mathematics education (the role and use of formulas and variables; how to plan lessons on this topic). Malle stressed that the education of teachers can, and should serve as a model for their future teaching.

The third speaker *Graham Jones* (Australia) highlighted the world wide phenomenon associated with the shortage of secondary mathematics teachers and focussed in particular on an Australian case study. The underlying thesis of Jones' paper was that the demand for mathematics by secondary students and hence the supply of potential mathematics teachers was largely contingent on the attractiveness of the school mathematics curriculum, the appeal of teaching and the attractiveness of the teacher education programs.

Evidence presented suggested that the small amount of time devoted to mathematics in school, the propensity to crowd too much into the curriculum, and the tendency to emphasize instrumental learning rather that relational learning, all contribute to a learning environment which fails to motivate many children. Indeed current mathematics curricula may alienate many capable students who might have been motivated to pursue a career in mathematics or mathematics teaching. Even among those students who achieved well and found mathematics attractive, there was a reluctance to enter teaching. Whether this reluctance stemmed from their own school experience, lack of enthusiasm by teachers to promote mathematics education as a profession, lack of appeal of teacher education courses, or negative attitudes of society towards mathematics, was difficult to isolate; however all factors appeared to have relevance to the current crisis.

Proposals outlined by Jones to resolve the problem included: political lobbying; financial and employment incentives for mathematics teachers, additional scholarships for suitably qualified persons in the work force; more flexible teacher education programs to accommodate capable students who had restricted their mathematical opportunities; and a more entrepreneurial stance by mathematics educators through the media to gain public support for mathematics and its teaching.

Preparation in both mathematics and mathematical education of prospective teachers of primary and middle grades

Chairs: *Heinrich Besuden* (FRG) and *Shirley Frye* (USA).
Secretaries: *Miriam Leiva* (USA) and *Jack Wilkinson* (USA)

Two groups discussed this topic. *Willoughby* (USA) and *Santos* (Brazil) reported concerns about the teaching of mathematics and advocated substantial changes in teacher education programs if children are to face the demands of a technological age. Both presentations called for a stronger orientation towards problem solving and opportunities for student teachers to engage in reflective and analytical activities. *Besuden* (FRG) described a methods course based on laboratory assignments incorporating manipulatives, background readings, and opportunities for students to evaluate their performance. *Lappan* (USA) emphasized the need for addressing the conceptual foundations of mathematical ideas and noted that the field components of a program should support this in real classroom settings. *Leiva* (USA) also stressed concept development in an integrated approach including manipulatives. Associated school based experiences modelling a consistent approach to mathematics learning was achieved through collaboration between teacher, teacher educator, and student peers. *Wilkinson* (USA) also supported an integrated approach to mathematics programs for elementary teachers but went further in advocating conceptual links across mathematics, science, social studies, and language. *Branca* (USA) discussed integrating mathematical content and pedagogy in a course characterized by extensive modelling of problem-solving processes, use of concrete materials, use of cooperative learning groups, and effective questioning techniques. *Frye* (USA) provided further support for the integration of content and pedagogy by presenting a sample of topics that addressed the content of the school curriculum in the context of its delivery to students of all ages.

In proposing a three-phase model for elementary mathematics teacher preparation, *Maher* (USA) conceived the preservice teacher first as learner, observer, and guide; then as designer, implementer, and evaluator and, finally, as philosopher. *Spiro* (Israel) also described a program in which student teachers in lower primary grades reflected on instructional techniques through observing their own actions in inductive–deductive problem solving situations. *Caughey* (Australia) described a program used by *Clarke* (Australia) in which preservice teachers use exemplary materials to present mathematical activities to children. Using a similar approach, *Childs* (USA) used the guiding principles from a model school mathematics curriculum to integrate content and teaching–learning processes. *Takata* (Japan) described a statistics course developed by himself and *Yokochi* (Japan) in which personal computers were used by preservice teachers to display and interpret real world problems.

While acknowledging that there is limited time available to increase the amount of course work in primary mathematics teacher preparation, both groups agreed that prospective teachers must be better prepared in both mathematics and mathematics education. The challenge for teacher educators is to generate more meaningful activity in the given time. In order to meet this challenge the groups proposed that mathematics teacher educators must: (1) integrate mathematics content with methods; (2) fully utilize available technology in the teaching–learning process; (3) plan preservice programs co-operatively with other faculty in order to maximize efficient use of learning time; (4) introduce preservice teachers to the profession through school-based experiences and involvement in professional activities; (5) break the chain of teacher-centered instruction by modelling exemplary teaching behaviours; (6) advocate continuing professional growth; and (7) be professionally active themselves.

The consensus was that if substantial changes are to occur in the teaching of primary and middle school mathematics, reforms must take place in the classrooms where preservice teachers are being educated. Teacher education programs must integrate methods and content with professional experiences that promote collaboration between teacher educators, school, and professional organizations. The challenge is to empower teachers to deal with the changing mathematics curriculum, the emerging instructional technology, and the forces which will affect mathematics education in the 21st century.

Preparation in mathematics of prospective secondary teachers of mathematics

Chairs: *Beth Southwell* (Australia) and *Fritz Schweiger* (Austria).
Secretaries: *Johnny Lott* (USA), *Rolph Schwarzenberger* (UK)

The contributors to these working groups were concerned with the issues of the mathematical content to be taught and how this content should be taught to prospective secondary mathematics teachers.

The students need greater confidence in secondary mathematics as well as some knowledge beyond the secondary school curriculum. Besides this, they need knowledge *about* mathematics, how to learn and how to practice it, and its relations to human thinking and the world outside mathematics. These latter aspects should have a higher priority than specific mathematics content. Courses should moreover adopt teaching methods which exemplify good practice in schools. *Swinson* (Australia) discussed the need for remedial mathematics courses to give prospective teachers the confidence necessary for their future career. *Dubish* (USA) pointed out that university level mathematics courses must provide insight into elementary mathematics. Many advanced topics become more meaningful to all students when they are related to topics previously studied. This implies a serious consideration of integration of content, pedagogy, and practice teaching. *Zhang Dian-Zhou* (China) described the development of a course to build bridges between contemporary and school mathematics. He urged the need for teachers to get help in understanding the stages in the development of ideas, for instance the concept of area working from rectilinear to curved shapes and finally to the concept of measure. *Reichel* (Austria) advocated an integration of "didactical aspects" into all mathematics courses, to provide a framework into which content can be placed.

Neubrand (FRG) stressed the need for teachers to be able to reflect upon their mathematics — its content as a school subject, its significance, and its relation to oneself. Topics from calculus are suitable for student reflections, and can stimulate them to talk about mathematics. *Movshovitz-Hadar* (Israel) described the use of mathematical paradoxes as a bridge between pure mathematics and its psychological and pedagogical issues. Besides its role in the history of mathematics, work with paradoxes also gives experience with cognitive conflict, logical challenge, concept formation and awareness of misconception, and problem solving. *Southwell* (Australia) emphasized the student's need to have not only a body of knowledge, but also a set of inherent processes. These processes should involve one's own experiences in reflection and lead them to teaching for, about and through problem-solving in their own classes. *Schwarzenberger* (UK) noted that students' experience with investigational approaches must be built on mathematical confidence. The students need guidance on how to link investigations with the rest of mathematics, how to link "finding of patterns" to proving and generalizing, and how to build these activities into a coherent theory. *Cheng Ping-Tung* (USA) insisted that to be a teacher in the year 2000, one must serve students as facilitator, tutor, consultant and friend in a computerized classroom. He suggested strategy games to motivate prospective teachers.

Tvete (Norway) applied classical geometry for a better understanding of physical space — giving examples from geography on the earth's surface. Such exemplifying use of mathematics is aimed at developing understanding. This same idea was mentioned by other contributors. *Marcucci* (USA) described a course where students had the opportunity to discover how secondary mathematics can be used to explain observable physical phenomena. Important educational features were discussed: working in small groups with problems; collecting, graphing and analyzing data; and conjecturing, verifying, and finally explicating the results.

Kepner (USA) suggested the use of statistical experiments, to provide a background to the use of statistics with real data and in informal nonaxiomatic settings with emphasis on exploratory data analysis. *Hennequin* (France) also encouraged teachers to consider including statistics, beginning with information from official sources of data, and continuing to experiment with probability using simulations and simple algorithms on small calculators.

Bush (USA) described the crucial nature of the personal understanding of, views on, and attitudes toward mathematics of future secondary mathematics teachers. Mathematics education programs must create courses which foster understanding, positive attitudes and accurate beliefs, in addition to the expected mathematical and pedagogical knowledge and skills.

In conclusion, there is a tension between the well defined aim of many mathematics departments (to get students to the research frontier as fast as possible) and the less well defined aim of teacher trainers (to develop learning and process skills together with content and to stress understanding rather than rote learning). The tension can be resolved only by joint efforts of mathematicians and mathematics educators. The program must ensure that students experience mathematics and not merely reproduce it.

Preparation in mathematical education and pedagogy of prospective secondary mathematics teachers

Chair: *Joop van Dormolen* (Netherlands)

Two subthemes were considered by this group: teaching practice and attitudes of prospective teachers. The presentations on teaching practice (*Price* (USA) and *Broekman* (Netherlands) stressed the need for teachers to observe the teaching–learning environment through activities such as listening, watching and helping children learn. *Egmond* (Canada) and *Backhouse* (UK) also examined the organization of practice and areas of responsibility and cooperation. The group suggested that the major goals of teaching practice were to: prepare students to survive their first years of teaching; enable them to meet the external and personal demands of teaching; and enable them to gain greater insight into their own abilities and the nature of teaching in order to decide whether to become a teacher or not.

Some aspects considered in relation to models of teaching practice included time allotted to practice where ranges varied from some lessons to a full year; the viability of school-based programs; the value of student teachers working together in groups of two or three; the responsibility of the school towards the student including the role of the head of department and supervising teachers; the responsibility of the college towards students during teaching practice including areas of cooperation with the school; monitoring of and feedback to the student teacher; training for the supervising teachers, particularly "master" teachers; and incentives for supervising teachers, both financial and professional.

Five categories of teaching tasks were identified for preservice teachers; teaching one-to-one as a preliminary activity; teaching small groups; teaching classes with student peers and the supervising teacher; teaching classes with the supervising teacher and student peers observing; and teaching classes alone with follow-up evaluation by the student teacher and the supervising teacher.

In summary, the teaching practice experiences should enhance prospective teachers' skills to reflect on their own goals, abilities, successes and failures in relation to classroom management and pupil learning. Moreover, by observing varied teaching styles and teaching–learning environments prospective teachers should be able to identify with those styles and strategies which best fit their own abilities.

In the discussion on attitudes, the contributors, *Farrel* (USA), *Knijnik* (Brazil) and *Milton* (Australia) considered experience and reflection as basic activities required for individual professional growth as a teacher, being especially important in relation to attitudes toward the nature of mathematics. Mathematics should not be seen as a discovery but as an invention and, for pupils, a re-invention. For many teachers this involves a substantial changing of ideas, especially since many of them are not even aware of their own beliefs or attitudes. In essence it requires a process of unpacking one's own ideas of goals, methods, and the nature of mathematics.

Use of observation in mathematics teaching

Chair: *Klaus Hasemann* (FRG). Secretary: *Janet Duffin* (UK)

The first group of papers dealt with models, methods, and strategies for preservice teachers to observe pupils and their teachers. (*Duffin* (UK), *Comiti* (France), *Jansson* (Canada), and *Putt* (Australia)) Discussion centered around the opportunity that observation gives for discussion and reflection about conceptualizations and procedures of teaching. Such opportunities are provided through students going to schools, through questionnaires on teaching procedures (instructional, interpersonal, and managerial), and through clinics for diagnosis and remediation. These possibilities can be further developed by using models of analysis based on theory or practitioner knowledge, using a checklist of observed facts, or through clinical interviews which follow up specific teacher or student behaviors.

The second set of papers dealt with subject — based observations through clinics, written evaluations and interviews (*Feldmann* (USA), *Johnson* (USA), *Hase-*

mann (FRG), *Kurth* (FRG), and *Olson and Olson* (USA)) Such activities should have a high priority in the development of a preservice teacher's understanding of the development of knowledge an behavior of the school-level pupil. Such activities serve as important steps for preservice teachers to grow in seeing the pupils constructing mathematics, individual levels of pupils' preknowledge, their interpretations, and their accommodation and assimilation patterns. Observation provides an avenue for observing pupil growth and self-reflection on the teaching process. This avenue should be used both as a base for observing students and our own growth as members of the mathematics teaching profession.

Research and teacher education

Chair: *Thomas Cooney* (USA). Secretary: *John Dossey* (USA)

Papers in this group were divided relative to their attempt to either influence or describe the interactions between the teacher; the pupil; or mathematical facts, concepts, procedures, or problem solving strategies:

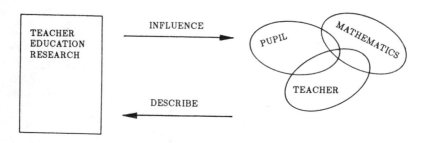

Eight papers dealt with influencing the interaction of the various facets to change the system and four dealt with describing the particular aspects of teaching or learning in an attempt to better understand and model mathematics teaching. *Abele* (FRG) described the use of video-tapes to examine the relations between methodological approaches and pupil behavior in small group learning situations. *Becker* and *Pence* (USA) outlined the results of a four-week long summer institute on teachers' knowledge of content, methodology, and leadership. The overall results showed that teachers placed less reliance on textbooks and grew in self-confidence as a mathematics teacher. *Kreinberg* (USA) reported on the EQUALS project's work with sex-related differences in student enrollments, career aspirations, and attitudes toward computers. One interesting technique asked students to describe a typical Wednesday when they are thirty years of age. Being a doctor or lawyer was one of the few occupations to receive a relatively equal proportion of the pupil's responses.

Carpenter (USA) described how teachers use knowledge from cognitive science in making instructional decisions. The approach provides teachers with knowledge about how children learn mathematics and then observes teachers' use of that knowledge in guiding and assessing their students. *Brekke* (Norway) outlined a similar project in which teachers are provided with a model of teaching and their use of the model is observed. Research questions focused on how mathematics teacher educators can learn about student cognitions, how research can change existing teacher behaviors, and how much time is needed to effect such changes.

Berthelot (France) described research that contains elements of both description and influence relative to the teaching of mathematics. His work with didactical variables deals both with the teaching situation and with student outcomes. This research is embedded in the notion of the didactical contract established between a teacher and the pupils. *M. Cooper* (Australia) reported on a survey of Sydney teachers which indicated that (1) some teachers have no interest in research, (2) many teachers are unaware of sources of information about research, and (3) when faced with problems of decisions, teachers turn to colleagues rather than to research. Efforts must be made by mathematics teacher educators to inject research findings into teacher education programs in productive ways. *Grouws* (USA) provided an overview of research on factors influencing teaching situations. His emphasis was on teachers as thoughtful decision makers and research that translates this conception into teacher education programs. He posed questions about (1) how new knowledge influences teachers' beliefs and conceptions, (2) how teachers come to appreciate perspectives other than their own, and (3) how teachers deal with competing paradigms for mathematics education?

The remaining four papers dealt with studies that attempted to describe particular aspects of teaching or teachers to better understand factors affecting mathematics teaching or learning. The first paper, by *Cooney* and *Jones* (USA), described the dualistic conception of mathematics held by teachers. Teachers tend to place "how" related activities ahead of "why" related activities. Research on teachers' meaning-construction requires humanistic approaches which help to understand teachers, their beliefs and the methods by which they are arrived at. *Ernest* (UK) discussed mathematics teaching and its relation to teachers' schemes, the social context of the teaching, and the teachers' levels of thought processes and reflection. Significant effects were noted for teachers' knowledge of mathematics and their beliefs concerning mathematics and its teaching and learning. *Kuendiger* (Canada) reported about relationships between beliefs preservice teachers established as learners of mathematics and their own teaching of mathematics. Questionnaires were used to build a learning history for the students and case study methods are being used to describe the relationships between these histories and classroom behaviors of these teachers. *Selkirk* (UK) outlined a study which focused on pupil reactions to student teachers' mathematics teaching. Teachers' appearance and habits were of great interest and their use of humor was appreciated by the pupils. Girls were quite aware of instances of sex-bias. The position was taken that knowing how pupils feel about different aspects of a teachers' work, can help improve teaching and teacher education.

Assessment of pupils and curriculum evaluation

Chair: *D. Downing* (UK)

Presentations and discussions in this working group indicated wide consensus that teacher education curriculum development has to be closely related to curriculum development for the respective school grades; is an integrating process which should involve mathematicians, mathematics educators, teachers, psychologists, administrators and others from related fields; must be considered as a continuing process and should not *primarily* be product-oriented; can involve teachers in didactic research and then foster their professional growth, as teachers possibly are the only individuals who can carry out the longitudinal and long-term classroom based research needed in mathematics education; should place much less emphasis on lectures, as teachers teach as they were taught; should have a stronger integration of mathematics and mathematics education; and should use new technologies as an agent for change and development.

McKillip (USA) reported on the development of new courses for the preservice education of middle school science and mathematics teachers. The project products were field tested in different schools. *T. Cooper* and *Smith* (Australia) outlined the development of an integrated course in primary teacher education which uses practical work to enhance the interrelation and critical analysis of content and methods. Experiences in mathematics education for student teachers which integrates the use of Logo graphic facilities were presented by *Ellerton* (Australia). In their second year the students have to tutor freshmen and to work with and observe children using Logo. Transition from manipulatives to computer representations was at the focus of *J. Olson*'s (USA) contribution. Prospective teachers were provided with many experiences illustrating this process, especially observing pupils in carefully planned lessons, to learn to appreciate the use of computers in teaching.

Bazzini (Italy) in her project on curriculum development for primary schools, besides many valuable outcomes, noted communication difficulties among various participants: different levels of theoretical knowledge, lack of a common language base for understanding, and lack of materials. *Downing* (UK) worked with groups of teachers for several years with the goal of enabling teachers to better assess student performance and simultaneously evaluate the curriculum. Teachers had opportunities to view themselves as researchers and to appreciate their crucial roles in transmission of curricula into the classroom.

Use of calculators and computers

Chairs: *Kenneth Travers* (USA) and *Jochen Ziegenbalg* (FRG).
Secretaries: *I. Weinzweig* (USA) and *Fong Ho-Kheong* (Singapore)

Beyond any doubt the computer will play a prominent role in the future learning and teaching of mathematics. Thus, serious attention must be given to the
various modes in which the computer can be used in and outside the classroom: as
a tool for numerical or symbolic calculations, as a teaching-tool (electronic blackboard, mathematical experiments, simulations and others), and as an individualized learning environment (from drill-and-practice to highly sophisticated computer
microworlds). These exciting opportunities necessitate an adequate education and
training of preservice teachers so that they will be competent to exploit the computer's capabilities. Beyond competency it is very important to develop positive
attitudes toward computer use in the classroom on the part of future teachers of
all grades.

Willson (Canada) reported on attitudes concerning the use of calculators but
many of her conclusions were very likely to hold in the case of computers as well.
One view was held unanimously — the use of computers has to be an integrated
part of the preservice teacher education. Preservice teachers should have extensive
opportunities to experience all kinds of computer (and calculator) use during their
tertiary education in mathematics and mathematics education. Such uses should
be accompanied by extensive discussions and reflections on methods, content, and
possible effects on the development of knowledge about one's own and pupils' use
of computers. In one way or another, most contributions to this working group
were concerned with the question of integrating the computer, its use and reflection
upon it in preservice education.

One principle guiding the education of teachers in computer use is to use
the computer in teaching mathematics to preservice teachers. The students can
thereby experience the power of the computer as a means for doing all kinds of
mathematics. Such experience will be the basis, and a viable model, for their
teaching mathematics with computers. *Ziegenbalg* (FRG) showed several examples
concerning interesting situations where computers help to avoid rather complex
analytic formulas and permit systematic search and experimentation as effective
solution methods. Since this is a style of problem solving different from the common "formula approach", preservice teachers should become well acquainted with
various aspects and general ideas involved in its use. The computer's graphic capabilities make it specifically suited to support the teaching of geometry. *Graf*
(FRG) reported on the use of Logo procedures in teaching geometric mappings
to student teachers and pupils in grade 8. The use of Logo as a language to enhance the learning of mathematics was supported by many participants. *Wilcox*
(Australia) and *Fong* (Singapore) presented examples of how to integrate Logo in
mathematics courses. Geometric configurations can be generated by many programming languages and related tasks offer many ways to integrate programming

and mathematical content. *Suzuki* (Japan) reported on BASIC programs written by students which generate many different geometric patterns.

The teaching of statistics poses many didactical problems which are very likely to be overcome solely by implementing the computer in the teaching process. Thereby the computer can be either a means for calculation or a means for demonstrating essential mathematical features of concepts involved. Contributors to this were: *Resek* (USA) (sampling, data analysis), and *Travers* (USA) (regression analysis on a spreadsheet). Nevertheless it is still open to debate whether one should use ready-made software packages or have the students write their own programs. *De Graeve* (France) reported on an introductory course on computer science (including Logo and Pascal) which was part of the preservice education of mathematics teachers. This course included algorithmics, numerical aspects and didactical consequences related to computer use.

Confronted with the large number of commercially available instructional software packages, teacher education has to address the immediate problem of using such software and specifically how to evaluate and to choose adequate products. Without proper guidelines the teacher will be in great trouble about when and how to use software. These problems were discussed by *Flake* (USA) and *Fong* (Singapore).

Though the discussion and contributions were quite optimistic, many problems and obstacles remain. *Johnson*'s (USA) contribution gave specific details of these. Points mentioned were: lack of computer equipment; absence of experiential background for educators and mathematicians with computers and the consequent lack of commitment and knowledge; problems of time (computer use is still very time consuming); and lack of adequate textbooks integrating computer use. Overall there is a strong need for more research on many questions: How does learning to program influence the learning of mathematics (e.g. the concept of variable)? Is knowledge acquired in a computer setting transferable to non-computer contexts? What are the effects of computer use on teaching styles and on the social structure in the classroom? What will be the effects of numerical and symbolic calculation software on curricula and methods of teaching?

Influence of the context on teacher education

Chair: *Cathy Brown* (USA)

The first set of papers (*Carss* and *Baxter* (Australia), *Hight* (USA), *Knight* (USA), *Nickson* (UK) and *Walther* (FRG)) focused on the influence of the society in which teachers learn and teach, e.g. prospects for employment, political atmosphere, and diversity of student population. This led to a discussion of mathematics education in schools as *production oriented* versus *individual development*. The conclusion was that most societies tend to emphasize the production model for school mathematics. The problem of diverse student populations (race, culture, language, socio-economic status, and ability) and its relation to teacher education was also considered. For teachers to effectively confront these issues, preservice programs must provide a greater number of practical experiences in school setting.

The second set of papers *Brown* and *Agard* (USA), *Dawe* (Australia), *Kvammen* (Norway), *Schultz* (USA), *Villar* (Uruguay) and *Wain* (UK)) focused on preservice teachers experiences with real-life mathematics. This is important background for motivating reluctant learners concerning the utility of mathematics. Traditional teaching has presented mathematics as an academic discipline and played down the role of mathematics in society and everyday life. This change in emphasis requires the development of "applied problems" and effective methods for studying them in classroom setting. The preparation of teachers cannot be isolated from either the broader needs of a society or the uses of mathematics in that society. If we view teachers as "providers of mathematics education opportunities" then we must prepare them to serve different pupils with varied contextual needs. We must also prepare them to teach with an appreciation and understanding of the discipline of mathematics that reflects its dynamic and useful nature.

Recruitment and selection of student teachers and training of mathematics teacher educators

Chair: *Larry Blakers* (Australia). Secretary: *Mary Lindquist* (USA)

The broad theme of this working group resulted from the combination of two planned working groups. However, several common considerations ran through the discussion of the range of issues. In the area of recruitment of preservice teachers, there was consensus that shortages would require the development of innovative and effective programs. One example of such a program (*Craven* (USA)) involved identifying capable students early in their secondary schooling and maintaining close contact with them through university. In another situation, a college changed its mathematics curriculum to ensure that it would meet the needs of the prospective teachers while not compromising its quality. *Campos* (Brazil) reported that only 25 % of the students in the program at her university completed the mathematics education course. They were addressing the problem through the establishment of a Center for Research in Mathematics Education. *Becker* (FRG) described a new program at his university which integrated theory and practice in a school-based setting. This allowed for a greater amount of time to be spent in the practical setting without changing the traditional coursework standards. Students have been quite satisfied with the program and the changes have aided in recruitment. *Linfoot* (UK) also reported improved recruitment through encouraging female graduates to take a one-year post graduate program in teaching. It has required careful placement and close follow-up the first year of teaching to guarantee the success of the graduates.

Discussion on enhancing retention of teachers who have already entered the profession was provided by *Yunker* (USA). He suggested that schools provide teachers with incentives to be professionally active, to share in departmental decisions regarding the teaching schedule, and to teach a wide range of courses throughout the totality of their careers.

Little time was devoted to the selection of preservice teachers. One particularly innovative suggestion was that potential candidates for such programs be interviewed as part of the acceptance procedure. Such a practice gives both the program and the individual a chance to reflect on the candidate's beliefs about mathematics and its teaching and learning.

Discussion of the training of mathematics teacher educators focused on the use of research information concerning beginning teachers. *Lindquist* (USA) suggested that training programs integrate knowledge about beginning teachers and teacher educators to assure smooth transitions to the tertiary classroom. *Lin* (Taiwan) reported on a program that retrains mathematicians to become mathematics educators. This program involves prospective teacher educators in studying international programs, working in research teams, and assisting in curriculum development. The overall consensus was that programs for teacher educators must involve both practical and theoretical aspects. Care must be taken that graduates are well trained in teaching, mathematics, research, and curriculum development before they assume their roles as teacher educators.

Supporting survey presentation:

Eric Ch. Wittmann (FRG): The Mathematical Training of the Teachers from the Point of View of Education.

ACTION GROUP 7: ADULT, TECHNICAL AND VOCATIONAL EDUCATION

Chief organizer: R. Strässer (FRG)
Hungarian coordinator: Zsuzsanna Berényi
Panel: Tommy Dreyfus (Israel), Ruth M. Rees (UK), Stephen B. Rodi (USA)

1. Organization of the Work of the Action Group

Before ICME 6 took place, the panel and chief organizer of Action Group 7 on "Adult, technical and vocational education" prepared a pre-conference paper on the topics to be discussed in the Action Group and planned the structure of the work at ICME 6. Contributors to the sessions were identified and guidelines for the presentations circulated.

At the conference in Budapest, the Action Group had four meetings of 90 minutes each. The 1st session was devoted to a short introduction (structure and topics of the Action Group, split into three subgroups) and work in parallel subgroups with subgroup 1 on "Vocational/Technical Education" (chair: *R. Rees*, reporter *G. Barr*), subgroup 2 on "Adult Education" (chair: *T. Dreyfus*, reporter: *J. Evans*) and subgroup 3 on "Distant Education" (chairs: *J. Adler – G. Knight*, reporter: *D. Visser*). In the 2nd session work in the subgroups was continued. For the 3rd session two subplenaries were scheduled: subplenary 2 on "Colleges/Polytechnics — pre-university tertiary education" (chair/reporter: *S.B. Rodi*). In the final session, short report from the subgroups and subplenaries were given, followed by a closing discussion.

Each subplenary or subgroup had a chairperson, who gave a short introduction to the topic of the meeting and acted as discussion leader. Short presentations by identified contributors followed the introduction in a way that these prepared activities did not exceed 45 minutes in all. The rest of the meeting was devoted to a discussion of the topics on the agenda, which was reported to the final session by the reporters.

At ICME 6, the work of the Action group was complemented by a survey lecture on "Teaching to Use Mathematics for Work and Cultural Development", which was given by *R. Rees* and *R. Strässer*. With pieces of research mainly from the United Kingdom and the FRG presented, the lecture ended up with an overview of current trends and a speculation about the relative lack of research in distant, adult and vocational mathematics education (for the full paper cf. *Rees/Strässer*, for the research reported cf. *Barr/Rees, Bromme/Strässer, Rees/Barr*. Part of the work of the Action Group will be documented more comprehensively in an issue of "Zentralblatt für Didaktik der Mathematik (ZDM)" on Adult Education to appear in 1989.

2. Topics of the Action Group

2.1 Mathematics in Technical and Vocational Education

Technical and Vocational Education "involves ... the study of technologies and related sciences and the acquisition of practical skills and knowledge relating to occupations in various sectors of economic and social life" (cf. UNESCO 1978. p. 17). Technical and vocational education normally takes place after general, compulsory education. The institutional system of technical and vocational education varies from full time technical colleges (e.g. France) to isolated and/or part-time activities in private enterprises or colleges (e.g. England for some levels of qualification).

With technical and vocational education related to occupations, mathematics used in employment was scrutinized: The mathematical procedures found on the job center around basic arithmetic, percentages and proportions (rule of three). In some vocations like hairdressing or plumbing, these techniques already seem to be all that is used, but "conceptual skills, such as spatial awareness, the understanding of orders of magnitude, approximation and optimization, are of equal or greater importance ... all mathematics which is used at work is related directly to specific and often limited tasks which soon become familiar ... it is possible to summarize a very large part of the mathematical needs of employment as a 'feeling for measurement'" (cf. *Cockcroft-report*, p. 15f and p. 24).

Research studies faced difficulties when trying to identify the professional use of mathematics: "Considerable differences ... were found to exist even within occupations which might be assumed ... to be similar. It is therefore not possible to produce definitive lists of the mathematical topics of which a knowledge will be needed in order to carry out jobs with a particular title" (cf. *Cockcroft-report*, p. 19). In a large number of cases, a close relationship between mathematics used at work and the vocational situation seems to render impossible the distinction between these two aspects of a given situation. If a solution to a vocational problem — be it mathematical or not — is found it will be used and improved as such. Usually, connections to related theoretical knowledge are cut off and forgotten.

Professional solutions tend to be transformed into mechanical procedures applicable by imitation rather than understanding — at least on the lower levels of the work hierarchy (for an example and a systematic elaboration of this transformation cf. *Damerow*, pp. 11–16).

Young employees or apprentices have difficulties to find the appropriate formula and/or technique to cope with a given vocational situation when learning mathematics. The result has to make sense for the vocational situation and may be most relevant to act in an appropriate way (cf. *Cockcroft-report*, p. 18f). Technical and vocational education (especially when organized along classroom lines) seems to overlook this need for learning to use mathematics in professional situations.

UNESCO formulated two major problems with technical and vocational education: First there is the widespread disapproval of technical and vocational education — compared to "general" education. Second there seems to be a problem with the adaptation of the social and economic process of production and distribution of goods and services and the preparation for this process in technical and vocational education. These problems materialize in the shortage of competent teachers and in the lack of adequate curricula and teaching/learning materials cf. *UNESCO* 1978a, p. 93ff).

2.2 Mathematics in Adult Education

Adult Education comprises all activities where people try to broaden their knowledge and/or skills while they are already mature members of a society. These activities are not necessarily related to earning a living, but undertaken e.g. for the search for promotion in a job, enrichment of leisure time or philosophical interest. Institutions, goals, learning/teaching strategies, learners, teachers and curricula in adult education are even more heterogeneous than in technical and vocational education — and so are differences between nations all over the world (cf. *Pengelly*, p. 86ff).

A certain classification of adult education activities concerning mathematics seems to emerge, distinguishing activities aimed at fostering adults' every-day use of mathematics as opposed to courses leading to formal certificates (cf. the *ICME-4-Proceedings*, p. 702ff). As for the subjects to be taught, the research study of the *Cockcroft-report* for adult education identified the "mathematical needs of adult life" as "ability to read numbers and to count, to tell the time, to pay for purchases and to give change, to weigh and measure, to understand straightforward timetables and simple graphs and charts, and to carry out any necessary calculations associated with these..." (cf. *Cockcroft-report*, p. 10).

The report goes on: "Most important of all is the need to have sufficient confidence to make effective use of whatever mathematical skill and understanding is possessed" (loc.cit.). Adult mathematics education often has to cope with the fact that its learner has already failed in mathematics or has the feeling of being a failure. In some contrast to this statement, one has to bear in mind that adult education in mathematics is also confronted with many adult learners who are really interested in and competent to do mathematics.

One way to cope with these problems may be the "comprehensive" community colleges to be found in the USA. Within these colleges, pre-university and beginning university training takes place in the same institution as technical and vocational education, with classes often taught by the same faculty. Frequently, theses institutions have articulated transfer programs with near-by senior colleges and higher level institutions, thus promoting mobility for students in vocational/technical programs by not "marking" them for one kind of education only. Remedial and developmental education is given emphasis by offering support systems students can turn to. The colleges are closely tied to their local communities by a board of local citizens governing the institution. Alliances with local industries establish training programs frequently taught on the industrial site.

Recently, mathematics educators have shown increased interest concerning learning problems in advanced mathematics, which expressed itself for instance in a series of conferences in the USA about the changing calculus curriculum. More material needs to be covered and — at least qualitatively — understood in less time, reaching as far as partial differential equations. An additional trend was an increase in cognitive research on advanced mathematical thinking. Such research concentrated in particular on problems of representations in mathematics (e.g. geometric vs. algebraic descriptions of the same concept or process), on abstraction, visualization, and on the possibility to use computers to help students overcome difficulties related to the problems mentioned above.

Most of the activities in adult mathematics education seem to be taught in close integration with other subjects such as reading and science. The different subjects are even more interrelated than in technical and vocational education, which is another difficulty for a discipline-oriented approach to adult mathematics education.

2.3 Mathematics in Distant Education

This topic was not explicitly discussed in the Action or Theme Groups of ICME 5, so one has to go back to ICME 3 and ICME 4 for a continuation of this subject within the ICME- frame. In 1976, *Pengelly* (following a document of the European Parliament) gave three characteristics of "distance education": "(1) spatial separation of student and teacher for all or almost all of the courses; (2) the use of teaching material in permanently recorded form, such as printed notes, films, cassettes, radio and television; (3) teacher control and goal directedness" (*Pengelly*, p. 87). Following this definition, "distant education" is a special organization or way of learning — and not so much a question of contents or age group.

From these characteristics, it is also evident that "distance learning is an unusual human activity, and for that reason it is difficult" (*Pengelly*, p. 89). Detailing this difficulty, *Pengelly* identified three major problems for the design of learning systems in distant education (cf. loc.cit., p. 98f): The designer of a learning system has to cope with "diversity" both coming from different, often quite distinct purposes of the course as well as from the student's background. He also has to cope "with a relatively small number of students spread over a geographically large area" and he has to choose a "method of assessing the student's attainment and of awarding certificates".

For the process of learning mathematics, three stages are distinguished — with certain media being more or less appropriate for these stages. The Open University in the United Kingdom distinguishes "exposure: the first meeting with new material" from "experience: in greater depth and detail" and "mastery: of either technique or concept" (cf. *Lovis*, p. 49) and seems to have a clear policy concerning the media used in the respective stages. For exposure, the Open University relies on television, which "must be linked with written material", whereas — besides written material again — "the audio-cassettes are particularly useful" in the stage of "experience", which is characterized by "structured questions and ... worked examples ... Mastery must mean doing, i.e. routine practice." In this stage, "tutorial radio programs seek to provide this" (loc.cit., p. 50f). *Knight* offers a similar pattern of stages, but a somewhat different use of media.

3. Key Issues

3.1 Impact of New Technologies

Computer-based information processing and communication technology can influence distant, adult, technical and vocational education in two possible domains: New technology — not only computers — can be used as a teaching aid. Communication technologies form new ways of transmitting information to new storage media (e.g. compact-discs) and a growing integration of the different media may affect the whole structure of training — on the job, in the classroom as well as in distant education.

Besides, there are changes stemming from the growing use of new technology in every-day life and vocational/professional situations. New demands on mathematics education are described as growing needs for graphics (especially in commerce and administration) and geometry (for technical drawing: coordinate and spatial geometry with the use of CNC- and CAD-machines), number systems, elementary algebra, optimizing, controlling, programming and algorithmic thinking, which is often accompanied by quests to train logical thinking and thinking in systems (cf. for instance *Fitzgerald* and *Strässer*).

As for new potentials, a changing relation between algorithms and holistic structures ("Gestalt") can be seen especially in business and administration. New possibilities of modelling, exploration and simulation heavily rely on the use of computers and adequate software and could compensate for the growing difficulties of training on the job because of potentially larger expenses of faults and losses of material, machines and time with training on the job. These new potentials are even looked upon as creating a new style of mathematical thinking — fostering a new sense of trial and error, exploration and "what–if"-type of mathematical practice and applications-oriented research. This may even lead to a different approach to teaching mathematics, creating the conditions within which the learner can proceed to more abstract mathematical concepts.

The use of new technology such as computers and video-cassettes seems to imply certain problems in distant education. Even in a relatively prosperous country like New Zealand, students do not have easy access to these new communication technologies. In addition to that, the great variety of standards used with these media and the incompatibility of the software is a major hurdle to use in distant education (cf. *Knight*, p. 72).

3.2 Developing Countries

It is only too obvious, that distant, technical and adult education are crucial issues for developing/rural countries. For a discussion in an Action Group however, there is a major obstacle: studies and papers concentrating for instance on technical or distant education in developing/rural countries are nearly non-existent. To cite a more general statement of the ICME-5-proceedings: "It is ... difficult to get substantial contributions in this domain" (loc.cit., p. 140; cf. also part 5 of this paper).

As *Broomes* put forward, "goal 1 of mathematics for rural development" is "the preparation of citizens as users of mathematics", specifying the "mathematical knowledge needed (as) ... arithmetic, algebra, statistics and logic" (*Broomes*, pp. 51–52). He points to the importance of the contexts mathematics is used in — thus confirming the above analysis of mathematics in working situations. The examples of *Gerdes* in a paper on "Conditions and Strategies for Emancipatory Mathematics Education in Undeveloped Countries" show in an impressive way that the topics named by *Broomes* must be reconsidered in the light of the needs of the society in question (cf. *Gerdes*, pp. 16–18).

Gerdes shows an additional relevant feature: A large number of nations have to alter the educational system which the former colonialist nation found appropriate. From their colonial history, a lot of countries inherited a system devoted to elite-education, which may have been working in west-european cultures of the 19th century. "It is quite another problem to build a system of mass education in the third world and embed mathematics education in both the school situation and the specific social and cultural contexts of that world" (*ICME-5-proceedings*, p. 143). "Ethnomathematics" (for examples cf. *Gerdes*, *Carraher et al.* and *Schliemann*) may be a way to determine what sort of mathematics education is in fact appropriate for a given society in a given situation. Not only in developing countries, the widespread separation of mental from physical labour has to be altered — together with the widespread hierarchical structure in the organization of work and educational institutions.

4. Contributions and Issues of Discussion

4.1 Technical and Vocational Education

In the subgroup on "Technical and Vocational Education", four presentations were given. *P. Bardy* spoke on "Mathematics Teaching in Preparation for Employment", starting with a review of the surveys which deal with the application of mathematics at work. After pointing to methodological problems in the surveys, he identified issues to be considered in curriculum planning for teaching mathematics in vocational education and ended with analyzing a vocational situation using mathematics as a tool to cope with the situation, turning the mathematical content into an algorithm.

In his presentation on "Assessment of Mathematical Skills in (Pre)Vocational Education", *G. Barr* discussed the curriculum development of vocational schemes at the City and Guilds London Institute. Being an examination board, City and Guilds does not run courses, but provides assessment packages with a trend towards assessment at the workplace.

B. Price presented experience with training programs for "Use of Mathematics at Work" coming from a larger US college. She described the way courses are devised and reformed at that college offering a matrix tying together mathematical competencies and vocational areas the competencies are needed for.

In the subplenary on "New Technology", *H. Abel* spoke on "Learning Mathematics for New Technologies in Metalworking". He gave examples how mathematics, especially coordinate systems, angles and vector geometry, is used when programming computer numerically controlled machines — showing the close integration of mathematical and vocational knowledge needed when using these machines. Computer assisted design tends to use even more geometrical knowledge, while technical knowledge has to be integrated when CAD-data are used to control for instance a turning lather.

R. Strässer detailed the alternative "Mathematics in Vocational Training — Tool or Scientific Knowledge?". Comparing the use of mathematical algorithms to produce numbers (the tool solution) to conceptually modelling a vocational problem by mathematics (the scientific knowledge), he pointed to constraints and potentials related to specific work situations and the classroom situation.

The main issue in the discussions was how to cope with the different way of using/applying mathematics at the workplace and in a classroom situation — with rote learning and skills more related to the workplace and understanding and scientific procedures/concepts linked more directly to the college. This difference was looked upon as a spectrum of possible goals of vocational training — with a choice to be made in every learning/teaching process.

4.2 Adult Education

The subgroup on adult education and the subplenary on colleges/polytechnics is reported together grouping the presentations into two categories: contributions mainly centering around the organization and environment of adult learning as opposed to those looking into special ways of teaching/learning mathematics in adult education.

R. Croasdale reported on a "Mathematics Degree Course at Newcastle Polytechnic" whereas *J. Searl* — on behalf of *R. Jordinson* — reported on the "Edinburgh Walk-In Numeracy Center". The Numeracy Center offered mathematics

learning for everyday use and entry tests to training courses as opposed to the polytechnics course which aimed at an approved certificate. This course was taught on
particular evenings with students tempted away to distant education when the
Open University began to offer courses. In contrast to that, the Walk-In-Center
is situated in a shop front in the center of the town offering two sessions a day to
go in and talk about any basic skill on demand. Having volunteers as tutors, the
center faced problems in terms of training and availability while funding problems
were common to both institutions — the center and the polytechnic.

In the subplenary on "Colleges/Polytechnics", *S.B. Rodi* briefly described the
various ways, vocational, technical and adult education is organized in two year
colleges in the USA including some basic demographic information and enrollment
figures. *L. Trivieri* detailed this overview by describing the variety of programs
available at a typical community college in the USA. Training versus education,
standards for adult students and specific learning styles of adults are some of the
issues discussed in the subplenary on "Colleges/Polytechnics".

Two contributions on the ways to teach/learn in adult education described
ways to foster student understanding by special ways to teach mathematics: In her
contribution on "College Algebra in Adult Education", *R. Moeller* exemplified the
use of special "devices" in the process of problem posing and reflection with the
phases of giving a problem, asking for examples, exploring properties and making
connections, possibly ending up in a formal definition. *J.-P. Drouhard* reported on
the "Discussion About Problems" — approach which starts from problems given
to groups of students. In a teaching session, the teacher asks for conjectures. He
writes these down — avoiding comments, in particular avoiding any indication
whether they are correct or not. With hopefully contradictory statements on the
blackboard, the teacher fosters a discussion oriented to bring to light the truth
or falsity of the statements. This procedure is meant to make the students speak
about mathematics and make them commit themselves to a position they must
argue for. The approach is aiming at learning and understanding of mathematics
which is at the same time intuitive (rather than formal) and abstract (as opposed
to concrete).

The contribution by *W. Schloeglmann* on "Justifying Mathematics in Courses
for Engineers" reported on experience with making engineers aware of mathematical proof, distinguishing the clarification of the range of applications of a result
from the justification of a result. Students showed different reactions for existence
and constructive proofs, preferring the latter. As could be seen in the two contributions described above, the main procedure again relied on asking students to
produce conjectures, using conflicting arguments as a motivation for proof.

The contribution of *C. d'Halluin/D. Poisson* in the subplenary on "New Technologies" reported on a different aspect. At a center of adult education, mathematics is taught taking into account the situation, patterns, "standard"-models and
formal theories to be studied. Numerical, algebraic and graphical representations
are used in this approach — with the use of computer produced function plotting

as a main help: "Computers Offer New Images" was the revealing title of the contribution. In the discussion, the possibility of simulation and access to higher order mathematical concepts — e.g. in statistics — were mentioned as major changes brought about by the use of new technologies.

4.3 Education at a Distance

K. Harrison spoke on "A New Initiative in International Distance Education" which brings mathematics teaching from the Australian Murdoch University to Malaysia by means of distant education. Apart from distance problems he reported on cultural difficulties — last but not least because extensive written material and the textbooks used are provided in English language. Regular contact of students with tutors seems to contribute to a high pass rate maintained by this initiative.

G. Knight reported on "Distant Education in New Zealand", where a faculty of mathematics offers written material and assignments marked by its lecturers every two weeks. New Zealand being relatively sparsely populated, education at a distance plays a major role in the educational system of the country. The students vary considerably in age, motivation to study and occupational background. With a majority of male students enrolled in a technical correspondence institute with entry requirements and female students forming the majority of enrollments in institutions with open entry policy, the figures reflect the nature of New Zealand society with clearly defined gender roles.

J. Adler spoke on "Basic Mathematics for Adults by Newspaper". The project funded by the *Sached Trust* was part of the many responses elicited by South Africa's racially discriminatory educational policy. Offering mathematics lessons as supplement to a cheap newspaper, the courses concentrated on enabling adults to re-enter the educational process to obtain a nationally accepted certificate. The contributor presented several examples of lessons and explained why algebra was chosen as main topic of the project. An evaluation of the project showed that the newspaper emerged to be most effective as a motivational medium for the mathematics education of adults in South Africa — the project being terminated in 1980 by the banning of the carrier newspaper.

J. Davis reported on "Learning Mathematics at the British Open University" showing the sophisticated and successful system unique in the amount of support to students. She mentioned that this constant support of students — necessarily including face-to-face contact with tutors — is directly linked to the high pass rate of the Open University. As a result of large student numbers, media other than written material — e.g. television, videos and audio cassettes — are used on a large scale.

Summing up the sessions on education at a distance, methods and success of this type of education was seen as dependent on

- characteristics of the student population — with confirming evidence in the contribution of *d'Halluin* and *Poisson* in the subplenary on "New Technology",

- the size of the country and the density of the population with its effects on student numbers and distance between students and tutors — as can be exemplified by the comparison of the United Kingdom and New Zealand,

- availability of media alternative to written material — with additional potentials from the use of graphical representations facilitated by computer based new technology and specific problems using television in mathematics learning at a distance mentioned in the final plenary session,

- the support given to the students, this was regarded as the most important factor.

The second part of *H. Abel*'s contribution in the subplenary on "New Technology" additionally showed that teaching the use of computers in metalwork can also bring up new problems in education at a distance. Experience of the simulation of processes or even production machines — including some type of interactivity — may be needed and call for face-to-face tuition complementing education at a distance.

5. Conclusion

In the final plenary, the question of how to essentially distinguish between adult and technical/vocational education arose. From the reports of the subgroups, it was apparent that vocational education tends to start from a skills approach whereas adult education from the outset tends to aim at understanding, may-be even proof. An attempt to detail this distinction referred to the fact that the main aim of vocational education is the functioning of a person in the vocational situation. The workplace situation often does not call for understanding, but habitualized use of algorithmic procedures. In contrast to that, the adult student would himself want to learn and understand — sometimes for very personal reasons, and tends not to be satisfied with a procedural, possibly mechanical presentation of mathematics. In addition to that, vocational/technical education usually has a shared context in terms of a more or less homogeneous workplace, while contexts tend to vary widely in adult education courses.

Adult, technical and vocational education in "developing" countries was not discussed in the Action Group for various reasons: Contributions specifically related to the topics of the Action Group could not be identified. A variety of contributions to the 5th day of ICME 6 on "Mathematics, Education, and Society" centered around this topic. It also became clear during ICME 6 that a better approach to the problems of discriminated and disadvantaged members of a society or nation is researching related problems in the country one is living and working in — regardless if it is looked upon as "developed" or "developing".

In the final plenary of the Action Group, a division between adult, technical/vocational education and education at a distance was suggested for future ICMEs. Education at a distance should be given an extra place in the program structure to foster an extensive discussion on problems and potentials of mathematics education at a distance — including education at other age levels, e.g. distance education in Australia at the primary level.

References:

Barr, G.–Rees, R.: Developing Numeracy Skills. Harlow (Longman) 1985.

Bromme, R.–Strässer, R. (1988): Mathematik im Beruf: Die Beziehungen verschiedener Typen des Wissens im Denken von Berufsschullehrern. Bielefeld (Occasional Paper No. 101 of IDM) April 1988.

Broomes, D. (1981): Goals of Mathematics for Rural Development. In: Morris (ed.):*Studies in Mathematics Education*, vol. 2. Paris (UNESCO) 1981, pp. 41–59.

Carraher, D.W.–Carraher, T.N.–Schliemann, A. (1984): Having a feel for calculations. In: Mathematics for All, Sydney (Macquarie University) August 1984 (Proceedings of the Theme Group "Mathematics for All" at ICME 5)

Cockcroft-report (Committee of Inquiry into the Teaching of Mathematics in Schools) (1982): Mathematics Counts — Report of the Committee. London (HMSO) 1982.

Damerow, P. (1984): Mathematikunterricht und Gesellschaft (Mathematics education and society). In: Heymann, H.-W. (Hrsg.) Mathematikunterricht zwischen Tradition und neuen Impulsen. Koeln (Aulis) 1984, pp. 9–48 (vol. 7 of "Untersuchungen zum Mathematikunterricht")

Fitzgerald, A. (1985): New Technology and Mathematics in Employment: A Summary Report. London (Dept. of Education and Science) 1985.

Gerdes, P. (1985): Conditions and Strategies for Emancipatory Mathematics Education in Undeveloped Countries. For the Learning of Mathematics, vol 5.1, 1985, pp. 15–20.

ICME-3-Proceedings (1979): New Trends in Mathematics Teaching — vol. IV. Paris (UNESCO) 1979.

ICME-4-Proceedings (1983): Proceedings of the Fourth International Congress on Mathematical Education (ICME 4) (ed. by Zweng et al.) Boston 1983.

ICME-5-Proceedings (1986): Proceedings of the Fifth International Congress on Mathematical Education (ICME 4) (ed. by M. Carss) Boston 1986.

Knight, G. (1987): Distance education in mathematics. In: Morris, R. (ed.): Out-of-school mathematics education. Paris (Studies in mathematics education, vol. 6, UNESCO) 1987, pp. 65–73.

Lovis, F. (1987): Broadcasting and the Open University of the United Kingdom. In: Morris, R. (ed.): Out-of-school mathematics education. Paris (Studies in mathematics education, vol. 6, UNESCO) 1987, pp. 43–64.

Pengellerey, R.M. (1979): Adult and Continuing Education in Mathematics. Paris (UNESCO) 1979, pp. 85–106. (chapter V of ICME-3-Proceedings)

Rees, R.–Barr, G. (1984): Diagnosis and Prescription. Some Common Maths Problems. London (Harper and Row) 1984.

Rees, R.–Strässer, R. (1988): Teaching to Use Mathematics for Work and Cultural Development. Survey Lecture to A7 at ICME 6. Bielefeld (Occasional Paper of IDM) 1988 (to appear)

Schliemann, A. (1984): Mathematics among Carpentry Apprentices: Implications for School Teaching. In: Mathematics for All, Sydney (Macquarie University) August 1984 (Proceedings of the Theme Group "Mathematics for All" at ICME 5)

Strässer, R. (1987): Neue Technologien und mathematischer Unterrich in gewerblich–technischen Berfusschulen. Bielefeld (Occasional Paper No. 89 of IDM) June 1987.

Strässer, R.–Thiering, J. (1986): Mathematics in Adult, Technical and Vocational Education. Report of the Action of ICME 5. In: ICME-5-Proceedings, pp. 124–132.

UNESCO (1978): Classification of information about technical and vocational education. Paris (UNESCO) 1978.

UNESCO (1978a): L'évolution de l'enseignement technique et professionel — Etude comparative. Paris (UNESCO) 1978.

Supporting survey presentation:

Ruth Rees (UK) – *Rudolf Strässer* (FRG): Teaching to use Mathematics for Work and Cultural Development.

THEME GROUPS

THEME GROUP 1: THE PROFESSION OF TEACHING

Chief organizer: Peggy A. House, USA
Hungarian coordinator: Judit Simon-Szász
Panel: Claude Comiti (France), William Ebeid (Egypt), Yoshitomo
Matsuo (Japan), Michael Silbert (Canada), Candido Sitia (Italy),
Derek Woodrow (UK)

Theme Group 1, "The Profession of Teaching", discussed issues related to the professional life and development of mathematics teachers. Subtheme groups were organized to examine more closely the issues of in-service education (organized by Derek Woodrow, UK); effective teaching of mathematics (organized by Yoshitomo Matsuo); the professionalism of teachers (organized by Peggy House, and Michael Silbert); the evaluation of teachers and teaching (organized by William Ebeid); and teachers and pupils in the classroom (organized by Claude Comiti). In addition, thirty other individuals from eight countries prepared short communications which were delivered and discussed within the subtheme meetings.

Issues in Mathematics Education: An International Consensus

From the papers submitted to Theme Group 1 it became apparent that, more than ever before, mathematics educators from around the world seem to be converging on a common set of issues that affect mathematics teaching in developed and developing nations alike. In her opening presentation to Theme Group 1, Peggy House identified the following themes that emerge across nations, as represented in the contributions to the T1 discussions:

1. It is widely accepted that teachers must emphasize *problem solving* and *higher-level thinking* as the most important goals of studying mathematics.
2. It is also generally agreed that teachers must stress *conceptual development* and *understanding* rather than the memorization of facts or the rapid performance of calculations.

3. Emerging theories of constructivism and the ways pupils build representations of mathematical concepts have implications for the ways teachers teach mathematics. In particular, teachers cannot be merely broadcasters of facts and theorems; they must, instead, provide classroom environments and activities that will facilitate the pupils' active construction of mathematical knowledge.

4. Emphasis on pupils' communication of mathematical ideas, both written and oral, stressing both precise communication and the unique personal expression of the student's understanding, is becoming a priority in most countries.

5. Presentation of applications that are relevant and interesting to the students is widely recognized to be important in motivating students in mathematics.

6. The use of concrete, physical manipulatives throughout the school curriculum is gaining acceptance as important to the pupils' success in constructing representations.

7. Likewise, it is recognized that mathematics teaching needs to employ multi-sensory approaches — that is, that pupils must not only listen to teachers talk, but that they also need visual representations, physical modeling, and acting out of mathematical ideas.

8. Cooperative group action is seen as important, especially in problem solving. This strategy of engaging pupils in small group activities is contrary to the long tradition of pupils working mathematics silently and alone, and it challenges the role of the teacher as the central figure in the learning process.

9. Estimation and mental mathematics are receiving increased emphasis in mathematics classes, especially when calculators are used to perform precise computations.

10. It is recognized that more attention needs to be given to spatial visualization and imagery in K–12 mathematics classes. Geometric as well as algebraic representations of mathematical ideas should be presented.

11. Technology will continue to have profound implications for the way mathematics is taught. Three quite distinct sets of concerns arise with respect to technology: (a) those concerned with the acceptance and utilization of calculators, especially in the early years of school; (b) those concerned with the use of computers as a tool to teach mathematics; and (c) those concerned with the potential of video technology, especially in the professional development of teachers.

12. Mathematics educators around the world are raising questions about what mathematics is to be taught and to whom. Additions to the curriculum of new content such as statistics, probability, and discrete mathematics, and movements to increase the amount of mathematics which students will study, place new demands on mathematics teachers.

13. Awareness that both pupils and teachers possess unique styles of learning and doing mathematics are leading mathematics educators to place more emphasis on helping teachers to develop a repertoire of teaching strategies that will better accommodate the needs of individual pupils.

14. Equity issues can be identified in many countries: in general, these are of three types. Questions of sex equity focus on the underachievement of females as compared to males. Ethnic equity concerns the differential performance by members of certain racial groups such as aborigines in Australia, blacks in the United States, and recent immigrants in several countries. Finally, geographic equity issues concern the inequality of opportunity afforded to teachers and pupils in countries or regions of countries where the density of schools is low compared to that of urban centers.

15. Testing and evaluation have become major concerns to educators world-wide. International assessment results published in recent years have led to criticisms of teachers and teaching practices in some countries and to a sense of competitiveness among some governments, their educational agencies, and the press. In addition, mathematics educators are aware of the divergence between contemporary goals, which emphasize problem solving and higher-level thinking, and traditional standardized tests, which largely involve factual recall and algorithmic performance. The need for assessment instruments compatible with modern program goals is a priority concern for many.

16 Considerable attention is being drawn to the question of what knowledge is necessary for mathematics teachers to possess. In particular, a tension arises between those who advocate the importance of mathematical knowledge and those who stress the importance of pedagogical knowledge. It is generally agreed that successful teachers integrate both mathematical and pedagogical knowledge, but the precise nature of that knowledge base remains open to examination.

17. The status and role of teachers has been changing in recent years. National reports on education have raised the level of public awareness and criticism of teaching in many countries and have resulted in attempts to restructure teacher education and professional development. Governments have mandated new curricula, programs of in-service education, testing of pupils and teachers, and more — and they have not necessarily consulted teachers on the changes that will affect them. Some teachers report a feeling of powerlessness to change their working conditions, while at the same time many nations continue to face shortages of qualified mathematics teachers, and they report difficulties in attracting top candidates into teaching.

18. Finally, there is widespread concern over the question of how to effect change in teachers. The process of change is a long and deliberate undertaking, but many report that teachers seem more interested in short-term solutions with immediate outcomes. The question of how to design, deliver, and sustain involvement in meaningful programs of in-service education continues to challenge mathematics educators the world over.

From consideration of the above common trends and issues, four general questions were posed to stimulate discussion in the working groups:

1. What do mathematics teachers need to know about mathematics that mathematicians do not need to know? How do teachers acquire that knowledge?

2. What is the necessary mathematical and pedagogical knowledge for *beginning* teachers? What is obtained in later professional development? When and how does this take place?

3. How do we recognize effective teaching of mathematics? How and by whom shall mathematics teachers be evaluated?

4. Assume that we accept the hypothesis that teachers need to change. How do we bring about this change? How do teachers themselves come to accept the need for change? How do we sustain their involvement in the process of gradual change?

Teachers as Workers; Teachers as Learners

In her survey lecture for Theme Group 1, Margaret Brown (UK) elaborated on two themes that touch the professional life of teachers: teachers as workers and teachers as learners.

Teachers as Workers:

A growing number of countries, especially Western countries, have recently imposed changes that decrease the professionalization of teachers. The call for accountability in education raised during the 1970s has continued throughout the 1980s. Industrial methods and mechanistic terminology increasingly are applied to education. Teachers and schools are expected to fulfill their primary objective by producing children with the skills required by the labor market, and their results are published for public examination. Thus teachers are losing control over their working conditions and working processes while coming to be regarded less as professionals and more as laborers. It is important, Brown concluded, to increase teachers' professional standing by helping them to become more expert, and more able to express their expertise, in the process of teaching and learning.

Teachers as Learners:

It has been said that most mathematics educators are constructivists. However, the same probably cannot be said about mathematics teachers. Yet, Brown pointed out, it would be contradictory to attempt to "teach" teachers to be constructivists. Rather, she called for teacher education that is itself inspired by constructivist assumptions both in method and in intended content. To this end, working collaboratively on activities that require reflection about mathematical learning is seen as a fruitful way to assist teachers to develop their professional expertise.

It is generally agreed, Brown noted, that to successfully implement changes in curriculum or in teaching practices, it is necessary to enable teachers to change their beliefs and attitudes. This is ordinarily a long and difficult process that is best accomplished by teacher educators working together with teachers to provide stimulating experiences that are likely to give rise to reflection and the reconstruction of beliefs. Thus, those who provide in-service education assume the roles of consultants, resource persons, and catalysts.

Brown's two themes, teachers as workers and teachers as learners, present, then, a fundamental conflict. On the one hand, external forces tend to reduce the professionalism of teachers, while, on the other hand, research reveals that children's learning is an enormously complex process which renders a teacher's job ever more demanding. The interaction between these two forces was further investigated during the subtheme discussions.

Teachers and Pupils in the Classroom

Contributors to this subtheme were Denise Grenier (France), Carolyn Maher (USA), Antoni Pardala (Poland), Karl Smith (USA), and Shlomo Vinner (Israel).

The teacher's principal role was assumed to be that of moving pupils from their initial conceptions to a state of mathematical knowledge. To do so, the teacher must take account of three interacting elements: the learners, the mathematical content, and the teaching situation. Thus, teachers must learn mathematics, and, although there was no consensus on what mathematics is necessary, there was general agreement that teachers must know mathematics at a high level in order to be able to perform epistemological reflection about the mathematical content that they will teach. However, some in the group disagreed, arguing that long, difficult studies of mathematics were not necessary for future teachers and were, in their opinion, the cause of mathematics teacher shortages in many countries.

If the group disagreed about mathematical content for teachers, they were quite unanimous about necessary pedagogical knowledge. Here necessary knowledge was taken to include the following: how children think and how children learn; how to motivate children to learn mathematics; how to create appropriate learning experiences relevant to children's conceptions and ideas; how to engage students in problem solving; and how to facilitate cooperative learning. Papers presented in the subtheme elaborated on these points.

Smith described a college classroom in which there was great diversity among the students, and he explained strategies that the teacher used to motivate the students using concrete and unexpected learning situations drawn from the students' daily lives. Pardala described an example of a teacher's action when dealing with students' mistakes and showed how difficult it is for a teacher to conduct a teaching activity while trying to deal with individual students' errors. Grenier pointed out how the teacher's interventions, choices, demands, and reactions affect the nature of student learning. To illustrate, she cited examples of pupils' responses when asked to construct precisely the axes of symmetry of certain geometric figures, and she discussed the discrepancies between what the teacher requested and what the pupils actually did.

Vinner argued that standard teaching procedures involve a problem-solving behavior that works by imitating the solutions of problems already seen and which, unfortunately, often leads students to correct results. He would prefer that teachers change their style of teaching and questioning from one that permits students to succeed by imitation to one that requires careful analysis of problems.

Comiti proposed a process for organizing the roles of teacher and students that is assumed to lead the students themselves to make sense of the constructed knowledge. In this model, the teacher is first responsible for organizing learning situations so that the chosen problem becomes the students' problem (called problem devolution). Second, the teacher is responsible for giving a status to the new knowledge which appeared during the learning situation and which will be used for further activity (called problem institutionalization).

Maher proposed a model of teacher development based on a constructivist perspective of teaching and learning. The model describes the teacher as learner, as observer, and as guide; the teacher as designer, implementer, and evaluator of problem solving; and the teacher as philosopher about the teaching of mathematics.

For the participants in this subtheme, good teaching is characterized by the following qualities: the flexibility to recognize and accommodate different children's learning styles; recognition that there is not just one "right way" of doing mathematics; incorporation of the richness of mathematics into the curriculum; and building meaning in mathematics for all learners. Teachers must have self-awareness of their own teaching, a quality that is enhanced by reflecting on their own learning as well as on that of pupils, and by discussions with other teachers. In addition, we must raise the image of teachers as professionals both by changing their own self-image and by enhancing society's image of teachers.

Effective Teaching of Mathematics

Papers in this subtheme were delivered by Derek Foxman (UK), Alex Friedlander (Israel), Barbara Jaworski (UK), Naomi Mitsutsuka (Japan), Judith Mumme and Julian Weisglass (USA), Linda Sheffield (USA), and Max Stephens (Australia).

Mumme and Weisglass initiated discussion by describing a K–12 site-based project designed to empower teachers to improve mathematics instruction by developing on-site leadership to provide both the impetus for change and ongoing support for teachers. In this model of change, children develop an understanding and appreciation of the beauty, elegance, and utility of mathematics; students construct their own understanding of mathematics; teachers respect children's thinking and understand their different learning and developmental needs; and females and males from different economic, ethnic, and racial backgrounds are ensured equal access to mathematical knowledge.

Stephens noted that among curriculum developers, teachers educators, and supervisors, exemplary curriculum materials are often seen as the key to improving mathematics teaching and to supporting reform of the school curriculum. His report described a national project in Australia that has developed exemplary teaching materials that challenge accepted definitions of school-worthy mathematical knowledge; that encourage teaching and learning processes which support active problem solving, applications of mathematics, and cooperative group work; and that foster a more inclusive curriculum which attends to the special needs of young people who have had limited access to or limited success in school mathematics.

Sheffield reported on the interaction between university education programs and public school programs for strengthening the quality of both preservice and in-service teacher education. Jaworski described an approach to working with groups of teachers to raise their awareness of issues related to the teaching of mathematics.

Foxman described the role of a mathematics coordinator who is a teacher to whom has been given the responsibility of developing a school's mathematics curriculum. Such a coordinator plays a crucial role in moving the curriculum toward desired goals.

Friedlander reported on a project that employed counselling as a model for in-service teacher education in urban junior high schools, while Mitsutsuka analyzed effective mathematics teaching from the viewpoints of (1) the structure of teaching materials and problem solving; (2) the structure of mathematics teaching; (3) general strategies for effective teaching; and (4) teaching based on those general strategies.

The Professionalism of Teachers

The subthemes dealing with the professionalism and the evaluation of teachers were combined for joint discussions. Presenting papers in this group were Raffaella Borasi (USA), William Bush and Vincent Altamuro (USA), Pat Costello (Australia), John Egsgard (Canada), Margaret Griffiths (UK), Charles Lamb (USA), and Patricia Tinto (USA).

Borasi, Griffiths, and Tinto all described approaches to developing teachers' professionalism. Borasi emphasized the importance of teachers' beliefs about mathematics and teaching and pointed out how such beliefs influence both teaching behavior and curricular and methodological decisions. She contended that the problematic nature of mathematics teaching defies hopes of identifying a non-controversial body of "expert knowledge" to be passed on to teachers, and that, therefore, teacher education should provide a "rich environment" to stimulate and encourage reflection, introspection, and dialogue.

Griffiths focused on strategies for guiding new teachers during the initial, probationary year of teaching. Activities employed in an experimental program in the UK include interviews with the new teachers, meetings with head teachers, supplying instructional materials needed by the probationers, releasing probationers to observe senior colleagues or to confer with their advisers, after-school courses at the Teacher Center, and small group seminars. Tinto described an experiment in school-university partnership in which university and school faculty members relate as colleagues to design and implement programs and to maintain a network for disseminating information. Experience has shown two types of activities to be most effective: symposia in which an outside speaker addresses a topic of mutual interest to all faculties, and workshops in which an outside keynote speaker is followed by parallel special interest sessions presented jointly by university, college, and school faculty members.

Costello broadened the consideration of professionalism by describing the Family Maths Project Australia which aims to develop good teaching and learning practices by assisting parent and teacher groups to coordinate activities and by encouraging partnerships between business, industry, and educational institutions. Ebeid offered a summary of guidelines for teacher professionalism that included membership in a professional organization which has a strong say in the preparation and standards of its members, and the development of a concept of teaching which enables the teacher to present learning activities in a pedagogically and psychologically appropriate manner. A professional mathematics teacher creates a positive affective climate conducive to optimal learning, maximizes class time, makes wise use of technology, adapts readily to new situations, works effectively as a member of a team, and regularly evaluates his or her own knowledge and teaching behaviour. Egsgard added to this by proposing a checklist for the evaluation of mathematics teachers based on characteristics that he assumed to be possessed by effective teachers. Debate ensued over whether such a checklist could be used effectively either by an outside evaluator or in self-examination. In particular, some participants questioned whether ineffective teachers recognized their own shortcomings and the extent to which the teacher's self-evaluation would be congruent with the evaluation by a mathematics educator.

This discussion led to the related question of teacher evaluation. Bush and Altamuro described the process of testing teachers which, in some states in the USA, has become a requirement for teacher certification and continued employment. They described four functions of teacher testing: admission to teacher preparation programs, certification, recertification, and promotion. Tests may be either written or performance-based, and they may focus on basic skills, on knowledge of the content area for teaching, on pedagogical and learning theories, or on some combination of the above. While most countries require some form of admission or certification test, few situations require recertification tests where those who fail are either placed on probation or denied recertification. Promotion tests are generally based on career ladder programs, and they usually include evaluations of teaching performance, extracurricular activities, and/or written assessments. Especially in the USA, there is a recent trend in favor of teacher testing with the possibility of creating a national certification examination. Subsequent discussion identified both positive and negative aspects of teacher testing and stressed the importance of not allowing testing to lessen the professionalism of teachers. As a specific example of these and related concerns, Lamb described curricular changes, teacher appraisal programs, and the planned abolishment, by 1991, of undergraduate degrees in education in the state of Texas.

Other discussions within the theme group focused on the need for mathematics teachers to be well-prepared in mathematics, but mathematics courses for teachers should be taught using the same instructional approaches that teachers themselves are expected to use. As one participant observed, "the 'bad' teachers I have had were bad not because of their knowledge of mathematics but because of their lack of knowledge about how to handle the classroom". The qualities of a good mathematics teacher identified by the participants in this subtheme group echo those already reported from other groups. In addition, discussion focused on the need for broad ideas about different aspects of mathematics and the links between them; about the foundations of mathematics; about methods of development and thinking in mathematics; and about the relationships of mathematics to other subjects, to the environment, and to every-day activities.

In-service Education

The final subtheme group concentrated on models of in-service education. Papers were contributed by Geoff Faux and Christine Shiu (UK), Paul Louis Hennequin (France), Harvey Keynes (USA), Donald Miller (USA), Susan Pirie (UK), John Savin (UK), Beth Southwell (Australia), and Ronald Wells (USA).

Miller, Wells, and Keynes summarized three projects in the USA. Both Wells and Keynes described models which brought together secondary mathematics teachers and university faculty for summer institutes during which they studied mathematics and participated in seminars to develop teaching materials. These were supplemented by follow-up workshops during the subsequent years which contributed to a high level of support and communication among the participants.

Miller's discussion elaborated on the problems of delivering in-service education in a rural area. The state of Nebraska has an area greater than that of Hungary and Austria combined, but its population is only about 75 percent of that of Budapest, with most of the teachers concentrated in two urban areas in the extreme eastern part of the state. The remainder are found in small and modest-sized schools separated by large distances, and often they work in isolation from other mathematics teachers. Thus, a model was developed in which selected teachers participated in a program that not only increased their own knowledge and skills, but also prepared them to assume responsibility for conducting dissemination activities for other mathematics teachers in their respective areas. In this way, a multiplier effect has been established to reach out to teachers who otherwise would not be well-served by in-service programs.

By contrast, Southwell and Hennequin described provisions for in-service education in Australia and France, respectively. Hennequin explained the role of the Institut de Recherche dans l'Enseignement Mathematiques (IREM) in France, and emphasized the value of clear structure and organization which the IREM provides. Southwell stressed the need for long-term development if one hopes to bring about changes in teachers' attitudes, and she emphasized that in-service education must be structured in a way that allows teachers to experience some of the changes that are envisioned so that they may reflect upon those experiences.

Shiu and Faux described an innovative form of in-service that uses teams of advisory teachers to work alongside teachers in one or more schools conducting "exemplar lessons", establishing resource centers, providing resources to teachers, introducing problem-solving investigations, developing calculator applications, and similar activities. Savin discussed the complex problem of evaluating these advisory teachers, especially since diverse groups of clients (schools, advisory teachers, local inspectors, politicians, etc.) have different interests and seek different information.

Pirie's paper summarized many of the difficulties one faces when designing in-service programs. For example, fundamental change is a long-term goal, but many teachers look only for immediate, short-term gains. A challenge is to provide sufficient short-term successes to sustain the effort of striving for longer-term outcomes. Other challenges relate to changing teachers' beliefs, to providing support during the threatening change process, and to modeling a process of active learning.

During discussions within the subtheme, four other questions arose: How should in-service activities be funded? What type of credit should be granted for such activities? What skills are needed by those who present in-service programs? How do we train in-service leaders? These questions need additional investigation. It was also noted that more discussion and examination are needed into the processes, as well as the structures, of in-service education.

Conclusions

As reforms in mathematics education are proposed in countries around the world, a common denominator of those reforms will continue to be the individual classroom teachers. The critical role of the teacher in the teaching and learning of mathematics cannot be denied. But as new content, new methods, and new technologies affect the goals of mathematics education and the means by which we try to achieve those goals, the key element in the improvement of mathematics education will continue to be the professional teacher. Theme Group 1 has aimed to further our understanding of the nature of the mathematics teacher's responsibilities and of the process of continued professional growth and development. The issues identified and discussed at ICME 6 provide an agenda for needed actions in the years ahead.

Supporting survey presentation:

Margaret Brown (UK): Teachers as Workers and Teachers as Learners.

THEME GROUP 2: COMPUTERS AND THE TEACHING OF MATHEMATICS

Chief organizer: Rosemary Fraser (UK)
Hungarian coordinators: János Kőhegyi, Jenő Székely.
Panel: James Fey (USA), Hiroshi Fujita (Japan), Hartwig Meissner (FRG), Josephine Guidy Wandja (Ivory Coast)

Introduction

Theme Group 2 gathered for a short plenary session to launch the seven areas of discussion chosen for this ICME 6 meeting.

Rosemary Fraser (UK) gave a short introductory talk highlighting two common questions:

(i) How can we move forward in understanding and sharing the wide range of possibilities computers offer teachers and learners?

(ii) How can we provide access to good teaching practices, involving technology, to the school classroom?

Comparison was made between the static nature of "paper media" to the dynamic nature of "computer media". Dynamic structures and visual images offer the student an exciting introduction to mathematics that is quite different to the sequence of a traditional curriculum. There is a need for teachers to grow with their students in order to understand how the developing mathematical imagery and thinking is influenced by the new learning environments.

A balanced use of available applications was called for. Four major areas of use (not necessarily mutually exclusive) were offered for consideration:

(i) Formal language interfaces — languages that help us to articulate and create mathematics both between ourselves and in machine-readable form.

(ii) Mathematical toolkits — such as graph plotters, spreadsheets, modelling systems, statistical packages etc. Such tools shift our efforts towards achieving skill in their use rather than developing traditional manipulative skills.

(iii) "Catalyst" software — materials that promote mathematical explorations and discussions and that help the building of conceptual understanding. This is a broad and varied set of materials including simulations, microworlds, supposers etc.

(iv) Tutorial modes — tutors, managers, AI systems etc.

Finally a balanced interaction between students, teachers and computers was discussed, emphasizing the need to provide the opportunities for classwork, small group work and individual work.

The focussed discussions of the seven working groups, reported below, reflect the need to consider the general issues that have been outlined.

Working group 2.1 reported on national initiatives and considered the classroom implementation of several curriculum packages.

Working groups 2.2, 2.3 and 2.4 considered the dynamics of the classroom and how interactions between teacher and students change in response to the new environments. The focus was on the demands made on young children, on older students and on teachers respectively.

Working group 2.5 explored the need for new mathematics curricula in the light of computer science etc.

Working group 2.6 challenged and reviewed new materials.

Working group 2.7 concentrated on reviewing the research and development of the use of LOGO in mathematics education.

Selected comments from the survey lecture are given at the end of this report.

As well as the summaries presented here each working group will contribute a full report of their work together with a selection from the contributed papers to a separate fuller Theme Group 2 publication. This publication will also include a theme group survey and a report from Topic Group 1 — Film and Video. Details may be obtained from E. Dubinsky (USA) or M. Emmer (Italy) or R. Fraser (UK).

Introducing Technology into the Classroom

Coordinator: Hans Brolin (Sweden)
Panel: Jerry Becker (USA), Guy Noel (Belgium), Carlos Mansilla (Argentina) and Kiyoshi Yokochi (Japan)

The aim of this group was to consider materials that are currently being used in classrooms and to report on their accessibility to teachers and students. The report describes some current initiatives.

Session 1. Learning environments — how do they change the teaching of mathematics

Chairperson: J. Becker (USA)
Reporter: G. Noel (Belgium)

This session started with a presentation by *Lars-Eric Björk* and *Hans Brolin (Sweden)* who talked about "The Role of a Mathematical Workshop in the Teaching of Mathematics". Here they described how the Swedish ADM-project "Analysis of the role of the computer in Mathematics Teaching" has implemented a Mathematical Workshop in the existing curriculum. In the normal curriculum the basic concepts of calculus are generally poorly understood, partly because too much time is spent on routine manipulations. With the Mathematical Workshop as a tool the students can be exposed to a much larger range of problems, encouraging a better conceptual understanding.

Next *Howard C. Johnson (USA)* talked about the Levy Apple Microcomputer Project. The purpose of this study is to create a learning environment in which students can acquire skills in systematic thinking in the domain of mathematical problem-solving. The computer allows students to investigate problems not usually attempted and increases motivation and enjoyment.

This first session ended with a paper by *Frank Demana (USA)*: "Using Technology to enhance Mathematics". He described two projects. The first is called "Approaching Algebra Numerically". This project has developed calculator based materials designed to prepare students for a successful study of algebra. The second called "Calculator and Computer Precalculus" has developed calculator and computer based materials for a year long precalculus course. These materials are used to strengthen student problem solving skills and improve understanding of functions graphs and analytic geometry.

Session 2. *Using computer software to support specific concepts and skills*
Chairperson: H. Brolin (Sweden)
Reporter: C. Mansilla (Argentina)

Adolf Ekenstam (Sweden), talked about "Programming and Understanding of Variables". Many teachers believe that training in computer programming makes students better at mathematics. Programming can increase the children's ability to solve mathematical problems, make concept understanding easier etc. This paper accounts for part of a Swedish investigation of the impact of computer programming on children's understanding of the variable concept. The purpose was to examine whether a carefully devised course in programming would increase understanding of variables by children in grade 8 (aged 14–15).

James Leitzel (USA) talked about "The Compulator and the Tablebuilder: Software that supports problem solving in the classroom". The usefulness and effectiveness of the hand-held calculator in mathematics instruction is well-established. However, young learners may still have trouble with certain aspects of its use. This talk described two programs that have been specifically designed with young learners in mind. The Compulator is a computer-based simulation of a hand-held calculator, which runs on any Apple IIe or IIc computer. It uses algebraic logic and allows students to enter, read and edit quantitative expressions and then evaluate them. The Tablebuilder program utilizes the power of the computer to help students structure, organize and do computations within a table.

Teri Perl (USA) talked about "Microworlds — Rich Environments for Problem Solving". She gave "Rocky's Boots" and "Robot Odyssey" as examples of sophisticated computer learning environments. In both environments users build circuits by connecting standard logic elements such as AND-gates, OR-gates and NOT-gates. These circuits power machines that perform certain tasks when activated. "Rocky's Boots" and "Robot Odyssey" provide good problem-solving environments for students who learn better in more structured settings and provide flexibility since teachers can construct and save partial or model solutions that are tailored to the needs of less talented program solvers.

This second session finished with a paper from *Brian Hudson (UK)* "Global Statistics". He has developed a database of statistics of 127 countries — in fact all the major countries of the world having a population of 50,000 or more. Twenty items of information are stored on each country which then can be used in the teaching of statistics.

Session 3. Calculators, computers, software packages
Chairperson: G. Noel (Belgium)
Reporter: J. Becker (USA)

This session started with a paper by *John Kenelly (USA)* "Calculators in mathematics placement testing: A near reality". He reported from the Calculator-Based Placement Test Program (CBPTP) Project. The tests produced by the CBPTP Project will permit active use of calculators in mathematics placement testing and should encourage both schools and colleges to incorporate calculators into their mathematical instruction. They believe that the result will be improved instruction and improved learning in mathematics courses.

Kiyoshi Yokochi (Japan) talked about "Spread of computers, and teaching of programming and geometry in Japan". At present 10–25 % of children at elementary schools and junior high schools have personal computers at home. At schools 2.1 % at elementary schools and 13.8 % at junior high schools had personal computers in 1985. Much more data was presented in the talk and a description of courses in programming for elementary school children and for Kindergarten teachers.

Y.L. Cheung (Hong Kong), classified broadly the teaching of mathematics into the teaching of 1. Skills, 2. Concepts and principles, 3. Problem Solving. He then gave examples showing how the use of calculators and graphic calculators could improve mathematics teaching in these three areas.

Session 4. National Reports
In this session national reports on the ways that different countries are using technology in the mathematics classroom were given. *Douwe Kok (Netherlands)* gave a report from The Netherlands and their NIVO project, *Jim Mayhew* (UK), *Guy Noel* (Belgium), *Bernardo Montero* (Costa Rica), and finally *Chantal Shafroth* (USA) compared the use of computers in mathematics education in USA to Japan.

53 delegates from 18 different countries took part. Written contributors to the group included S. Sarode (India)

The Dynamics of the Classroom: considering the young students (12 years old)

Coordinator: Jan Stewart (UK)
Panel: David Kressen (USA), Connie Widmer (USA)

This group considered how the new learning environments with their strong visual images, dynamic feedback, structured databases etc. will affect the young student's view of mathematics. The role of the teacher and the role of the student will change and evolve in response to these developing environments.

Session 1. Exploring the potential and problems of using calculators and interactive video

Chairperson: J. Stewart (UK)
Reporter: D. Kressen (USA)

This session concentrated on the current use of calculators and interactive videos in school. Countries discouraging the use of calculators with young children still exist but are rare. However the calculator is still not widely adopted. Research into the benefits of calculator use have declined rapidly over recent years. The current projects represented here by *Sheila Sconiers (USA)* from the University of Chicago School Math Project (UCSMP) and *David Pagni and James Wiebe (USA)* from the Calculators and Mathematics project, Los Angeles (CAMP-LA) mainly emphasized the development of materials for pupils and teachers. Both also felt that workshops to help teachers understand the potential of calculators were the best means of producing change of practice. The unchanging curriculum was, however, widely lamented - text books and guidelines scarcely recognizing the calculator's existence. Here above all else seems to be the major place for action if such a readily available and cheap tool is to become more widely adopted.

Connie Widmer (USA) alone represented the work being done on introducing interactive video into school. The work is very experimental and cost prohibitive. This has prevented wider adoption and accompanying research in the field. The group felt that, at this stage, some of the best applications they had seen in the field were for training, as opposed to education. The possibilities for the latter remain largely unexplored — but having seen computer software at similar stages in the early days we were optimistic that, not only was there a lot of work to do, but that it might eventually move into more exciting domains.

Session 2. Using a computer as a tool

Chairperson: P. Bender (FRG)
Reporter: J. Stewart (UK)

Large tutorial systems formed the major part of our discussions. Two specific ones were considered, an integrated learning system, introduced by *Joanne Weiss (USA)* of WICAT and "Shopping on Mars" presented by *Sara Hennessy and Ann Floyd (UK)* of the Open University. The two systems were widely different — the first covering the whole curriculum and totally managing the learning taking place (from giving reports to setting homework); the second planned for inclusion in a classroom where many activities are taking place — some with and many without the micro. "Shopping on Mars" is also specifically aimed at encouraging pupils to explain and discuss the algorithms they invent and use.

The group showed a general suspicion of the "all-embracing" systems. Some thought it dangerous to encourage a belief in teachers that computers can replace them — or, at least, do everything for them. Others, having tried such systems, reported that they made many teachers feel insecure by emphasizing the wide range of differences between their pupils and offering only computerized materials to help. They reported that creative talented teachers could cope but that less able teachers adopted very worrying strategies — the two most popular being making the class all work on the same exercise at any one time or asking brighter pupils to help slower ones holding back the former until the latter caught up. All agreed that feed-back on pupil performance from all such systems could provide a management nightmare for teachers with too much information on pupil performance and little support for what to do with it.

The second session also included looking at computer tools the pupil could use. *Jo Russell (USA)* of TERC, had reviewed and developed materials for existing software. Her aim was to encourage pupils to work with real data and move beyond the common practice of simply making graphs to analyse and interpret their data in a range of ways. She felt that children should use a variety of tools — not just those on the computer — for doing this and pleaded for more flexibility in this genre of software.

Session 3. Environments to Explore
Chairperson: C. Widmer (USA)
Reporter: B. Blakeley (UK)

This session was intended to look at "Microworlds" though in the end it was dominated by a discussion on LOGO. We were fortunate in having LOGO enthusiasts working with all age groups — *Pat Campbell (USA)* with K–1 grades, *Miriam Holsen (USA)* with 2–5 grades, *David Kressen (USA)* with all ages and *Claudia Turco (Italy)* with the 12 year-olds. The talk stimulated much discussion. Concerns were expressed about the paucity of the use of manipulatives in the early years in some countries and the tendency to work on computers rather than with real materials. Claims of LOGO enthusiasts that LOGO promotes real problem solving as an autonomous activity generated deep philosophical discussion and disagreement. The hope of the group was that no program or technological tool should be promoted to the exclusion of others. The child needs a wide variety of experiences and an emphasis on balance — of curriculum, types of software, materials, technologies and experiences would be the recommendation.

Session 4. Young children from different countries
Chairperson: D. Kressen (USA)
Reporter: C. Widmer (USA)

This session was a pot-pourri of experiences from around the world plus an opportunity for members to cement the links made in the sessions into positive contact over the next four years until our next ICME conference. A range of software was shown — *Bernard Tressol (France)* demonstrated materials for basic skills; *Barry Blakeley (UK)* with software for dynamic geometry for children to explore; *Flans van Galen (Netherlands)* showing "Packaging" for open mental computational activity and *Donna Berlin (USA)* concluding by discussing generally suggestions for integrating technology into the teaching of mathematics. We left feeling we had seen much to stimulate and encourage us in the sessions but, equally important, we had made many new contacts and friends in the field.

31 delegates from 12 different countries took part. Written contributions to the group included papers from: Joost Klep (Netherlands), Antonio Fernandez Cano (Spain)

The Dynamics of the Classroom: Students older than (12 years)

Coordinator: Ed Dubinsky (USA)
Panel: Tommy Dreyfus (Israel), Serge Hocquenheim (France), Robert Tinker (USA)

This group focussed on the implications of using and designing materials. How is the older student's view and understanding of mathematics affected? Again the role of the teacher and learner will need to respond continuously to the opportunities the new environments offer. Each session had a topic that was introduced by three 10–minute presentations; then the group split into two subgroups for discussion. In this report the four sessions are taken separately but within each session themes that were dominant will be stressed rather than speakers and what they said.

Session 1. The computer as a tool
Chairperson: J. Leitzel (USA)
Reporters: J. Cesar (France), O. Legare (USA)
Presenters: Andee Rubin (USA), David Green (UK), Stefano Finz Vita (Italy)

A first theme that was discussed was the question of modelling. To what extent can the computer programs, e.g. simulations, represent the real world, e.g. in statistics, and how is the connection to mathematics established. It was suggested that this connection could be strengthened by asking "What happens if ..." questions. For example how does the mean or median change if we add values to a set of data.

The visual aspects of computer graphics were mentioned as the most powerful aspect of computer use as a tool, but examples were given where visual representations led to misconceptions. It was concluded that extreme care is needed in the generation of such representations. Different representations may be needed for different but related concepts.

Serious doubts were expressed with respect to the usefulness of programming as a tool for learning mathematics.

Session 2. Didactic engineering and computer learning environments

Chairperson: T. Dreyfus

Reporters: M. J. Winter (USA), B. Jensen (USA)

Presenters: Carolyn Kieran (Canada), Ed Dubinsky (USA), Baruch Schwarz (Israel), Rafi Nachias (Israel) (Note: Nachias presented in session 4, but his contribution fits better here)

The central theme here was the interplay between theory and intuition in software development. Whereas some claimed that a rather complete theoretical analysis of the learning processes with respect to a given concept must precede the development, others felt that intuition was a far better guide. Two learning environments were used as examples in this discussion. The tentative conclusion was that, at present, theories are insufficiently developed, and that possibly they never will be, but that software development should generally be accompanied by prior and concurrent efforts to build theory and by concurrent and posterior (formative) evaluation, i.e. lots of classroom trials.

Session 3. Graphical Work with computers

Chairperson: S. Dugdale (USA)

Reporters: A. Arcavi (Israel), B. Jensen (USA)

Presenters: Michele Artigue (France), Henry Kepner (USA), Peter Giblin (UK)

Two themes were central to session 3. The first was the degree of specificity of a piece of software which is necessary to achieve given learning objectives. Can commercial software be used or does each curriculum development project need to develop its own software? In the first, teachers often find specific features which they need at a particular point are lacking. Specific cognitive difficulties can rarely be addressed.

The second theme ranged from considering unexpected events on the screen to the necessity of proof. Difficulties with scaling, the off-screen part of a graph etc., as well as surprising or confusing results on the screen should be exploited didactically to lead students to argue about what happened and to deal with error and misconceptions. Such arguing may be done graphically (limit-concept for instance). Participants were divided as to whether graphical argument could suffice as proof and to what extent proof should be given.

Session 4. Proving and intelligent tutoring

Chairperson: W. Barz (FRG) (Israel)

Reporters: A. Reuben (USA), B. Schwarz

Presenters: Janice Flake (USA), Gerhard Holland (FRG)

Proofs were discussed again: problems are epistemological (is the "truth" known to the teacher or to the computer system?), connection to content (is there a unique proof?); can we teach proof techniques?

It was difficult to discuss intelligent tutoring systems, because we only had one incomplete example. One problem is, that such systems are necessarily logic-based, and we need to ask ourselves to what extent mathematics should be formalized for the student. Problems of implementation in the curriculum were addressed: "What is the educational value of intelligent tutoring systems?" Maybe at ICME 7, 8, 9 ... we will be in a better position to discuss this topic.

53 delegates from 15 different countries took part. Written contributions to the group included papers from: W. Barz/G. Holland (FRG), P. von Blokland (Netherlands), B. Hughes(USA), F. Terada (Japan), M. K. Heid (USA), T. Dreyfus and T. Halevi (Israel), D. Clegg (UK), J. Sasser (USA), P. Perham (USA), B. Waits and Demena (USA), R. Allen et al (France), A. Arcavi (Israel), G. Bitter (USA), T. Dick and G. Musser (USA), I. Feghavi (USA), B. Jensen (USA), D. Resek (USA), S. Sigurdson and S. Churchat (USA)

The Dynamics of the Classroom — focussing on the Teacher

Coordinator: Nurit Zehavi (Israel)
Panel: Klaus Graf (FRG), Janine Rogalski (France), David Tall (UK)

The aim of the group was to consider the demands made on the teacher to respond to different learning environments. The role of the teacher and the role of the student will change considerably in the next decade.

Session 1: What are the theoretical changes in the teacher's role?
Chairperson: K. Graf (FRG)
Reporter: J. Rogalski (France)

The contributions highlighted different aspects of the changes appearing, or needed, when computers are introduced in the classroom.

C. M. Fernandez and M. Pinilla (Spain) discussed changes in the relationships between teacher, students and knowledge. The computer allows the teacher to move from the role of a giver of information to one as a counsellor and coordinator. S/he becomes a coordinator instead of an informer, and acquires the role of an expert. Moreover his/her role of researcher is now less hidden than in traditional ways of teaching.

Shoichiro Machida (Japan) emphasized the modification of the teacher's role when computers are used as human teacher's aids, in parallel with other hand manipulated material. This situation allows the teacher to improve individual student's creativity by assistance and adapt interventions to the variety of student's attainment and interests. Machida presented as an example how computer can be used with 14 year old students in order to fill teaching goals oriented toward exploration, simulation, elaboration of conjectures and collective discussion between students themselves about how to prove these conjectures.

Janice Flake (USA) devoted her talk to exemplify the fact that changing the teacher is a condition for introducing computers in the classroom. Using software in pre- or in-service training allowed teachers themselves to adapt lessons for their own students and to ask specific questions about computer use; this appears also as a possible way to pay more attention to reflective abstraction by students.

Janine Rogalski (France) presented the dimensions related to the control the teacher can have on the relationships between students and knowledge, to time management of the classroom in short-term as well as in long-term, and to the design of didactical situations , introducing new concepts or operating on recently presented notions. Depending on the mode of computer use : as a "dynamic blackboard" , as a "common resource" for teacher and students or as an "individual partner" for students, teacher's activities may be more or less deeply changed.

Joao Ponte (Portugal) stressed that each way of using computers puts its own demands on the teacher and Ponte analysed the pressure on teachers for a deep change of their role in the classroom, the need of new competencies and the consequences for training. The teacher becomes the one who orients the learning activity, the one who has experience in learning how to learn (expert at a metacognitive level) and s/he is able to face with new and challenging situations (as a researcher).

During the discussion an important point was underlined : a lot of changes in teacher's role presented above were not specific to the use of computers in the classroom ; coordination of students' work, controlling group dynamics, development of problem solving , creativity and reflective thinking by students are examples of roles not always fulfilled in teaching, but not directly related to computer use. Nevertheless some important modifications may be induced by the change of students' attitude towards mathematical content when they are using computers as a tool in classroom situations, as well as by the development of new expectations towards the teacher.

Session 2: *What is the reality of use in typical classrooms in different countries?*
Chairperson: D. Tall (UK)
Reporter: N. Zehavi (Israel)

Martin Perkins (UK) gave a summary of a survey into the use of computers in the United Kingdom in which 33% of the teachers in the survey never used the computer, whilst 39% used it very rarely; only 12% used a computer more than once a week.

Ulla Kurstein–Jensen (Denmark) reported that a new mathematics curriculum introduced in 1988 will make the use of computers compulsory. Most teachers have had short in-service training courses in the use of computers, but many do not feel confident when using them in class.

Sou-Yung Chiu (Taiwan) reported that there are many computers available in the schools but only exceptional teachers use them in mathematical instruction.

Dick Lesh (USA) reported a large commercial project in the U.S.A. involving 700,000 students in 400 schools. The experience is that the material is covered at a greater speed, with substantial increases in test-scores. The time saved is used to introduce more "real-world" problem-solving in which mathematics is used as a tool.

In discussion the results of a 1987 survey across the USA were given: in this survey 70% of teachers reported having access to the computer, of which nearly half (approximately 30% of the sample) did not use the computer in mathematics teaching.

The working group split up into three smaller groups for interchange of ideas from people from different countries.

Session 3: What practical research is being undertaken on the role of the computer in teaching?

Chairperson: J. Rogalski (France)

Reporter: K. Graf (FRG)

Michael Shaughnessy (USA) discussed a paper "from data to theory" for teaching probability and statistics starting with a problem-solving situation, proceeding from guesses through experiments, then computer simulations, leading into theoretical models. Teachers took time to become confident and benefitted from being able to try out the ideas cooperatively.

Dominique Guin (France) reported research into teaching situations where the teacher presented ideas relating to various representations of the notion of function using LOGO and set the children appropriate tasks. This again required considerable preparation on the part of the teacher.

Alan Schoenfeld (USA) described one student's difficulties using computer software to find the formula for a line through given points and showed how the teacher needs to be aware that what happens in the student's head may be very different from what is on the screen — or, as he said — "every silver lining has a cloud".

Harm Jan Smid (Netherlands) discussed a project focussing on the use of microcomputers for teaching mathematics at high-school level which closely followed the text-book; he reported that teachers did not use the computer because it seemed to make the curriculum more complex and did not relate to the goals of the current curriculum.

Discussion concentrated on the need for teachers to take account of the changes which are occurring because of computer technology. This will involve changes of teaching style, for instance, a move towards a more flexible problem-solving attitude, with teachers working in cooperation with students. There are evident difficulties for teachers in attempting to keep pace with new technology, however, it is essential to take account of long-term technological changes in society.

Session 4: What needs to be done to help teachers' professional development?

Chairperson: N. Zehavi (Israel)

Reporter: D. Tall (UK)

The session was opened by four short presentations. *Rosemary Fraser (UK)* asked that we be very open-minded without closing off any possibilities. She proposed various roles that could be played by the student, the teacher and the computer and also suggested that in research we distinguished between four levels of activity ranging from the individual learner via a single classroom, and several classrooms, to the large curriculum project. *Mary Jean Winter (USA)* spoke of four different levels of professional development, from the entry level, via the convert level, the implication & instruction level to the reflective level. She proposed different activities appropriate for each. *Willem Van Der Vegt (Netherlands)* spoke about the questions that were being asked about the use of the computer in the classroom and how these were being presented to teachers in in-service and pre-service courses. *Edward Dickey (USA)* talked about his work using MuMath to get teachers involved in the use of computer algebra systems.

This was followed by a group activity conducted by *Diana Resek (USA)* in which each participant was asked to write down what s/he thinks is the most important activity needed by teachers. She reported some of her own observations:

1. It takes time to develop taste as to what is good software.
2. Appreciating good software requires much time.
3. Integrating software into the curriculum is difficult.
4. Choices must be made in the curriculum.
5. Teachers need to recognize their own worth.

There followed a general discussion where individuals reported what is going on in different countries.

50 delegates from 17 different countries took part.

The written contributors included: J. Maldonado Bernal (Spain), Tang Rui-Fen, Wang Ji–Qing, and Wang Yu (China).

The Effects of Technology and of Computer Science on a Mathematics Curriculum of the Future

Coordinator: Bernard Cornu (France)
Panel: Bernard Hodgson (Canada), Audun Holme (Norway), David Johnson (UK)

Session 1. New topics: Which new topics should be included in a mathematics curriculum by the year 2000?
Chairperson: B. Cornu (France)
Reporter: B. Hodgson (Canada)
Introductory talk by *Bernard Cornu*

We have to take into account the fact that mathematics evolves and that society's need of mathematics evolves as well. Evolution of curricula should follow these evolutions.

There are many questions: Do we need to teach computer science to mathematics students? Which mathematics should be taught? How can we design a continuously evolving curriculum? How can we support teachers so that the implemented curriculum and intended curriculum move closer together? How will the order in which topics are taught change? Rather than following a mathematical sequence should we encourage a constructive or experimental development?

Tony Ralston (USA) stressed the centrality of technology as *the* force for curriculum change, this has impact on educational research and calls for widespread teacher training initiatives. He raised questions: what traditional mathematics is no longer appropriate? What new mathematics should be introduced? What is the role of manipulative skills? In what order should topics be learned? Can everyone learn mathematics?

Some elements of answers appeared in other contributions : *Ralston:* A framework developed by a task-force of the Mathematics and Science Education Board (NSEB) of the National Research Council (NRC), *Falcone (Italy):* An experimental curriculum in Italian universities, *Galada and Kockin (Czechoslovakia):* Teaching of engineers, *Goebel (USA)* presented six themes for the fourth year of high school: the computer as a tool; modelling; application of functions; data analysis; discrete phenomena; numerical algorithms, *Herget (FRG)* stressed the importance of the process of mathematizing and the need to choose topics suited to the use of computers and to make use of computer science theories, *Niman (USA)* presented a technology based programme for elementary school teachers.

Session 2. Consideration of an algorithmic approach to the mathematics curriculum — its effect upon the learning process and the content of the curriculum

Chairperson: D. Johnson (UK)
Reporter: A. Holme (Norway)
Introductory talk by *David Johnson*

The study of algorithms and procedures represents both an extension of mathematics and a new way to view the current school curriculum. Pupils designing algorithms to implement mathematical ideas and to solve problems provides the learner with dynamic representations of mathematical concepts and relationships. The algorithmic approach gives more power to the understanding: concepts become usable.

Lazarev (USSR) gave three trends: teach algorithms, creation of algorithms, using algorithms to solve real problems. He spoke of a set of two hundred problems that he had designed. *Giard (Canada)* presented activities for an active and pupil orientated curriculum. *Haguel (Canada)* reported on an experiment using a procedural approach to introduce a mathematical concept (limit of functions).

Many written contributions gave examples of an algorithmic approach (Cohors Fresenbourg (FRG), Forcheri/ Furinghetti/ Molfino (Italy), Hurley (USA), Mascarello/ Scarafitti (Italy), Olivier (South Africa).

In the discussion was raised: what mathematical topics, new and old are particularly amenable to this approach? Given full technological support, how would the way pupils study be changed?

Session 3. Presentations of new techniques for problem solving, both with and without computer resources, based on the theories and practices of computer science

Chairperson: A. Holme (Norway)
Reporter: B. Cornu (France)
Introductory talk by *Audun Holme*

It is necessary to take a fresh look at traditional techniques in the light of the developments of computer science theories, for example, constructive mathematics and algorithmic approaches. Presenting traditional mathematics on a computer can revitalize images and approaches. However, we must not base the mathematics curricula of tomorrow on the computer science techniques of yesterday. We should revise approaches to topics such as practical arithmetic, everyday accounting, classical elementary geometry, algebraic equations, and so on. We must also evaluate the effect on the student's mathematical abilities.

Many examples were presented in the short talks and the written contributions: *Dagdileldis (France)* — the use of the spreadsheet to develop the understanding of algebraic concepts, *Laborde (France)* — Cabri geometrie, a new tool for classical geometry, *Marconi (Italy)* — trigonometrical and periodic functions, *Tufte (USA)* — introducing calculus, *Guzicki ( )* — examples of problems to be solved with the computer, *Czuba (Poland)* — MIZAR, a new method of teaching.

The discussions showed the link between programming and learning concepts, but we noticed the need for a model of learning in order to implement effective learning strategies. After speaking at length about the teacher and the role of the computers in teaching we should not forget that the main priority is the pupil's learning. How can we help the pupils to be creative users of the computer in full command of their own abilities to generate ideas?

Session 4. Demonstrating the "state of the art" of symbolic manipulators and experiments of their use in educational environments

Chairperson: B. Hodgson (Canada)
Reporter: D. Johnson (UK)
Introductory talk by *Bernard Hodgson*

While Symbolic Manipulation Systems (SMS) have been used for quite a while in research, it has only been recently that experiments of teaching in actual classrooms have been done. Many more studies need to be done, to assess how much time, if any, can be gained by their use; what, in present curricula, could be topics of diminished importance, topics of increased importance, or indeed new topics? How can SMS turn the computer into an "object to think with" (S. Papert) or a device to help in formulating or testing hypotheses. In opposition to the view of many users of mathematics, SMS should not result in a reduction of the level of mathematical prerequisites, but it should provoke a shift from purely computational skills to more complex interpretive skills which need a strong theoretical mathematical background.

It was shown in the presentation of *Beilby (UK)* how standard programming languages can be used, among others, in symbolic manipulation. *Heid (USA)* reported on three studies that she made about the influence of SMS in mathematical modelling and conceptual understanding. Her studies tend to show that SMS help in acquisition of concepts. *Muller (Canada)* gave an example of the use of SMS in a first year calculus course. Laboratory exercises were presented related to concept acquisition and to exploration. *Zorn (USA)* presented interesting examples of problems to be solved with SMS. He noticed that learning the syntax of the system is not an obstacle for the students and that these activities helped the students to understand better the concept of a function.

81 delegates from 25 different countries took part. The written contributors included: R. Anderson (USA), D. O'Shea (USA)

Presentations and Reviews of Current and Future Developments

Coordinator: Herbert Loethe (FRG)
Panel: Klaus Menzel (FRG), Thomas O'Brien (USA), Richard Lesh (USA)

Session 1: Advanced computer use in elementary schools
Chairperson: H. Loethe (FRG)
Reporter: R. Lesh (USA)

As an opening address the chairman Herbert Loethe took the role of an advocate of the silent majority of mathematics educators who are not yet convinced about the benefits of computers in learning and teaching of mathematics. He questioned: Are there current software developments which are ideal or perfect or at least exemplary? The chairman asked the presenters to try to answer questions like: Why do we need this piece of software? What does it help, what does it upset? What does it encourage, what discourage? Does it fit into the system of the curricular goals of mathematics education? The reason for asking these questions is not to get arguments against an innovation but to make sure that mathematics educators feel in some sense comfortable with an innovation. Therefore, innovators always should express their difficulties, their uncertainty, their scruples and second thoughts about their product.

Tim O'Shea (UK) claimed that the standard arguments for applying computers in mathematics instruction are based either on the adaptivity of computer tutors or the expressivity of computer programming languages. The adaptivity argument is based on the notion that a computer tutor can incorporate a student model derived from an analysis of the standard errors that pupils actually make. The expressivity arguments usually steer around drill and practice and focus instead on the value to the pupil of expressing their own mathematical understandings as computer programs. The speaker then illustrated these ideas by giving examples out of "The Math Galaxy": Shrink-a-cube, a simulation of the Dienes blocks, Shopping-on-Mars, a collaborative adventure game, etc. In the discussion the statement of the chairman and the presented examples of O'Shea gave rise to controversy. Especially the teaching of mathematics hidden in adventure games was considered to be not appropriate in general.

Session 2: Exemplar uses in school
Chairperson: T. O'Brien (USA)
Reporter: K. Menzel (FRG)

The second session demonstrated the influence of the different school situations in the US and the UK. This can serve as model for the general influence of national educational politics even on computer innovations. *Richard Lesh (USA)* (co-author:*Joanne Weiss)* from WICAT presented an approach to Computer-Assisted Instruction using integrated learning systems. The activities of the project are oriented to the learning and teaching of the basic skills mainly in mathematics and takes into account that the school system has to deal with teacher shortage and economic cutbacks. The lessons are designed for a network of personal computers and they belong to comprehensive on-line curricula in mathematics, reading and language arts. The same system has integrated computer-adaptive testing programs, learning management facilities to route students through the courseware. One main aspect is the use of computers to motivate students to deal with mathematical skills at all; significant positive results were especially found in individual learning progress of deprived students: "They learn better, faster and more". The discussion clarified the situation of teachers who are learning content better and methods of teaching by using the system. The student - computer relation is nearly 1 to 1 and a major aspect is that students often start teaching each other.

The *ATM (Association of Teachers of Mathematics)* group from the UK presented examples of a total problem solving oriented approach to mathematics learning. This means that the teachers "set a problem and look at what the children are doing". The computer is only used if the objectives of teaching cannot reached in another way. Two problems were demonstrated by working with the audience and the related computer aids were shown.

The discussion dealt with the possibility of the generalization of the demonstrated examples, with the use of other materials, with teacher interaction and with the role of the computer during the problem solving process. It was claimed that the computer influences the learning process also by its computer -game-like motivation, which need not fit with the problem solving process.

Session 3: Computer environments for problem solving
Chairperson: R. Lesh (USA)
Reporter: T. O'Brien (USA)

In the third session *Judah Schwartz (USA)* claimed that mathematics teachers all over the world ask children to learn ideas that have been constructed and perfected by other people. Making mathematics should, in contrast, be the task of children in classrooms. What does it mean to "make mathematics"? To make and to explore conjectures about mathematical objects — numbers, vectors, matrices, categories etc. Schwartz then demonstrated the use of "The Geometric Supposer", software he developed for classroom use. The "Supposer" is a powerful arena for the exploration of relationships and conjectures involving shapes.

Marge Kosel (USA) discussed single context programs and educational tools. She then demonstrated "The Factory" and "The Super Factory" software program in which pupils deal with shapes and transformations. She also demonstrated "Blockers and Finders" and theme groups participants solved several complex problems. Single context software such as this, Kosel claimed, causes complex processing by children usually involving generalization, and often having more than one solution.

David McCarthy (USA) described an NSF-supported project involving the development of an intelligent tutoring system and tools for mathematical learning. An "initial teaching system" involves modelling and coaching. He then described in detail the operation of an algebra tutor which teaches children to solve equations.

Session 4: Computers as a tool
Chairperson: K. Menzel (FRG)
Reporter: H. Loethe (FRG)

In the final session *Sharon Dugdale (USA)* presented a package for manipulating polynomial functions and their graphs. An objective is to develop a "qualitative sense and an intuitive basis for theorems on polynomials" and the software should motivate student investigations of mathematical ideas. She demonstrated the building up of polynomials by adding terms of x in a formula and in the graphics representation in parallel; on the advanced level a given graph is approximated step by step using the same process and making the decisions about the exponents and coefficients by intuition concerning symmetry, number of extrema and so on. In the discussion the question was raised whether it makes sense to teach this stuff since all polynomial operations are done by computer systems in reality.

Herbert Moeller (FRG) reported about "Elementary Analysis" which means an introduction to calculus by using "Lipschitz-restricted" concepts of limit, continuity, differentiation and integration. In consequence of these concepts all definitions can be visualized and most results can be discovered geometrically. That means that the geometric representation and the basic concepts are very near together and a strong completeness theorem allows the construction of many algorithms.

49 delegates from 17 different countries took part.

LOGO and Mathematics Education

Coordinator: Celia Hoyles (UK)
Reporters: Laurie Edwards (USA), Tom Kieren (Canada), Uri Leron (Israel), Richard Noss (UK)

The sessions within the theme group focussed on four key strands within the field of LOGO and Mathematics education.

Strand 1: LOGO and teachers. The four presentations in strand 1 raised a number of common themes in addressing the use of LOGO in teacher education which ranged from a consideration of the personal experiences and attitudes of individual teachers to the social and institutional contexts of computer use.

Joao Matos (Portugal), in describing the MINERVA education project for secondary school teachers, emphasized the importance of creating for teachers the opportunity for personal mathematical development. This was also evident in the Inset LOGO course for primary teachers presented by *Candida Moreira (Portugal)*, and in the Microworlds Project, described by *Celia Hoyles, Richard Noss*, and *Rosamund Sutherland (UK)*. The Microworlds Project involves research and evaluation of a 30 day course for secondary mathematics teachers in which significant shifts in attitude to pupils and to classroom practice have been detected. As in the Inset programs described by Matos and by Moreira particular attention is paid to the social context of the teachers' learning experience — the importance of discussion, cooperation, reflection and debriefing in the process of building new LOGO/maths activities. The social context of teacher-education was also a focus of the Sunrise School project (*Liddy Nevile, Australia*) which is investigating the nature of effective teacher interventions in a technologically rich environment for inner city children, and attempting to address the question of the influence of social setting on learning.

In summary, the session brought to light a number of issues important in mathematics education generally: the need for time and commitment to both personal and professional teacher development, the influence of the social context of learning, and questions of pedagogy related to teacher intervention and learner autonomy.

Strand 2: LOGO and geometry. Three common themes of the second strand were (a) extending Turtle Geometry (TG) beyond its usual familiar domain, (b) using turtle geometry to explore geometrical transformations and (c) linking LOGO to school geometry.

Jean Cesar (France) presented a study in which tools for constructing 3D solids were supplied. Pupils can construct solids by specifying their faces using ordinary plane TG, then assemble these using special new procedures. *Johnny Lott (USA)* reported using list processing to implement a matrix-based approach to transformational geometry. Using these tools, students can experiment with transformations and, in particular, obtain all transformation as products of reflections. *Medhat Rahim (USA)* applied LOGO was concerned, *Laurie Edwards (USA)* presented a microworld for transformation geometry and reported a teaching experiment using this microworld in which the pupils were involved in studying a single transformation, invariants, group properties (composition, inverse) and creative use of the transformations. *Elizabeth Gallou–Dumiel (France)* presented research on the learning of reflection and point symmetry. She explained how a structured LOGO environment was helpful for the learning of these two transformations, and how it modified pupil's mathematical representations in these two cases. *Joel Hillel (Canada)* described an environment for learning about the properties of several geometric properties, particularly those in an isosceles triangle, and reported how pupils learned to use the tools given them to come to understand the relations and geometric concepts embedded in the tasks.

Strand 3: Recursion and recursive thinking. In recent years, considerable interest has been focussed within the field of LOGO and Mathematics on the idea of recursive processes and the kinds of thinking which are generated by the construction and use of recursive LOGO programs. These issues formed the basis for the third strand.

Tom Kieren (Canada) began the discussion by pointing out that although LOGO offers children and adults the opportunity to represent mathematical ideas by sophisticated multilevelled recursion there is also a tendency to represent these same ideas through concatenative recursion and systematic replication — both essentially iterative in nature. *Trevor Fletcher (UK)* proposed the need to build bridges between LOGO activities and existing mathematical curricula, and took as an example, a LOGO procedure for a binary TREE and a traditional piece of combinatorics at grades 11 and 12 focussing on binary arithmetic. *Rosamund Sutherland* and *Teresa Smart (UK)* reported the results of a teaching experiment to investigate the nature of grade and 12 pupils' mental models when approaching a problem with a linear recursive solution. Their approach hinged on encouraging pupils to develop a "recursive frame" which pupils could then initiate when presented with the task.

Al Cuoco (USA) presented an example of his work with advanced high-school students; he reported the activities of two students who worked on the combinatorial problem of determining the number of ordered partitions of a finite set. *Blagovest Sendov (Bulgaria)* outlined a new version of LOGO which has been enlarged by the inclusion of primitives to solve problems in plane geometry (such as CIRCLE, POINT, and SEGMENT). Finally, *Uri Leron (Israel)* addressed the fundamental question — 'what makes recursion hard?'. He pointed to a 'complexity gap' between recursive procedures and the processes they generate and argued that unless we specifically wish to understand the behaviour of the computer (as in debugging), it is often best to suppress process and move directly to the LOGO procedure as a description of the product.

Strand 4: the LOGO mathematics learning environment.

De Corte et al. (Belgium) reported on a research project "Computers and Thinking" in which sixth graders were systematically taught a planning strategy within a LOGO learning environment. The researchers identified a strong relationship between the quality of the planning and the time needed to achieve a positive outcome. *Hoyles and Noss (UK)* reported two studies based around a parallelogram microworld. They found a mismatch between pupils' initial intuitions and formalized definitions, which was at least partially resolved as a result of the LOGO activities. They also pointed to the importance of synthesis between visual and symbolic modes of representation, and described instances of the computer offering "scaffolding" for pupils to make sense of partially comprehended geometrical relationships. *Bruno Vitale (Switzerland)* argued for the importance of mastering the tools of informatics as a way of enriching learning, by introducing procedural thinking through exploration of the strengths and limitations of a formal language. Vitale also described a psycho-cognitive analysis of pupil's perspectives of what the computer can offer.

In conclusion, the theme group illustrated the rich variety of research, teacher-education and curriculum development that currently characterizes the international LOGO mathematics community. In particular, clear progress has been made in a number of key areas: — in understanding the potential of LOGO tools for the construction of mathematical representations; in clarifying the role of turtle geometry as offering not only a context for geometrical activities, but also as a medium for expressing and exploring a range of algebraic and more general mathematical ideas; finally, in identifying the significant features of the pedagogical context in which children's LOGO work is situated.

56 delegates from 20 different countries attended the meeting.

Supporting survey presentation:

Jim Fey (USA): Computer and Teaching of Mathematics.

The speaker was asked to review some key issues with respect to this Theme. Selected passages from the full paper are given here. The greatest challenge and opportunity in mathematics education today is the revision of curricula, including teaching methods, to take advantage of electronic information technology. Developments in this decade alone have presented us with inexpensive and powerful hardware and software tools that challenge every traditional assumption about what we should teach and what we can teach.

There is no shortage of speculative writing on the promise of revolution in school mathematics following from the application of various calculating and computing tools to teaching, learning, and problem solving. Since ICME 5 in Adelaide, there has been a profusion of conference reports and position papers outlining potential technology-based innovations. An exciting array of experimental projects have begun demonstrating the prospects for mathematics classrooms that use calculators, computers and videodisks to change the goals of curricula at all age levels and longstanding patterns of teacher/student interaction.

The reports from ICME 5 itself include hints of nearly every proposal and project reported since 1984. However, it is very difficult to determine the real impact of those ideas and projects on the daily life of mathematics classrooms; there is very little solid research evidence validating the boundless optimism of technophiles in our field.

There are several possible ways to impose order on the array of technology-motivated ideas in mathematical education today. One is to look at the major tasks involved in designing school mathematics — selection of content and process goals, organization of teaching and learning environments, and assessment of achievement — and to describe the impact of technology on each. There are many suggestions and active development projects working on each of these dimensions.

1. Content/Process Goals — The most prominent technology-motivated suggestions for change in these focuses on decreasing attention to those aspects of mathematical work that are readily done by machines, and increasing emphasis on the conceptual thinking and planning required in any "tool environment". Another family of content recommendations focus on ways to enhance and extend the current curriculum to mathematical ideas and applications of greater complexity than those accessible to most students via traditional methods.

2. Teaching/Learning Styles – Many mathematics educators have looked at the new information processing tools and envisioned a striking change in the traditional teaching/learning patterns of mathematics classrooms. They see teachers shifting their roles from expositor and drill-master to tasksetter, counselor, information resource, manager, explainer and fellow-student, while students engage in considerably more self-directed exploratory learning activity. The most common strategy for creating these new kinds of classroom interaction is the provision of some kind of computer microworld. In a microworld, mathematical or real world objects, relations and operations are represented electronically in a way that permits controlled exploratory manipulation and observation of properties by the student searching for abstractable mathematical principles.

Unfortunately, while there is a certain naive logic in sorting ideas by their focus on content or pedagogy, very few projects or proposals can be categorized in this way. Nearly every development program has as agenda of goals that imply changes in both content and pedagogy of school and university mathematics. To focus attention on technology prospects and the implication of those prospects for school mathematics, I have chosen to approach the survey task by looking at the principal capabilities and characteristics of the technology and describing the implications of those conditions for both content and pedagogy in mathematics.

I distinguish:

1. Numerical computation, 2. Graphic computation, 3. Symbolic computation, 4. Multiple representations of information, 5. Programming and the connections of Computer Science with Mathematics curricula, 6. Artificial intelligences and machine tutors.

In each area there are a number of interesting studies, described in the full paper. However, the potential for using technology to extend the range of human mathematical learning and problem solving is only beginning to be tapped by research and development projects, much less in the day-to-day life of typical mathematics classrooms. While some may choose to wait until a clearer picture of the "best" response emerges, the situation right now offers impressive opportunities for progress.

Effective use of computers in instruction can permit the kinds of teaching/learning environments that most teachers long for, while they struggle with the constraints of the traditional classroom and curricular conditions. Revision of curricular goals to acknowledge that computers and other electronic information technologies are now standard tools for problem-solving and decision-making in science, business, government and industry will lead to significant changes in what we ask and empower students to learn. There are many important questions to be answered, but we really have no choice but to tackle those questions and to bring school and university mathematics into the electronic information age for which we are ostensibly preparing our students.

THEME GROUP 3: PROBLEM SOLVING, MODELLING AND APPLICATIONS

Chief organizer: Mogens Niss (Denmark)
Hungarian coordinator: Tünde Kántor,
Panel: Leone Burton (UK), Mary Kantowski (USA), Ian Isaacs
(Jamaica, p.t.), Zbigniew Semadeni (Poland), Kaye Stacey (Australia)

Introduction

For the first time in ICMEs Modelling and Applications, on the one hand, and Problem Solving on the other, were merged together to constitute one theme and one theme group. At the same time the closely related area "Mathematics and Other Subjects" was made an independent theme to be dealt with by another theme group (T6).

Since the theme of T3 is rather broad a few working definitions may prove useful.

A (mathematical) *problem* is a situation giving rise to certain open questions that are intellectually challenging to somebody who is not in immediate possession of direct methods/procedures/algorithms etc. which can answer the questions and solve the problems. Problems are thus different from exercises. Problems can either be *pure*, i.e. imbedded in a purely mathematical universe, or *applied*, i.e. characterized by defining situations belonging to some extra-mathematical universe. *Problem solving* then simply refers to the process of dealing with problems with the aim of solving them.

A (mathematical) *model* is a collection of mathematical objects and relations selected to represent and reflect aspects of a given extra-mathematical area (called "reality"). The process of constructing a mathematical model is called *modelling* or *model building*. In this term we include all stages in the modelling process, from initial investigations into the extra-mathematical context, through the translation from "reality" to mathematics (*mathematization*), to the validation of the model leading to either acceptance, modification or complete rejection of it. In the latter case a new model may be constructed and the whole process repeated.

Models may be built with different purposes. If a model is built as part of an attempt to solve problems selected from the extra-mathematical area being modelled, the modelling process takes the shape of applied problem solving. However, models may also be built in order to describe or better understand segments of "reality" with no direct aim at solving specific problems.

Whenever mathematics is activated towards an extra-mathematical area we are faced with an *application* of mathematics. The term "application" is thus a very general one. It is possible to deal with applications without performing modelling, but the application of mathematics always presupposes a model (or several models). However, applications may be dealt with without making the underlying model(s) an explicit object of study or treatment.

So, there are very good reasons for bringing Problem Solving and Modelling and Applications together, such as: (i) they both deal with students operating with some degree of creativity, often in less than well-defined situations; (ii) the psychological (cognitive, affective, emotive) processes involved in solving problems, in performing modelling and in applying mathematics have much in common; (iii) much problem solving addresses applied problems requiring modelling and models for their tackling, and, dually, a major aim of performing modelling and applying mathematics in given situations is to solve problems of one kind or another.

Although the two components of the theme, Problem Solving, and Modelling and Applications, thus *do* overlap in important respects, they are not simply equivalent, neither in content nor in sociology. Central features of problem solving are focussed on purely mathematical problems in which applications and modelling in extra-mathematical areas are absent. Conversely, applications and models may be made objects of study without involving problem solving activities in the sense which the term problem solving carries today. This difference in content is accompanied by a difference in sociology: The communities of people engaged in problem solving research and development, and in modelling and applications research and development, are not the same, a fact that manifested itself clearly with the actual T3 attendance.

The organizers of the Theme Group saw it as an important aim and challenge to accommodate, in a balanced way, both communities in the programme, but also to foster closer interaction and exchange of experiences and views between the two communities. The Theme Group programme was, therefore, designed to emphasize and encourage active participation of those attending in the working sessions.

In order to facilitate this, the element of oral presentations was limited in order that small group working sessions devoted to the discussion of specified subthemes (as described in the sequel) could take place.

Unfortunately it has to be said that for a number of reasons that intention was not brought into reality to a satisfactory extent. The foremost reason for this was the large Theme Group attendance, with originally about 500 attenders having a diversity of backgrounds and interests, in combination with the difficulty of accommodating so many people in rooms suitable for small group discussion.

In addition to the input provided by the relatively few oral introductions and presentations, invited background papers (not presented orally) of different categories were made available to Theme Group attenders (in limited numbers):

Pekka Kupari (Finland): Problem Solving, Modelling and Applications in the Finish school mathematics: Some observations and trends.

Ian Isaacs (Jamaica/Australia): Using Problem Solving to encourage mathematical thinking in Jamaican secondary school students.

Gabriele Kaiser–Messmer (FRG): Survey of the present state, recent developments and important trends of modelling and applications in the Federal Republic of Germany.

Kaye Stacey and Susie Groves (Australia): The teaching of Applications, Modelling and Problem Solving in Australia: 1984–88.

Zalman Usiskin (USA): Applications and Modelling in the University of Chicago School Mathematics Project (UCSMP) Materials. A Summary.

Invited papers available in a few copies only were prepared by

H. J. Burscheid and H. Struve (FRG): Applying mathematics by pupils from an epistemological point of view.

Milton Fuller (Australia): Mathematical modelling in distance education — A challenge for teacher and learner.

Solomon Garfunkel (USA): Survey of Applications and Modelling in US high schools and colleges.

A. O. Moscardini (UK): The identification and teaching of Mathematical Modelling Skills.

Alan Rogerson (Australia): Background paper for ICME 6 Theme Group 3: Problem Solving, Modelling and Applications.

Some of these papers together with papers given in the oral presentations in Session 2 (see below) will be published in a joint volume of contributions to the themes "Problem Solving, Modelling and Applications" and "Mathematics and Other Subjects" edited by Chief Organizers Werner Blum (FRG, T6)and Mogens Niss (Denmark, T3) with the collaboration of Ian Huntley (UK).

Theme Group specific posters were prepared, on invitation, by

Morten Anker (UN, New York)

Wendy Caughey and Max Stephens (Australia)

István Drahos (Hungary)

István Ábrahám (Hungary)

Janet Jagger (UK)

R. N. Mukherjee (India)

Finally it should be mentioned that outside the Theme Group work proper a joint survey lecture of the Themes relating to Theme Group 3, "Problem Solving, Modelling and Applications", and T6, "Mathematics and Other Subjects", was given by Chief Organizers Werner Blum (T6) and Mogens Niss (T3).

Theme Group Organization

The main part of the Theme Group sessions was spent on discussion in small working groups. To accommodate the variety of interests and backgrounds with Theme Group attenders, working groups were formed according to the matrix below. For each cell of the matrix the chairperson(s) of that cell is(are) indicated. For cells which attracted a large number of participants several subgroups were formed. Group members were asked to stay with their group for all three working sessions (1., 3., 4.), but as was expected this happened only to a certain extent.

major student interest age groups	problem solving in a mathematical context	applications and modelling
ages 5–12	no division chairpersons: Leone Burton (UK) and Zbigniew Semadeni (Poland)	
ages 13–19	chairperson: Kaye Stacey (Australia)	chairpersons: Ian Isaacs (Jamaica/Australia) and E. L. Kantowski (USA)
tertiary education	chairperson: Mogens Niss (Denmark)	chairperson: Susie Groves (Australia)

Originally, it was planned to include a fourth row in the matrix: vocational and adult education, but at the formation of subgroups in Session 1. it turned out that almost no attenders chose that option. It was, therefore, cancelled.

Session 1.: Foundational Issues

Chairperson: *Mogens Niss*, Chief Organizer, Denmark

The session opened with a 30-minute plenary with simultaneous interpretation. The first 10 minutes contained welcoming and opening remarks by the Chief Organizer, Mogens Niss, who presented the organizing panel, recapitulated the Theme Group programme briefly, and asked for participants' choices as to the matrix options described above.

In the remaining 20 minutes panel member *Kaye Stacey* gave an introduction to the Theme Group work, with particular regard to the (sub)theme of Session 1. In the Programme Statement the following foundational issues were pointed out:

- clarification of *concepts, notions* and *terms* involved in the theme, e. g. "problem", "problem solving", "modelling", "model", "applying", "application", and their interrelations;
- identification of the *purpose, role* and *position* of problem solving, modelling and applications work in different mathematics curricula;
- discussion of balances between *process* and *product* in different types of problem solving, modelling and applications work;
- identification of *research* and *development contributions* of importance to the field.

Kaye Stacey, in her lecture, suggested the following specific discussion questions to be dealt with by the subgroups in the remaining 60 minutes of Session 1.:

Regarding *clarification*: What are the important characteristics of problem solving, applications and modelling for your matrix cell? How do you communicate them to others, in particular teachers? How do you assess them? Examples?

Regarding *purpose, role and position*: Do problem solving, modelling and applications with a relatively low mathematical demand have a place in the curriculum? Where? For whom?

Extended investigations take time. What should be omitted? Where does the balance lie?

Is it appropriate to spend mathematics teaching time on the non-mathematical aspects of applications?

Regarding *product and process*: Is it possible/helpful to try to separate content and process in teaching and assessment respectively?

Regarding *research and development*: What fundamental questions about problem solving, applications and modelling are most urgently in need of (further) research? How could such investigation best inform practice?

Report on subgroup discussions in Session 1
Clarification of concepts, notions and terms

Subgroups worked in general accordance with the working definitions proposed in the introduction of the present report, but various additional points were made.

Several subgroups devoted some time to classifying different *types of problem*. Even if "exercises" — characteristic by the straightforward activation of newly introduced methods, procedures or algorithms to non-complex situations — be left out it might be debated which of the following kinds of activity should be included in the concept of a "problem". It seems that the demarcation line is drawn differently in different languages and in different countries.

The following types were identified:

(a) *"dressed-up" exercises*: activities for which a very brief verbal description of a situation is given and a few specific questions are posed, the answering of which is an exercise as defined above. In dressed-up exercises neither an interpretation of the background situation nor a search for methods are really needed.

(b) *"advancement" exercises*: situations or activities for which one or more specific questions are given to guide or assist the development of a new concept or a new method.

(c) *"classical" problems*: activity for which the questions asked are well-defined and clear from the outset, where, however, the methods to answer them are not immediately available but have to be found or devised.

(d) *"neoclassical" problems*: activities for which the questions asked are general and open-ended rather than specific and well-defined, such that precise questions to coin the general ones have to be formulated, and solution methods have to be sought or devised accordingly.

(e) *open problems*: activities for which not only the questions asked are open-ended and general, but where also background situations are described in general terms, such that additional information has to be collected, decisions have to be made etc., before solution methods can be sought or devised.

(f) *investigations*: situations described only vaguely or in rather general terms and with no questions posed from the outset. On that background students are invited to explore the situations on their own in such a way that they themselves formulate questions, seek information and look for methods to deal with the questions formulated.

Any of the above six kinds of situation and activity may deal with a purely mathematical universe as well as with an extra-mathematical universe whether real-life or artificial.

There was general agreement that it is not an intrinsic property of an activity to be a problem or an exercise. Whether one or the other is the case depends on the knowledge and experiences of the person(s) engaged in that activity. What to one person is a problem may be an exercise to another. The distinction between a problem and an exercise is thus a *relative* rather than an absolute one.

It was pointed out by many subgroups that problems may be classified in several different ways, for instance according to their educational aims, their content of extra-mathematical substance, their objective or subjective relevance to students, their relevance to the world outside mathematics. The "applications and modelling" subgroup addressing the secondary level (ages 13–19) suggested that for the lower secondary level (13–15) *"action" problems* (aiming at making things happen) might be the most appropriate, whereas for the upper secondary level *"believable" problems* (i.e. other people's problems with which students can identify) might be more preferable, partly because of the relative scarcity of proper action problems for that age level. Thus the classification given above should not be thought of as a canonical one.

As regards *modelling* there were several attempts to distinguish between different types of models and modelling:

– One type consists of models built to deal with isolated situations without much attention being paid to connections to other models or to theory. Terms such as *empirical* or *ad hoc* modelling were suggested as possible labels for this type of modelling.

– A second type consists of modelling which draws on or is related to established theory of the field to which the object/phenomenon area being modelled belongs. *Classical* or *theoretical* modelling were proposed as possible labels for this type of modelling.

One subgroup characterized the difference between modelling and *applications* by saying that in modelling it is not clear — in principle, if not necessarily in practice — from the outset what model the modelling process will lead to, whereas application is rather a matter of selecting from a library of established models.

Purpose, role and position

One issue on *problem solving* and *modelling* was predominant in the discussion in most subgroups regardless of which matrix cell they belonged to:

Should problem solving/modelling be given in *separate* courses to allow for emphasis to be put on the process and/or extra-mathematical components in the activity? Or should problem solving/modelling be integrated within the ordinary mathematics courses, for instance with the purpose of enriching or assisting the acquisition of mathematical concepts and methods, a type of approach which in the ICME 5 proceedings of T6 was called *the mixing* approach.

Participants held different views on this issue, but there was overall agreement that the two kinds of approach serve different purposes and needs, and that both lead to highly desirable results of independent value, if successful. Problem solving and modelling should be viewed as arts in themselves which are unlikely to be satisfactorily accommodated within the boundaries of courses aiming mainly at developing mathematical content.

The difference in participants' viewpoints correlated strongly with the age level of their target groups.

For the *primary* and *lower secondary* levels (ages 5–15) the mixing approach was advocated. It was considered crucial that the problem or modelling situations dealt with should be (made) interesting to pupils on independent grounds, so that they may see a point in activating mathematics towards the situation in question. One subgroup phrased this as follows: situations should be extracted from pupils' real worlds. Some participants found that it would be ideal to take a problem solving/modelling situation as a starting point and then develop the mathematics needed accordingly, but for various reasons this was considered less realistic, primarily because it is too time consuming and demanding to both pupils and teachers.

For the *tertiary level* participants' views differed, but the overall trend was to express the need both for courses exploiting the mixing approach and for separate problem solving/modelling courses.

Other points made under the present heading include:

There might be a paradox in the fact that in many places applications are used to assist less resourceful students in gaining from their mathematics instruction, while at the same time it is widely recognized that applicational activities are more demanding and difficult than "pure" mathematics.

Process-product balance

There was general consensus amongst participants that the process aspects of problem solving, modelling and applications are essential in themselves. There should be no dichotomy between process and product. If such a dichotomy seems to exist it is mainly due to external strain on mathematics instruction exerted by curriculum authorities, colleagues, parents, receivers of graduates etc. Some subgroups agreed that the modelling process must dominate the mathematics to be used, not vice versa. Others, however, felt that this is easier said than done, because mathematics programmes are subject to different types of forces and interests, some of which are conflicting.

It was pointed out that as soon as problem solving, modelling and applications are *taught*, i.e. are made objects of study for students — which is for instance the case when heuristics, strategies, approaches etc. are taught explicitly — those *processes become content* themselves, though of a different order than the first order content belonging to mathematical or non-mathematical areas.

Research and development

Instead of attempting to identify research and development contributions (as suggested in the Programme Statement) most subgroups addressed the question of urgent research needs, raised in Kaye Stacey's introductory lecture.

The prevailing concern to several subgroups was that it is difficult to *prove,* in a classic strict sense, that knowledge and skills gained from problem solving, modelling and application activities within mathematics instruction can be transferred to new situations (whether inside or outside mathematics). This is important when we have to convince colleagues, educational authorities, students etc. of the value/necessity of including problem solving, modelling and applications in the curriculum. Although there *is* research (for instance by Alan Schoenfeld) which gives evidence for some degree of transferability, the field is largely unexplored. This is due to several factors, such as the fundamental complexity of problem solving, modelling and application abilities, and the fact that genuine problem solving and modelling activities are still not firmly established as a practice in most curricula.

Most Theme Group participants, however, shared a belief — based on their own experience with courses and students as well as on general reflection — that problem solving, modelling and application skills can be taught so as to be transferable. In any case it was agreed that evidence abounds to the effect that pure mathematics instruction alone is not sufficient to generate the kind of problem solving, modelling and application abilities which we are after.

Another research and development issue raised some discussion. Many participants expressed the need for suitable problems, modelling and application situations for different purposes and different levels. Especially for the upper secondary level it was considered difficult for the average teacher to find modelling and application situations of an appropriate level of difficulty for students: It is easy to find too simple cases; it is easy to find too difficult cases, but is not easy to find cases inbetween.

Others felt that in recent years a wide variety of different sources to this end have been made generally available. This was also recognized in the discussion but it was commented that it is one thing to have published "banks" of suitable cases, but a different thing to find or devise original modelling and application cases to fit into a specific instructional situation and purpose. Therefore *methods* for finding and devising *new* situations were asked for and proposed as an item needing development.

The issue of *assessment* was a major concern to the subgroups. The problem is that even if it is accepted that problem solving, modelling and applications can be taught and learnt, it is difficult to create forms of assessment and testing which reflect the spirit, content and complexity of the field, and pay due respect to the higher order knowledge and skills it involves. It is even more difficult, if not impossible, to use the assessment and testing frameworks usually applied in educational systems. Various attempts to devise assessment forms were reported by participants (e.g. from Denmark and the UK) but they are mainly of an experimental character, and not suitable for large scale implementation. In some countries certain aspects of *applications* of mathematics are tested in examinations to a modest degree, but so far this is only rarely the case with problem solving and modelling. So it was urged that substantial research and development efforts should be invested in assessment and testing relating to problem solving, modelling and applications, both as regards the evaluation of students individually or in groups and the evaluation of programmes/teaching.

Finally, a number of miscellaneous points were raised, including the following research questions:

- Do students learn/understand the mathematics more deeply when they are taught problem solving, modelling and applications?
- Do you need to know specific mathematics content before applying it?
- How to develop an appropriate taxonomy of problems?
- Would it be an idea to compare the expert problem solver with the average problem solver in order to gain insight into the crucial factors of problem solving ability?

Session 2.: *Presentations*

No subgroup discussions were scheduled for this session which was devoted to a number of oral presentations of different sorts, the aim of which was to provide an opportunity for Theme Group participants to become acquainted with valuable work in the field, and with experiences and conditions in other places, in order to widen the background for further discussion in sessions 3 and 4.

The presentations were given in four parallel blocks each having a specific heading. Participants were invited to choose blocks according to their own wishes.

Four presentations were planned for each block, but in Block 4. two of them had to be cancelled due to the absence of the presenters from the congress.

Block 1.: Problem solving with an emphasis on research aspects

Chairperson: *Kaye Stacey*

(1) Zbigniew Semadeni (Poland): "Verbal problems with missing, surplus or contradictory data as means for instruction".

(2) Jerry P. Becker (USA): "Cross-cultural research on student strategies and difficulties in problem solving, performed by the Japan-United States Joint Study Group for Mathematical Problem Solving".

(3) Tetsuo Matsumiya, Akira Yanagimoto, Yuichi Mori (Japan): "Mathematics of a lake — problem solving in the real world".

(4) Francine Grandsard (Belgium): "Problem solving for first year university students".

Block 2.: Presentations of national situations in problem solving, modelling and applications

Chairperson: *Mogens Niss*

(1) Samson O. Ale (Nigeria): "The position of problem solving, modelling and applications in mathematics curricula in Nigeria in a societal perspective".

(2) Nicos Klaudatos (with *Stavros Papastavrides*) (Greece): "The actual and potential role — and problems related to it — of problem solving, modelling and applications in post-elementary mathematics education in Greece".

(3) Paulus Gerdes (Mozambique): "Using local situations and conditions to generate mathematical problems for mathematics instruction, with examples from Mozambique in particular".

(4) Kirsten Hermann (Denmark): "Recent trends and experiences in applications and modelling as part of upper secondary mathematics instruction in Denmark".

Block 3.: Curricular aspects of problem solving

Chairperson: *Ian Isaacs*

(1) Akira Takata (with *Kiyoshi Yokochi*) (Japan): "Applied mathematical problem solving of ages 5–12 with special regard to Japanese experiences".

(2) Stephen Krulik (USA): "Content and implementation strategies for including problem solving in the school mathematics curriculum".

(3) Erkki Pehkonen (Finland): "Possibilities of low-attainers in mathematical problem solving".

(4) Tünde Kántor (Hungary): "How to know mathematical structures by problem solving at ages 15–16".

Block 4. Applied modelling

Chairperson: *Leone Burton*

(1) Julian S. Williams (UK): "Teaching modelling in classical mechanics to high school students using practical approaches".

(2) Wolfgang Schlöglmann (Austria): "Experiences with students' mathematical modelling of problems from industry".

As mentioned in the introductory section it is planned to publish some of these papers in a volume edited by Werner Blum and Mogens Niss.

Session 3. : Content, Form and Resources

The entire session was devoted to discussion in (matrix) small groups of the following points listed in the Programme Statement.

- *types of content* (in terms of reality as well as of mathematics) which should be represented in problem solving, modelling and applications work;
- the *impact* of problem solving, modelling and applications on the *content of mathematics curricula*;
- *teaching and study forms* suitable as vehicles for problem solving, modelling and applications work in different mathematics curricula;
- available or desirable material and immaterial *resources* for problem solving, modelling and applications work in different curricula.

Subgroup discussions concentrated mainly on the content issues and dealt a little with resources, whereas teaching and study forms were only touched upon in passing.

Types of content, impact on content

As was already touched upon previously in this report, content in problem solving, modelling and applications encompasses much more than mathematical content in its narrow meaning. For modelling and applications, characteristic features of the area being subjected to mathematical treatment is content too, to the extent that it is explicitly dealt with. Process aspect are essential in any problem solving, modelling and applications work. If they are made objects of teaching or study they constitute content in themselves.

In addition to general abilities such as reflecting, creating, generalizing, reporting etc., more specific process components mentioned by Theme Group participants include: conjecturing, simplifying, looking for related problems, identifying variables, collecting data, representing data, generating data-matching functions, estimating parameters, validating results.

Participants recognized that for process aspects to be taken seriously in mathematics curricula a lessening of "classic" mathematical content is likely to be necessary — given a limited amount of time at the disposal of mathematics instruction. We should be honest but not apologetic about this. Rather we should insist on the value of the content associated with the inclusion of problem solving, modelling and applications in mathematics curricula.

Some subgroups discussed what mathematics content might be left out to make room for problem solving, modelling and applications. The answer to this is different in different countries. In the USA, for instance: complicated manipulations and calculations. In Australia: Euclidean geometry, a suggestion which was vigorously opposed by other participants to whom geometry is an area par excellence for problem solving.

Finally, miscellaneous different points were made by subgroups. They include:

- The question of how realistic problems, and applications and modelling situations should be was discussed in several subgroups. It was agreed that at the elementary level most problems, but not necessarily all, should be extracted from children's everyday life and immediate surroundings. However, imaginary and artificial situations may have educational value as well if they can be made interesting to the children. At the secondary and tertiary levels it seems that the appeal to students of a particular problem, application and modelling situation is mainly a question of personal variables, one of which might be gender.

For tertiary level students the scope of problem solving and modelling may be widened by involving students in problems that occur in industry. This approach (which is being practiced in some places) teaches students to ask questions and pose problems, to discuss their work with others, verbalize their ideas etc., in addition to solving the problems formulated.

- Problem solving, applications and modelling would be facilitated if cases requiring a minimum knowledge of substance-related theory are chosen. An example of this is Linear Programming problems.

- The teaching of many mathematical topics (e.g. differential equations) would gain if their historical origins in modelling situations were taken into consideration more often than is the case.

- Are there differences, for a given educational level, between what should be taught in problem solving, modelling and applications to all students and what should be taught to only a minority of more resourceful students?

Resources

It was generally recognized amongst participants that a large variety of *source materials* — mainly books, but also computer programs, video-films etc. — are now available at all educational levels. This does not imply, however, that all these materials are widely known, but they can be disseminated into actual instruction through key persons closer to the front line of development than the average teacher. There is still a great need for source materials, in particular flexible ones which can be modified for use in specific instructional situations. And as pointed out earlier in this report there is a great need for *methods* which make it possible for teachers to find or generate suitable problem solving, modelling and applicational situations themselves. Anyway, it is now no longer the lack of materials that prevents the adoption of problem solving, modelling and applications activities in mathematics instruction.

Still, the *teacher* is the crucial resource for work in the field. Therefore emphasis should be put on encouraging and preparing teachers, in their pre-service education as well as in in-service courses, to teach problem solving, modelling and applications.

However, it was a general concern that the traditional components and instruments of mathematics curricula (such as *syllabus* and *assessment*) should be exploited to foster work in the field rather than being viewed as only providing obstacles. To this end a lot of research and development work is needed.

Session 4.: Fundamental and Practical Obstacles

The obstacles to giving problem solving, modelling and applications a reasonable position in mathematical instruction, and how to overcome them, was the theme of the last session. Subgroups were invited to focus their discussions on formulating policy statements and recommendations, if possible. Subgroups worked on this for 60 minutes. The session, and the entire Theme Group programme, finished with a 30-minute *plenary session*, chaired by *Mogens Niss*, in which subgroups reported on their work.

In the discussion it became clear that most of the important obstacles materialize in one obstacle, mentioned again and again in subgroup reports: *lack of time* — in the curriculum, for students, for teachers, for examination authorities etc.. There are, of course, obstacles which cannot be subsumed within this category. This is the case with psychological and cultural barriers, such as preconceived notions of what mathematics is all about, educational backgrounds and traditions, feelings of uneasiness and insecurity towards unknown or blurred situations, lack of civil courage etc..

In many respects the obstacles are, of course, *real*, both from an objective and a subjective point of view. There are external and internal pressures on courses, exerted by authorities, by other institutions, by receivers of course graduates further on in the educational system or outside it. There are class sizes, students' commitment (problem solving, modelling and applications are more demanding than traditional courses), and financial constraints.

That the obstacles are real does not imply that they are a matter of *principle*. It was a generally held opinion amongst participants that most of the obstacles encountered are rather of a *practical*, sometimes even of an imaginary, nature. Attempts to overcome them are therefore more likely to be successful if they operate on a practical level also ("demonstratio ad oculos") and not on a theoretical level only. In this connection it was suggested that separate courses in problem solving, modelling and applications might serve to open a gateway for an inclusion of such activities in mathematics courses proper, thus exercising a pressure on the general mathematics curriculum.

The session finished by the presentation of a few additional policy statements;

- Problem solving, modelling and applications should be made part of the mathematics curriculum for *any* pupil or student in the educational system from school to university.
- Leadership in research, curriculum development and teacher training is needed to make this happen to a larger extent than is presently the case.

Concluding Remarks

It seems natural to conclude this report by looking a little at the development which has taken place in the field since ICME 5 in 1984.

In many respects the situation in ICME 6 was not very different from that in ICME 5. Many issues, concerns and points made were the same. So was the considerable amount of uniformity in views held by participants working at different levels of the educational system. Also in both congresses a large proportion of those attending the relevant Theme Group activities had not many personal experiences with problem solving, modelling and applications work. They attended the Theme Group in order to obtain information about current developments in and contributions to the field. This fact tends to focus subgroup discussions on fairly basic issues, whereas progress is displayed in presentations and papers.

The relative stability in the field over the last half decade is probably a reflection of the trend described in the joint T3/T6 survey lecture by Werner Blum and Mogens Niss (cfr. (5)) that the rate of innovation at the frontier of problem solving, modelling and applications seems to be decreasing, concurrently with (and related to) a reduction of the distance between the frontier and the mainstream of mathematics instruction. That distance is, however, still considerable. Thus research is awaiting a larger number of ordinary curricula to implement problem solving, modelling and applications, so that research on general practice may be carried out. There is reason to believe that this is going to be a major object of research in the years to come. The results of such research are likely to be points of major interest in future ICMEs.

Yet there are things which have changed since ICME 5. Of course, the fact that the distance between the frontier and the mainstream is being reduced is in itself an indication of wider inclusion of problem solving, modelling and applications in mathematics instruction at various levels. But it is possible further to detect changes in emphasis. The different components in the theme, "Problem Solving", "Applications", "Modelling", have come closer together: they are becoming increasingly unified. In ICME 5 the main emphases were on "Problem Solving" (then T7), and on "Applications" (then T6). In ICME 6 "Modelling" was much more the focus of attention. There is reason to predict that this trend will gain much more momentum in the years up till ICME 7 in 1992.

References:

The following list mainly contains books and articles published after ICME 5, 1984. For earlier references see the proceedings of ICME 5 ((8)).

[1] J. S. Berry – D. N. Burghes – I. D. Huntley – D. J. G. James – A. O. Moscardini (eds.): Teaching and Applying Mathematical Modelling (Proceedings of ICTMA 1), 1984, Chichester (Ellis Horwood)

[2] J. S. Berry – D. N. Burghes – I. D. Huntley – D. J. G. James – A. O. Moscardini (eds.): Mathematical Modelling Methodology, Models and Micros (Proceedings of ICTMA 2), 1986, Chichester (Ellis Horwood)

[3] W. Blum: Anwendungsorientierter Mathematikunterricht in der didaktischen Diskussion, Mathematische Semesterberichte, 1985, Band 32, Heft 2, pp 195–232.

[4] W. Blum – J.S. Berry – I.D. Huntley (eds): Applications and Modelling in Learning and Teaching Mathematics (Proceedings of ICTMA 3). To appear 1988/89, Chichester (Ellis Horwood)

[5] W. Blum – M. Niss: Mathematical Problem Solving, Modelling, Applications and links to Other Subjects. State, trends and issues in mathematics instruction (Joint survey lecture of ICME 6, Theme Groups 3 and 6) To appear in several versions

[6] W. Blum – M. Niss – I. D. Huntley (eds.): Modelling, Applications and Applied Problem Solving: Teaching Mathematics in a Real Context. To appear 1989, Chichester (Ellis Horwood)

[7] H. Burkhardt – S. Groves – A. H. Schoenfeld – K. Stacey (eds): Problem Solving — A World View. Proceedings of Problem Solving Theme Group ICME 5, 1985, Nottingham (The Shell Centre for Mathematical Education, University of Nottingham)

[8] M. Carss (ed.): Proceedings of the Fifth International Congress on Mathematical Education (ICME 5) 1985, Basel; Boston; Stuttgart (Birkhäuser). Reports on "Applications and Modelling" (T6), pp 197–221, on "Problem Solving" (T7), pp 212–226.

[9] Contemporary Applied Mathematics series 1987, Providence (Rhode Island) (Janson Publications)

[10] M. Cross – A. O. Moscardini: Learning the Art of Mathematical Modelling, 1985, Chichester (Ellis Horwood)

[11] H. P. Ginsberg (ed.): The Development of Mathematical Thinking, 1983, New York (Academic Press)

[12] HIMAP modules, 1987, Arlington (MA) (COMAP Inc.)

[13] A. G. Howson – B. Wilson: School Mathematics in the 1990s. ICMI Study Series 1986, Cambridge (Cambridge University Press)

[14] G. Kaiser-Messmer: Anwendungen im Mathematikunterricht, Band 1–2, 1986, Bad Salzdetfurth (Franzbecker)

[15] J. de Lange : Mathematics, Insight and Meaning, 1987, Utrecht (OW and OC, Rijksuniversiteit Utrecht)

[16] L. C. Larson: Problem Solving Through Problems, 1983, New York (Springer Verlag)

[17] M. Niss: Aims and Scope of Applications and Modelling in Mathematics Curricula. Plenary lecture delivered at ICTMA 3, Kassel (1987). To appear in (4)

[18] M. Niss: Applications and modelling in the mathematics curriculum — State and trends. Int.J.Math.Educ.Sci.Techn., 1987, Vol 18, No 4, pp 487–505

[19] M. f. Rubinstein: Tools for Thinking and Problem Solving, 1986, Englewood Cliffs (NJ) (Prentice-Hall)

[20] A. H. Schoenfeld: Mathematical Problem Solving, 1985, Orlando (Florida) (Academic Press)

[21] E. A. Silver (ed.): Teaching and Learning Mathematical Problem Solving: Multiple Research Perspectives, 1985, Hillsdale (NJ) (Lawrence Erlbaumer Assoc.)

[22] UMAP modules. Tools for Teaching. Annual publications, Arlington (MA) (COMAP Inc.)

Supporting survey presentation:

Werner Blum (FRG) and *Mogens Niss* (Denmark) presented a joint survey lecture for Theme Groups T3 and T6. Selected papers of the speakers in these groups are planned to be published in a separate volume.

THEME GROUP 4: EVALUATION AND ASSESSMENT

Chief organizer: **David F. Robitaille** *(Canada)*
Hungarian coordinator: **Júlia Szendrei**
Panel: **Antoine Bodin** *(France),* **Raimondo Bolletta** *(Italy),* **Desmond
Broomes** *(Barbados),* **Toshio Sawada** *(Japan)*

Thirteen sessions were scheduled for Theme Group T4 over a four-day period, and 47 papers were accepted for presentation by scholars from 13 countries. Unfortunately, some papers which had been accepted were not presented because the speakers were unable to attend. Summaries of those papers have not been included. These proceedings contain brief summaries of most of the presentations for Theme Group T4, with the exception of those few where the speaker had not prepared a paper in advance.

The work of the theme group was partitioned into three major subthemes, and a set of papers on each sub-theme was presented on each of the four days scheduled for theme groups at ICME 6. The three sub-themes dealt with Large-Scale Evaluation Projects, Teachers and Teaching Processes, and Testing Practices. It had been hoped that a number of papers dealing with evaluation of teachers would be presented, but no submissions on that topic were received.

A fourth sub-theme dealt with data analysis. This activity was designed to enable participants to gain experience with computerbased techniques for analysis of data using the data bank from the Second International Mathematics Study which was conducted in some 20 countries in the early 1980s by IEA (International Association for the Evaluation of Educational Achievement). The organizers of ICME 6 made available a number of microcomputers for this session, and the workshop leaders brought a number of software packages with them along with the data. Unfortunately, the initial session in this group was poorly attended, and the periods which had been tentatively scheduled on subsequent days to provide participants with time to work on projects of their choosing had to be cancelled.

THEME A: Large-Scale Evaluation Projects
Leaders: Raimondo Bolletta (Italy), Júlia Szendrei (Hungary)

On the first day, the papers in this sub-theme dealt with results from the Second International Mathematics Study. Several such papers had been presented at ICME 5 in Adelaide, Australia where a number of speakers discussed preliminary results from that project. At ICME 6 the authors were able to present conclusions based on more complete analyses of the results of that study.

Identification and Description of Opportunity to Learn and Growth in Achievement: R. G. Wolfe (Canada)

IEA has, for some time, employed the concept of opportunity to learn as an important variable in seeking to account for students' achievement. Professor Wolfe illustrated, with many examples, the difficulties inherent in attempts to describe and analyze relationships between variables concerning opportunity to learn and those concerning growth in achievement from pre-test to post-test as was done in the longitudinal versions of the Second International Mathematics Study.

Such analyses are further complicated by variations among countries in the structure of their school systems; and, for that reason, the analyses presented in this paper were limited to findings from one country and one population of students: 13-year olds in the United States. Results showed that most gains in achievement were on items dealing with lower level competencies and this may be due to the fact that a great deal of the teaching of mathematics at this level focuses, to a large extent, on the development of computational and algorithmic skills, rather than on applications and problem solving.

Participation and Opportunity to Learn as Functions of Structural and Organizational Factors of Educational Systems: E. A. Kifer (USA)

The paper dealt with the effects of early streaming or teaching of students in different levels of courses in mathematics. It was reported that systems which teach students early profoundly affect the opportunities students have. A contrast was drawn between the United States, where students are tracked early, and several other systems where either no tracking occurs, or it occurs much later.

Papers presented on the second, third, and fourth day in this sub-theme were grouped under the general heading of National Initiatives in Education. They dealt, as one might expect, with a fairly wide variety of topics.

The Curriculum of the Scuola Media since 1979: R. Bolletta (Italy)

Raimondo Bolletta presented a description of the national curriculum in mathematics of Italian "scuola media" students between the ages of 13 and 14. A new and innovative curriculum had been proposed for these schools in 1979, but not officially prescribed. Given the innovative character of that proposed program, and the lack of prescription it was considered necessary to conduct a survey to assess the extent to which the program had been implemented. Results showed that, although a large majority of teachers is in agreement with what constitutes a core program in mathematics at this level, they tend to eliminate many topics which they find to be too innovative or difficult for most of their students.

Statistical Indicators of the Condition of Science and Mathematics Education in the United States: R. M. Berry (USA)

Results from a number of recent studies in the United States and internationally have raised serious concerns in the United States about the teaching of science and mathematics. The concerns may be summarized as follows:

1. Achievement of students in mathematics and science is inadequate.
2. Too few students are taking academic courses, particularly in mathematics and the sciences.
3. The curriculum presented to students is inadequate in many respects.
4. The teaching force is inadequately prepared.

Brief mention of several steps which are being taken to ameliorate the present situation was made.

Communication of Evaluation Results to Students and Parents: J. Provost and A. Gagneux (France)

The authors of this paper described a number of ways of presenting test results to students and their parents. They described how, by making use of Bloom's Taxonomy of Educational Objectives in the preparation of tests and teaching units, it is possible to gather a great deal of useful information about the general academic profile of each student.

The Feasibility of a Scheme for National Attainment Targets and a Related National Assessment System: B. Denvir and M. Brown (UK)

The authors presented the results of a 10-month feasibility study on the educational implications of setting attainment targets for 11 year old students in England and Wales. Particular attention was devoted to describing the methods employed for the empirical study on the basis of which some possible targets were formulated, as well as to the necessity of achieving a homogeneous standard at the national level.

Replication of the Second International Mathematics Study in the State of Virginia, USA: E. L. Edwards Jr. (USA)

The paper dealt with the results of a replication of the Second International Mathematics Study in Virginia. The replication demonstrated the feasibility of conducting a large-scale survey in order to compare local results and performance levels to national and international results. In addition, the survey was found to be very useful for purely local purposes.

Replication of the Second International Mathematics Study in the Province of Ontario: W. Beevor and H. Russell (Canada)

Several school districts in the Province of Ontario, Canada have conducted replications of the Second International Mathematics Study and the authors reported on their general findings. It was seen to be a highly useful experience for the school boards involved, and it is predicted that more such surveys will be conducted in the future.

Curriculum-Evaluation Instruments: A. M. Artiaco and E. Guala (Italy)

The authors described a major teaching experiment which was conducted in a large region of Italy with students between the ages of 10 and 14. A number of instruments were specially developed to gather information on the development of students, abilities and didactical processes. Particular attention was given to evaluation of higher-level cognitive processes.

The Graded Assessment in Mathematics Project (GAIM):
M. Brown (UK)

GAIM is a classroom-based scheme for continuous and progressive assessment of students aged 11 to 16. The goal is to record what each student knows, understands, and is able to apply in order to assist teachers. A bank of profile statements or objectives describing 15 levels of difficulty in 6 topic areas has been produced. A pilot project was instituted in September 1988.

Inspecting Mathematics: J. H. Mayhew (UK)

Mayhew reported on the work of Her Majesty's Inspectors in the schools of the United Kingdom. He pointed out some of the benefits which this program has provided to the schools of the country.

APU Monitoring Surveys 1987: Preliminary Results: D. Foxman and G. Ruddock (UK)

The sixth national monitoring surveys of 11- and 15-year olds in England, Wales, and Northern Ireland were carried out in 1987. As in previous surveys, a light sampling technique was used in which each pupil involved took only a fraction of the total assessments used. Modes of assessment relating to new technology and small-group problem solving were included for the first time in addition to the modes utilized in previous surveys. Initial analyses of the age 11 survey data show a similar picture to 1982 in the overall pattern of results, although there have been some shifts in detail.

Findings from the Fourth National Mathematics Assessment in the United States: J. O. Swafford, E. A. Silver, and C. A. Brown (USA)

Results from this survey show that there has been some improvement in students, achievement in mathematics, although most of the progress has been in the domain of lower-order skills. There has also been an increase in participation rates in mathematics courses at the senior secondary school level, and this is taken as a positive sign. Large-scale revision of the content of the mathematics curriculum is seen as an urgent need.

THEME B: Teachers and Teaching Processes
Leaders: Antoine Bodin (France) and Desmond Broomes (Barbados)

The papers in this group dealt with a variety of topics concerning the link between teaching practices and achievement, with classroom processes in mathematics, and with the evaluation of students progress in mathematics. A total of ten papers was presented over the four days.

The Teaching of Selected Concepts and Procedures to 13-Year Olds: D. Robitaille (Canada)

As part of the longitudinal component of the Second International Mathematics Study, teachers were asked to complete questionnaires dealing with the teaching practices they employed in their teaching of mathematics. These classroom process questionnaires were specially developed for the study, and were designed to collect highly specific information on the teaching of a number of topics from the curriculum for 13-year olds in the participating countries. These included some aspects of arithmetic, algebra, and geometry.

Results showed that, in most countries, teachers spend an inordinate amount of time re-teaching material which should have been completed in earlier grades. Moreover, despite the large amounts of time that are devoted to the teaching of algebra at this level in many countries, students' performance levels are low and this raises the question of whether students of this age are capable of learning such abstract content. The situation with respect to the teaching of geometry is extremely complex because of the differences that exist among countries with respect to the geometric content which is included in the curriculum at this level.

Content Representation in Mathematics Instruction: Characteristics, Determinants, and Impact: T. Cooney and V. McKnight (USA)

Using data from the Second International Mathematics Study, the authors studied the classroom practices utilized in the teaching of fractions, and the effects of certain classroom processes and characteristics of teachers on students' achievement. The research strategy employed was to identify three countries in which there were major differences in the profiles of classroom features, and then to compare the relationship which emerged between classroom variables and students' achievement in each of the three countries. The three countries selected were France, New Zealand, and the United States.

Classroom Processes as Predictors of Success in Mathematics: A. R. Taylor (Canada)

A survey was conducted in a large school district in the Province of British Columbia in which a pre-test was employed and the design of the project permitted teachers to be linked to their classes for purposes of analysis. Results showed the importance of the affective domain in curriculum design, and of teaching prospective teachers the fundamentals of classroom organization and methodology. In addition, a number of teaching practices were identified as being particularly effective.

Gender Differences in Students, Attitudes and Performance: E. Kuendiger (Canada)

The paper dealt with an analysis of gender differences in a number of countries which had participated in the Second International Mathematics Study. It was reported that, in several countries, older students were more likely to stereotype mathematics as a male domain than younger students. No link between the development of such attitudes and achievement was detected. The degree of gender stereotyping of mathematics seems to be independent of achievement and of other characteristics of the school system.

Evaluation in Mathematics: The Quality of Students, Knowledge: A. Bodin (France)

In the author's opinion, most of the work on evaluation has shown that evaluation is generally not well done because of a lack of understanding of its proper role and function. Teachers need a lot of specialized training and assistance to help them evaluate better. Some approaches which had been tried under the auspices of the IREM de Besancon were described and discussed.

Causal Models of Concept Development and Skill Acquisition in Measurement and Arithmetic: K. Pedersen and C. Tayeh (USA)

The purpose of the study was to determine causal models for studying patterns of students, learning in measurement and arithmetic. Data was drawn from the results of tests administered as part of the Illinois Inventory of Educational Progress for a random sample of 7200 students. The study resulted in a "best fitting" model for males and in two "best fitting" models for females. Conclusions include gender differences in knowledge of mathematical concepts at Grade 4 and in the effects of performance in measurement and arithmetic at Grade 8.

School-Based Assessment: D. Bain and J. Sunter (UK)

The introduction of the General Certificate of Secondary Education (GCSE) has resulted in a greater emphasis upon the evaluation of coursework. As from 1991, all candidates will receive a substantial proportion of their marks from coursework. School-based approaches to the assessment of coursework have been developed in response to demands emanating from outside as well as within the school. Concerns have been expressed about a number of difficulties which teachers face in attempting to implement such schemes.

Toward Greater Efficiency in the Teaching of Mathematics: J. Porte (France)

The author is a teacher of mathematics at the secondary school level and an educationist associated with the IREM. For several years she has been working on an approach to evaluation which provides students with the tools they need to evaluate their own work. This approach, she feels, would work equally well at the tertiary level.

Issues in Research on Estimation: H. Schoen (USA)

The estimation procedures used by students in Grades 5 through 9 as they responded to computational estimation test items were examined. Interview-based process descriptions were cross-validated using large-group test data. Students demonstrated a strong mental set to round numbers to the nearest leading power of ten even when the items required other estimation processes. Performance differed according to item format, types of numbers and operations in the items, and grade level of students.

Using Professional Judgement in Evaluation M. M. Moustafa (Egypt)

One of the most important professional responsibilities of teachers is that of allocating and reporting marks or grades as measures of students' performance. After assigning marks for different aspects of a students' work, the teacher usually combines them all into one to produce the final mark or grade. Usually, teachers have a predetermined procedure for combining these grades using nominal weights known to both teachers and students.

The author reported results of a study conducted with 45 elementary school teachers in Bahrain who received marks under four testing procedures. Analysis of the results showed that the informed judgment of instructors was a necessary component in the production of valid final measures of students, success or failure.

THEME C: Testing Practices
Leaders: Derek Foxman (UK), Toshio Sawada (Japan)

The papers in this group were partitioned among four major topics. These were Evaluation for Diagnostic Purposes, Evaluation of Problem Solving, Evaluation of Assessment Programs, and Program Evaluation.

Evaluating Students, Performance on Spatial Tasks: J. Izard (Australia)

The author presented the results from a number of studies which have investigated the performance of children aged 10 to 12 years on spatial tasks. In some cases the classroom instructional strategies varied and consideration was given to the influence of instruction on the predictive power of some factor-referenced tests. Other studies involved manipulation of the types of stimulus material in the test instruments and the extent to which manipulation of solids used in the testing situation contributed to success on the test.

An Informal Diagnostic Instrument for Algebra: D. Edge (Canada)

Professor Edge has developed an instrument for use in diagnosing students' understanding of algebra, particularly in the area of ratio and proportion, and he conducted a study in which the main focus was on students' ability to explain their work in writing. About 100 students aged 12 to 14 participated. It was found that students described what they did with varying degrees of success. These ranged from responses such as, "I guessed", to written sentences. However, none described why a particular strategy had been chosen. It was concluded that this approach had some advantages over a paper-and-pencil test, but that it could not replace interviews as a diagnostic tool.

Methods of Diagnostic Assessment at the Primary Level: B. Denvir (UK)

Three studies concerned with the strategies used by 7 to 11 year olds in tackling mathematical tasks were described. The first study involved lower-attaining children and was designed to find a framework for describing their acquisition of concepts, to develop a diagnostic instrument, and to design, carry out, and evaluate a remedial program. The second study investigated the extent to which a group assessment would yield reliable diagnostic information about individuals. A third study described the strategies children used in a practical investigation. The importance of setting tasks in a meaningful context and one which is challenging was stressed.

Assessment of Mathematical Modelling: J. Gillespie (UK)

The session dealt with the development of classroom materials and assessment procedures in the Numeracy through Problem Solving project. The goal was for students to pose, solve, and evaluate their solutions to extended real-life problems. The work has shown that nearly all students can appreciate and employ strategic and technical skill in problem solving. Moreover, students' previous experience of engaging in this work enables them to achieve a higher standard than those without such experience.

Diagnostic Testing Improves Performance for College/University-Bound Students: J. Caballero (USA)

The California State University and the University of California jointly created the Mathematics Diagnostic Testing Project to detect weaknesses in students, college preparatory mathematics. A comprehensive analysis leads to more accurate placement, remediation of students, weaknesses, and modification of curriculum. Several effective techniques for strengthening weak areas have emerged, and studies indicate that there is a high correlation between test scores and performance in the course.

Assessing Investigational Work and Extended Projects:
S. Pirie (UK)

With the advent of the CSE teachers were required to assess investigations of extended projects, where previously they had virtually no involvement in the assessment of their students. There are problems about what to assess in coursework, but the dilemma for the teacher is to reconcile the roles of facilitator of learning and examiner of pupils. The answer appears to lie in the teacher's exercise of confident professionalism, placing assessment firmly in the learning environment. As to grading extended work, mathematical processes and thinking should be the foci of the assessment.

Assessing Problem Solving in Small Groups:
D. Foxman (UK) and L. Joffe (Australia)

Small-group problem solving was included in the 1987 mathematics monitoring survey of the Assessment of Performance Unit in Britain. About 90 groups of 3 pupils, aged 11 or 15, of the same gender and similar attainment participated. Each group took one task administered by a specially trained, experienced teacher. The assessment schedules provided three sets of data ratings of aspects of performance, categorization of specific activities undertaken, and detailed observations. A factoranalysis of the results produced two main factors identified as cognitive and attitudinal. Amount and type of cooperation were task dependent, as were the correlations between assessors' evaluations of the outcomes of the pupils' problem solving and pupils' written test scores.

Assessment of Open-Ended Work in the Secondary School: D. William (UK)

A distinction was drawn between conceptual and procedural knowledge. In order to generate mathematical activity it is necessary to utilize tasks which reduce the effects of procedural knowledge and present them to students so as to engage them. Approaches to assessing such tasks included "cognitive demand" and heuristic-based schemes stemming from the work of Pólya. These latter schemes have tended to regard the cognitive demand of the task as of secondary importance; what is required is a scheme which considers demand with the degree of difficulty of the task. A model was presented which characterizes this degree of difficulty using two factors: the structure of the search space of the problem, and the complexity of the mathematical relationship between the variables.

An International Comparison of Examination Papers in Mathematics at GCE "A" Level and Equivalent: R. Croasdale (UK)

This paper dealt with a comparison of the syllabus content of written examination papers in mathematics at, or near the level of the British General Certificate of Education (CGE) Advanced Level, taken in 20 countries in 1979. Comparisons were made for all the papers on the basis of a broad system of classification. A subset of 12 papers was subjected to a detailed comparison, employing a more refined classification devised specially for this purpose. Cluster Analysis was used to identify groups of papers that exhibited certain similarities.

Widening the Perspective of Program Evaluation: D. Nevo (Israel)

Evaluators must continue to seek evidence about the impact of evaluation on educational programs. However, it is important that evaluations include activities directed toward the assessment of the rationale for the program and its strategy and process of operation. Tools are available for this task. They include the use of observational techniques, content analysis, experts, opinions, and the recently developed methods of meta-analysis for quantitative synthesis of research literature.

Analysis of the Effectiveness of Assessment Instruments: E. Guala and A.M. Artiaco (Italy)

The paper addresses what the authors term a complex problem. In the evaluation of learning processes is it possible to reconcile the requirements connected on the one hand to the research on the learning process and, on the other hand, to the collection of observations and reliable data? Moreover, is it possible to reconcile those requirements with the involvement of teachers and the development of their spirit of professionalism?

Curriculum-Linked Assessment: A Model Based on the Second International Mathematics Study: K. Travers (USA)

The author reviewed the model employed in the Second International Mathematics Study and described how that model had been employed in a number of jurisdictions to replicate the study. A number of features of the model were highlighted.

Evaluation of an Innovative Graduated Assessment Scheme for Low-Attaining Secondary School Students: G.S. Close (UK)

Results of this study showed that a graduated assessment scheme, based on externally provided tests and marking schemes, administered by the pupil,s teacher could be operated by a wide range of specialist mathematics teachers and other teachers. Overall, the evidence suggests that the components of a system of graduated assessment can bring about a substantial improvement in both the mathematics standards and the attitudes and motivation of low-attaining pupils.

Evaluation of a New Program: H. Schuring (Netherlands)

In 1985 a new mathematics program was established for the last two years of pre-university education in the Netherlands: Mathematics A, with emphasis on the applications of mathematics; and Mathematics B, pure mathematics. On the whole the programs have been favorably received. The paper describes the kinds of evaluation techniques that have been used, and the kinds of items that have been included on final examinations for this program.

Supporting survey presentation:

Thomas A. Romberg (USA): Evaluation: a Coat of Many Colors.

THEME GROUP 5: THE PRACTICE OF TEACHING AND RESEARCH IN DIDACTICS

Chief organizer: Nicolas Balacheff (France)
Hungarian coordinator: Sarolta Pálfalvi
Panel: Fred Goffree (Netherlands), Jeremy Kilpatrick (USA), John H. Mason (UK), Anna Sierpinska (Poland), Gerd Walther (FRG)

Introduction

This theme was introduced during the two previous ICMEs with the development of research about the theoretization of mathematics teaching. The particular and fundamental relationships between didactical theories, the experimental work related to them and to the practice of teaching (what is happening in real classroom situations), is precisely the focus of many contemporary approaches of mathematics teaching. These relationships appear to have a two-fold character: on the one hand, the kind of effect and influence research and its results may have on teaching practices, and on the other hand, research itself, which has to take into account practice as an object of investigation.

The survey lecture associated this theme group was organized by its two authors, André Rouchier (France) and Heinz Steinbring, (FRG) in close relationship to these two directions.

- First, from a rather formal point of view, the survey showed that systematic interconnections between research and the practice of teaching are necessary. Consequently, in relation to the problem of mediating mathematics as theoretical knowledge, they analyzed the difficulties of this particular transposition, arguing for the development of a didactical literacy as an integrative culture to help teachers to assimilate theoretical knowledge.

- Secondly, the relations between theory and practice are set up in real situations which occur in teacher training sessions as well as in experimental research.

The relation between the researcher and the teacher is then understood as the product of a theory–practice negotiation.

Then the survey considered basic facts which play a central role in the relations between didactical research and practice of mathematics teaching: research objects, the impossibility of abolishing the fundamental tension and dialectic between theory and practice, the need for specific mediators, the relative autonomy of both actors involved in these relationships...

All these issues underline the work done in the theme group, and in fact were presented to the participants through the initial questions (which we present at the conclusion of the working groups' report). The report will try to give an account of the important contribution to the theme group of all the participants and their commitment in gaining more understanding of the problems offered for discussion. The activity of the theme group was organized to try to allow a fruitful exchange between researchers, teacher educators and teachers with respect to our specific topic. For this purpose two sessions were devoted to discussions in small groups based on the presentation of short communications. A plenary session was then organized to report on the result of the discussions. About 40 short communications were presented. It is not possible here to present them in detail, because each of them contribute to several aspects of our topic. The list of contributors is given at the end of the text. In order to help us to reflect on our activity and its related outcomes, we asked five people to react to them. These reactions were given during the final plenary session and outlines are presented as a conclusion to this report.

Report on the working groups' discussions

We have organized the report on the activity of the working groups around two main topics:

- what can teachers expect from research and how?
- the relationships between teachers, theory, and practice. *What can teachers expect from research and how?*

Let us start from the following remark: the first goal of a teacher is to improve pupils' learning, so the teacher will be chiefly interested in information which can produce an immediate effect on their teaching.

A first category of such results consists of those which allow a better understanding of children learning, and thus enable a coherent interpretation of the behaviour of children, especially where pupil behaviour very often seems incoherent from the point of view of the teacher (at least at first glance). This includes:

- information of all kinds about learning processes, or about the kind of mental image of a given problem which the pupils construct;
- information about pupils' misconceptions, and links with the errors they produce. Of particular importance here is the possibility of an "epistemological theory" which could link the mathematical knowledge and its historical development with pupil responses, and thereby provide a coherent and useful framework to help teachers to understand pupils' behaviour. But this raises

an important problem: *the teacher can be faced with different and sometimes contradictory theories.* This is a normal situation in the field of research, and especially of "human science" research. *How can the teacher cope with such a situation?*

A second category of immediately useful research results for teachers concerns ways of teaching which can facilitate to effective use of the former information in the his/her classroom. The communication of these results appears to be far more difficult than in the case of information about children's cognitive behaviours. Some solutions have been suggested:

- the teacher can be a member of the research team, discussing and collaborating with the researchers to prepare classroom experiments. Such a collaboration can be viewed as one of the most efficient ways to organize professional development for the teachers.

- the teacher can experiment in class with "tools" made available by the researcher (intellectual, or material, including specific classroom organization). But in this case we have to verify that such tools, and the way in which they are presented, can be used without any further help from the researcher, and without specific knowledge about the theoretical basis of the proposed tool.

Whatever the approach, one problem of great importance arises: *that of classroom observation,* which is a question for both the researcher and the teacher who intends to use the classroom setting proposed by researchers. It appears that the failure to establish a satisfactory standard method of observation in classrooms is one of the main reasons why didactical research has not been notably successful. Several different methods of classroom observation were presented and discussed: ethnographic research, observation related to an *a priori* theoretical analysis, and methods for the analysis of transcripts, or of video tapes. As a means of provoking fruitful discussions of the principles of classroom observation, a workshop was organized within Working Group 4: participants worked in subgroups at a variety of mathematical activities. Attached to each group was an "observer" who watched how the group responded to the activity and to the person who presented the activity and who was acting as a teacher. The observers were then invited to tell the group what had been "observed", and in turn, the group was able to respond to these observations. Discussions showed that, during the workshop activity, many of the issues related to observations, the way in which learners work together, and the relations between learners and the teacher emerged in a way very similar to that of school situation.

A fundamental problem arises which concerns especially the case in which the teacher uses "tools" or results proposed by research without a sound knowledge of the background of research: *how to deal with the conceptions or implicit theories (theory in action) of the teacher which might interfere with the researcher's proposals?*

A key issue, in relation to this question, appears to be that of teacher training, including pre-service teacher training. Beyond the differences between national and even local "paradigms" of didactical research and theoretical knowledge, there seemed to be a common starting point for implementing didactical theory in teacher training courses: it was the context of practice, whether in the form of pre-planning for an actual mathematics classroom, or in the form of "theoretical" reflection on data related to mathematics classes. From our discussion about this problem there emerged a great need for mutual exchanges of such ideas, methods, and theories. Theoretical knowledge and theory-fragments (local theories) should be seen both as *objects* of study and as tools for helping the teacher or student-teacher to work in a professional way.

The relationships between teachers, theory, and practice

The survey lecture pointed out that "*theory* is in the field of *reflection* and follows special rules of development" whereas "*practice* is in the domain of *action*, hence it contains its own laws for functioning". While support for untested theory is often sought *via* empirical investigation, research often generates or extends theory.

Research findings reach the learning situation indirectly, by way of the teacher. This may occur through the teacher's interpretation of a syllabus, curriculum, or textbook; or it may happen by teacher-directed use of other didactic materials which result from research. Research findings and the elaboration of theory may come to the teacher's attention by a number of routes.

It has been demonstrated that teachers rarely turn to research journals or academics for information about developments in the learning–teaching process. Instead, they refer to the syllabus, textbooks, teaching journals, "consultants," or colleagues. It was believed that articles published in research journals were generally too technical and unrelated to applications–researchers writing for other researchers, but not for teachers. Research articles published in teaching journals, on the other hand, although more palatable for teachers, were far too rare. Hence, many teachers relied on untested methods and materials with which they felt comfortable rather that seeking other, generally tested approaches, except when necessary.

Other contributions described research projects conducted jointly by researchers and teachers. As well as resulting in the production of curriculum materials, these preservice/inservice collaborative activities served to develop teachers professionally and helped researchers to become more aware of the real teaching situation. In each case the researcher-teacher contract negotiated at the outset involved both parties accepting modifications made necessary by the outlook and situation of the other.

One of the American projects focussed on the development of a problem-solving unit, and was conducted by means of a collaborative inservice programme in an elementary school. The research framework, the method of implementation and a three-dimensional "training model" were all derived, step by step, from research findings.

In Italy work was generated by the implementation of new syllabus units on probability and on mathematics and computer science. Here, teachers and re-

searchers (and occasionally mathematicians) collaborated in developmental programmes which eventually resulted in the nationwide distribution of booklets and
other curricular material.

From France we heard an elaboration of how theory is used in a unit on
fractions and decimal numbers, in a teacher-education course. It was explained
specifically how the *theory of didactic situations* is used to support the teaching
process, how the theoretical notion of *obstacle* assists in the process, and how the
didactic contract idea helps the teacher to help the learner.

Finally, in the Scandinavian context, some distinctive features of mathematical
education as *critical education* were illustrated by means of an example, and ways
of handling a theory–practice relationship in this context were demonstrated.

On the question of providing models and theoretical frameworks for the guidance and support of teacher activity, it was argued that such provision depends on
the researcher's efficacy in embedding research in practical classroom activities. In
addition, a distinction was drawn between theory and models.

It appears that the application of research results to classroom practice is
made more difficult in some countries by factors such as isolation, the low general
educational level of teachers and the paucity of research. Some ethnomathematical
projects are carried out. In these, institutionalized mathematics is introduced via
ethnomathematics in such a way that the learners' previous and culturally dependent knowledge is respected. Their existing mathematical concepts are discovered
and coded in terms of their buildings, plantations, hunting, fishing, body paintings,
crafts, myths, and so on, and later decodified and introduced in teaching modules.
This turns out to be a meaningful way in which learners can incorporate formal
basic mathematical concepts, but it also preserves their way of seeing the world.

From discussion in groups, it was generally agreed that didactical theory could
certainly have practical value, especially if researchers and teachers work together.
For one thing, such theory can make the objectives of teaching interventions more
explicit. Furthermore, if a teacher comes across a theory which produces dissonance
with a previously-held belief or practice, the teacher, after reflection, may thereby
be better able to change. Moreover, when a theory supports a teacher's long-
held beliefs or practices, the teacher's self-confidence can improve, and a greater
willingness to try new strategies can follow.

Research results and didactic theory can be communicated to teachers in a
number of ways. A "linking teacher", or consultant, can be a useful vehicle for
dissemination of information, provided, of course, that the person involved *is* aware
of such theory and research. Researchers could report *directly* to groups of teachers.
More articles, written in teacher-friendly language and emphasizing applications,
should appear in teachers' journals. It was believed that, whatever the mode of
presentation, theoretical material and research findings should always be advanced
in a way calculated to *inspire*, and that one should make an attempt to weave
specific material into the broader fabric of which it is a part.

One speaker emphasized that there is a need for pure research as well as the more common applied variety, stressing that Dewey recognized this sixty years ago. Building on this comes "practical" research, one form of which is so-called "action" research, which involves both researchers and teachers working collaboratively. Indeed, it was generally agreed that research and theory could well make optimum input into the improvement of teaching by way of this sort of activity.

Here, the French notion of the *research contract* is important. The relationship between the two parties must be a dynamic one. The teachers, for their part, can help the researchers to be more conscious of the problems because the teachers may be more realistic than the researchers when it comes to classroom practice. On the other hand, the researchers can assist the teachers in the construction of the theoretical frameworks, etc. Once a comfortable working relationship has been attained, it is essential that the teachers have an active role in the research work.

Conclusion

Organizing the activity of the working groups, we offered the participants three sets of questions related to our topic. These questions were intended to structure the discussion and to focus the short communications. The discussions reported above contain tentative answers, which should actually be seen as possible new questions for future work on the relationships between research and practice in the field of mathematics teaching.

Set 1

What is the role of a didactical theory? Has it any practical value?

There is a consensus that the answer is YES, provided that teachers and researchers work together.

What could a teacher gain by knowing didactical theories or research methodologies?

Which theories produce research results that have any value for real practice? How are they useful for teachers in their everyday classroom activities?

On the whole, participants were *not* prepared to identify particular theories. The identification of possible different theories even appears to be quite difficult, a fact which reflects the problems of communication. Beyond that, external constraints on teachers (e.g. national constraints) are likely to play an important role in the effectiveness of the use of theories in everyday classroom activities.

Set 2

How can research results, and their theoretical backgrounds, be made understandable for teachers? How can research results become of practical use?

What are the possible means of dissemination of research results?

Set 3

How are the relations between research and practice institutionally organized in the different countries (depending perhaps on school levels)?

How are research results taken into account in pre-service and in-service teacher-training?

What are the means and the organization of research results in different countries?

This set of questions has been left almost unanswered, mainly due to lack of time available for our activity. Nevertheless we report it here, thinking that these questions must be considered in the future.

For further discussions: addresses from the reactors

Paolo Boero (Italy)

Why are so many teachers so suspicious and skeptical about didactical research? I think that the analysis here was not sufficient to explain it; I think that we should also keep in mind the effects of:

– teachers' experience of many "research" products in the last 20 years (modern mathematics, taxonomies, general curriculum theories, perhaps constructivism...)

– teachers' confusion about the differences between general educational research, developmental psychology, didactical research...

– teachers' involvement in a context of school traditions, parents' constraints, personal interests in the traditional way of teaching and learning ("I speak, you listen, you study... I am in control": an *economical* process for pupils *and* teachers!)

– lack of visible, widespread positive consequences of learning results while adopting "manners of thinking" and behaviours derived from didactical research.

We should also answer the following question (which remained unstated in the discussions): why are so many researchers in the field of mathematical education so suspicious about a *unified* perspective of didactical research? It seems to me that some motivations are:

– lack of an established methodology and key-word system

– reference to other, well-established fields of research (developmental psychology, epistemology, sociology of education...), also due to career constraints

– national traditions and needs

– lack of contact with, or excessive involvement in, the school system

– unilateral, unintegrated development of research: coming from the traditional conceptions of mathematical education ("teacher who transfers mathematical knowledge into the heads of pupils"). We are exposed to the danger of *separate* approaches (research of developmental psychology, focussing on the pupil; research of sociology of education, focussing on the teacher; research into the history and/or epistemology of mathematics, concerning mathematical knowledge), which in fact contradict the reasons for the existence of didactical research.

About the questions: "Who needs the results of didactical research? And what results are needed?" I agree, as a reactor, with many things which have been pointed out, concerning, in general terms, a better awareness of didactical processes. On the contrary, *in the present situation* I don't agree with the perspective of a *large-scale* involvement of teachers. I think that, *before that*, we need to create an extensive, international community with some common frame of reference and key word system, and we need to involve project designers and teachers who experiment with projects, "innovators", and so on. Indeed, the crucial problem is, in my opinion, how to overcome the manner of functioning of the innovative process synthesized in the following scheme:

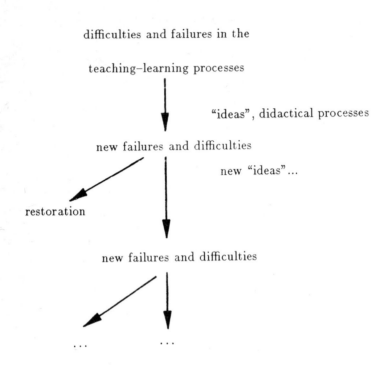

According to my experience, perhaps, one manner of breaking this frustrating process is to realize innovative "prototypes" theoretically framed in an explicit way.

Guy Brousseau (France)

   We were invited to discuss theories, their role, their use, their tractability and the related institutional structures. The proposed questions for theme group T5 allowed us to speak about all kinds of theories without real distinctions. This is the rule of the game. But the answers depend on the type of theory which we are considering, and more precisely on what we think a theory is. Our discussion made some progress with the question of knowing what constitutes a didactical theory, or rather, which type of theorization is related to a given practice and what role is given to the person responsible for the communication of knowledge.

   We began to understand that a didactical theory is a theory about the relationships between the person responsible for the teaching, and the knowledge to be transmitted. There are some difficulties with this idea because, in my opinion, it encounters two pairs of epistemological obstacles. I would like to propose here a systematic analysis as is normal in didactics: we characterize the objects by their function within a system.

Pair A

A1: "Mirror Theory": the "theoreticians" borrow the "spontaneous" concepts of teachers. They try to rationalize them, they "sacralize" them by means of experimental manipulation, and then return them to the teachers as "scientific" knowledge.

A2: "Supermarket Theories", the teacher is considered as a consumer who "buys" theories from mathematics, psychology, epistemology, sociology, etc. The theoretician proposes "results" but leaves to the teacher the responsibility for making a synthesis and for deciding about possible contradictory conclusions.

Pair B

B1: The fusion of research and teaching. Some researchers seem to be willing to fuse theory into practice and vice-versa, as is the case with action-research.

B2: A complete separation of the relationships between research and practice.

These four ways (A1, A2, B1, B2) of dismissing both the problems of theorization, and of serious examination of didactical practice appeared in our debate as various methods of intimidation of speakers, and as objections that we have to eliminate: especially the tyranny of the obligation for immediate usefulness, and the requirement for naive compatibility with any external theory.

We have too many *a priori* requirements which are presented as justification for not doing anything serious. Perhaps because our community is too small, we try to escape our responsibility by transferring some of it to the pupil, the teacher, knowledge itself, or to other sciences.

We have too strong a tendency to be happy with vague ideas, to look for new fashionable ideas as idols which we then destroy without retaining anything.

Our studies are too local, they do not give an account of the complexity of the phenomena we study (as in ecology, or in economics). Our studies are still too superficial. There are very few studies which *a posteriori* allow our community to reconsider the same facts. We very seldom find *a priori* analysis: fundamental place for theoretical analysis, and key means for communication with practice.

We must approach classroom phenomena respectfully and with humility, in the manner that psychology teaches us about the concerns of the child. The object of our study is the didactical situation. Most work attempts to take as the object of study either pupils or their knowledge. In this way the researchers lose or miss the very question they research.

Our theoretical work aims to ensure the consistency of the instruments of our study. Teachers can recognize the relevance and the usefulness of a theory, but not necessarily its validity: we have responsibility for this question. Finally I would like to remind you that the purpose of didactical research is the study of what is specific to mathematics in the relationship between a pupil and his/her milieu.

All didactical study is a study of the functioning of a piece of mathematical knowledge. Very often this epistemological dimension is missing in work that is done.

In conclusion: what is the use of a theory? It aims at recognizing, describing, explaining a given class of phenomena by means of a not too large set of concepts. It also aims at designing elements of "didactical engineering" (avoiding absurd and unrealistic attempts), and classifying, describing, and predicting their effects. So a theory allows one to organize the relationships between several parts of the didactical social organization.

Jeremy Kilpatrick (USA)

My reaction to the work of the theme group as it relates to my own perspective can be summarized in the following points:

– There has been a striking increase in collaborations between teachers and researchers since ICME 5, and such collaborations should now be taken as objects of study.

– In the same way that theory can be seen as a tool for researchers — making their work more powerful — research can be seen as a tool for teachers.

– The practical value of didactical research is not so much in its results as in its constructs. Efforts to summarize what research says on an issue should be supplanted by efforts to bring teachers into a professional conversation.

– There may be some problems ahead in our eager embrace of qualitative research methodologies. Although attractive to both teachers and researchers today, such methodologies may demand much more technical skill then they have received from mathematics educators.

– The widely accepted view of children as constructors of their own mathematical knowledge has not been accompanied by a similar view of the teacher's research knowledge. Constructionism has its merits, but the extreme view that "learning is not a matter of collecting but a matter of constructing" should be moderated to "learning is both a matter of collecting and a matter of constructing".

– If teaching is a science, a strong grasp of theory and research ought to be essential to its effective practice from the beginning. If teaching is an art, both theory and research are perhaps less essential until one has reached a level of mature practice. I think teaching is both. Therefore some creative synthesis of these views is needed.

Stieg Mellin-Olsen (Norway)

I myself feel at home with the tendency to choose a position closer to the practice of teaching aspect, thinking in terms of developmental research. From this position I am very happy with all those reports about *collaboration* and *cooperation* between researchers and teachers. The concept of research contract is helpful, because I see it as vital that the teacher relates to the researcher in a similar way to that in which the student relates to the teacher: as one whose knowledge is appreciated, not so much on an external level of motivation, but really as a dynamic source for development. But, in this cooperation between teacher and researcher, there are a few snags and I think we have to elaborate the problems of cooperation before we suggest new methods of innovation.

I have learned from the psychology of organization, in which "model-strong" and "model-weak" participants in cooperative projects have been distinguished. Model-strong persons are often leaders who have a high status in contrast with model-weak participants, who mostly have a low status and want to be just one among the other participants. The latter sometimes may have important experiences and suggestions for innovation, but are not always heard because they have to be presented in terms of the strong model. In the end this might lead to resistance, because no-one wants to be cooperative in an innovation without being allowed to be creative. Model-strong leaders in industry can have a negative influence on development. Model-strong researchers may meet similar resistance in education.

We can learn from the anthropology of medicine as well. Take staff-meetings in a hospital, in which doctors are model-strong and nurses (and other caretakers) are mostly model-weak. It happens now that nurses have been involved in two different kinds of "discourse". The staff-meeting discourse, in which doctors dominate with their theories and concepts, and the (what I call) corridor-discourse, in which particular practice and patients are discussed. Of course problems occur when these two discourses meet together. I leave the parallel between this medical situation and the researchers in classroom situations to the imagination of the audience.

Back to mathematics education, and the cooperation between researchers and teachers.

In several working group sessions I found the "resistance" issue reflected. In my point of view the cooperation between researchers and teachers, the discourse and the possible resistance have to be investigated in depth as an object of study.
Anna Sierpinska, (Poland)

One question for this theme group concerns the difficulties of communicating research outcomes to teachers and the possibilities of their misinterpretation. It was remarked via one of the working groups that researchers have a tendency to write their papers for other researchers rather than for teachers.

In this congress we ourselves have experienced genuine difficulties of communication on several levels.

One level, the most explicit one, was the level of the natural language — the language in which we presented our papers or ideas and in which we discussed. This was English. But, as English is not the mother tongue for all of us, interventions were difficult both to make and to understand, especially when the paper was read. Moreover, people for whom English is a foreign language, sometimes feel too shy to take part in a discussion although they may have interesting things to communicate. They are not heard in the congress. And therefore, our idea of what the general opinion in a working group was can be biased by this fact.

Another level of communication difficulties is the level of educational terminology. This can be more serious. Translating a term from one language to another may not really give the idea of what is in the speaker's mind. By changing the language we often change the research paradigm. Paradigms differ from one country to another and this makes the communication extremely difficult.

This difficulty concerns the vocabulary related to a local system of education. Out of this context these words are meaningless or mean something quite different.

One of my Polish colleagues here, after having listened to a report on an American project designed from the point of view of teacher training said that she understood nothing: "It was just words and round sentences, no example was given to hang on to." Another colleague, from Mexico, probably fairly well acquainted with this kind of project, mentioned the report as a particularly interesting one, containing useful ideas and methodological tools.

At the plenary session an English gentleman asked for an explanation of this French idea of "didactical contract". Fortunately, the author himself was there to explain. How many other people in other sessions simply hadn't had the courage to ask?

"Didactical contract" is a fairly specialized term. "Research" is more common. Yet, do we all understand it the same way? Translate it into your own language. What does it denote then?

In referring to the kind of research that is being done in France, the late Mme Krygowska would use the French word "recherche" and not the Polish word for it. "Recherche" for her meant research that focuses more on "what is there" in mathematics teaching rather than on "what should be?". It meant research aimed at constructing some knowledge on how an individual learns mathematics, what obstacles he/she has to overcome to develop his/her mathematical knowledge, what are the mechanisms of mathematics classroom functioning, what phenomena does it involve, what are its paradoxes, etc. This knowledge has to be founded on a sound theoretical basis which has to be developed specifically for these purposes and not just borrowed from other fields like statistics, psychology and pedagogy. It is not the task of this kind of research to establish norms and ideals from which recipes for the practice of teaching could be deduced.

Some time ago, before I saw Nicolas Balacheff in this panel, I thought that it is even the ambition of research French style not to be too close to practice — to underline its pure scientific character.

In other countries, people called mathematics educators (in their own language, of course) and working in mathematics departments may be expected to express opinions, establish norms, curricula; to design teaching material, write textbooks, train teachers. And, if they want to do fundamental research and be appointed for that, they are invited to apply for a position in a pedagogical or psychological institute.

"Research" does not mean the same for everyone. Doesn't the same apply to "didactical theory"?

It may well mean:

− a set of principles that the teacher should observe;

– or a set of methods of teaching a particular topic: what kind of introductory exercises and stimulating questions, what definitions, what theorems, what exercises for developing skills, what problems, what applications, etc. What do French educationalists mean by a "didactical theory"? I have also heard at the T5 survey lecture, of a *"real* didactical theory". What is this?

So, how do we want to be understood by teachers if we do not even understand each other?

I would like to end on an optimistic note:

It is true that our understanding of words and our points of view may differ a lot. But we have something in common: we all probably like mathematics and are interested in teaching and learning problems. Why don't we spend more time on doing mathematics and educational research together in these congresses instead of giving incomprehensible talks? Here, in Budapest, I have taken part in a workshop organized by Daphne Kerslake where we were doing exactly this. I must say I enjoyed it immensely and was sorry it was so short.

Contributors

Working group 1: On the relations between teachers and research Chair: M. Cooper (Australia)

R. Berthelot (France), M. Cooper – M. Kumar (Australia), G. J. Cuevas (USA), R. P. Hunting (Australia), M. Reggiani (Italy)

Working group 2: Theory–practice relationships. Chair: M. Fasano (Italy)

T. Cooper – R. Smith (Australia), M. A. Farrell (USA), M. Fasano (Italy), O. Skovsmose (Denmark)

Working group 3: Research and teacher training. Chair: G. Walther (FRG)

D. Clarke – C. Lovitt (Australia), J. A. Dossey (USA), P. Lanier (USA), G. Lappan (USA), J. Weber (France)

Working group 4: Observation of teaching practice. Chair: D. Kerslake (UK)

J. Adda (France), R. Dekker (Netherland), M. Eberhard (France), D. Kerslake (UK)

Working group 5: Learning processes and teaching design (1). Chair: G. Arsac (France)

C. Chasiotis (Greece), G. Ervynck (Belgium), J. M. Fortuny (Spain), K. Hasemann (FRG), H. Mansfield (Australia), R. S. Millman (USA)

Working group 6: Learning processes and teaching design (2). Chair: J. Confrey (USA)

J. Ainley (UK), T. H. F. Brissenden (UK), R. Douady (France), L. C. Hart (USA), R. Hershkowitz – S. Vinner (Israel), S. Maury (France)

Working group 7: Research–practice relationships: the case of algebra. Chair: E. Filloy (Mexico)

J. B. Dubriel (USA), E. Filloy (Mexico), D. Miller (USA), D. R. S. Moncur (UK), H. Okamori (Japan)

Working group 8-1: Research–practice relationships, some case studies. Chair: J. Mason (UK)

K. Schultz (USA), S. Yanagimoto (Japan)

Working group 8-2: Teaching mathematics in developing communities. Chair: S. Pálfalvi (Hungary)

E. S. Ferreira (Brazil), J. Suffolk (Zambia)

Supporting survey presentation:

Andrè Rouchier (France) and *Heinz Steinbring* (FRG): The Practice of Teaching and Research on Didactics.

THEME GROUP 6: MATHEMATICS AND OTHER SUBJECTS

Chief organizer: *Werner Blum (FRG)*
Hungarian coordinator: *Anna Racsmány*
Panel: *Ubiratan D'Ambrosio (Brazil), Rudolf Bkouche (France), David N. Burghes (UK), Dilip K. Sinha (India)*

1. Background of the work of the group

1.1 Mathematics and the real world

From its very beginnings, mathematics has been both the most *esoteric* and the most *practical* of human creations. There have been and there are close *relations between mathematics and everyday life, the world around us and other sciences.* Problems in the real world have inspired and stimulated the development of mathematical concepts and theories, and theoretical achievements in mathematics have contributed essentially to solve practical problems.

In the last few decades an enormous *extension* of applicable mathematical topics as well as of disciplines related to mathematics has taken place. Many sciences such as biology, economics and sociology have become more and more *mathematized* (see e.g. Pollak 1979,1988 or Jaffee 1984). This is also and especially due to the rapid development in the field of *computer science.*

As is well-known, there are many simplified models for the complex *interrelations between mathematics and the real world* (for a synopsis, see Kaiser–Messmer 1986). By "real world" we mean the "rest of the world" outside mathematics, i.e. everyday life, the world around us, other disciplines and especially *other school or university subjects.* We choose the following diagram (taken from Blum 1985) as a concise illustration:

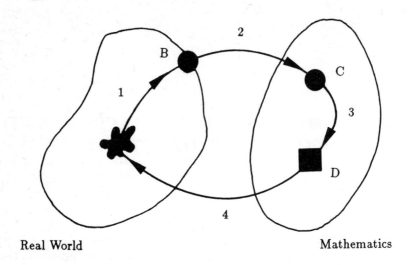

A) Real problem situation   1, Specifying, idealizing, structuring
B) Real model   2, Mathematizing
C) Mathematical model   3, Working mathematically
D) Mathematical results   4, Interpreting, validating

The starting point is a "real problem", i.e. a situation in the real world with some open questions. This situation has to be simplified, idealized, structured and made more precise by the "problem solver" according to his/her interests. This leads to a "real model" of the original situation. The real model has to be mathematized, i.e. its data, concepts, relations, conditions and assumptions are to be translated into mathematics. Thus, a "mathematical model" of the original situation results. Then, by working within mathematics, certain mathematical results are obtained. They have to be re-translated into the real world, i.e. to be interpreted in the original situation. In doing so, the problem solver has also to validate the model, i.e. to establish whether he/she can use it for his/her purposes. When validating the model, discrepancies of various kinds can occur which lead to a modification of the model or to its replacement by a new one, i.e. the problem solving process may require going round the loop in the diagram several times. Sometimes, however, even several attempts do not lead to usable result.

The use of this *model conception* of the relationship between mathematics and the real world, especially between mathematics and other subjects, is often very helpful for an adequate solution of a given applied problem. There are, however, also disadvantages. For, by strictly separating mathematics from the rest of the world, inseparable links in content — as they have grown up in many centuries especially between mathematics and physics — are examined in a merely formal manner, i.e., artificial distances between a real situation and its mathematical description are created, e.g. in the case of natural laws.

Besides such complex processes there are also abbreviated and restricted links between mathematics and the real world, especially other subjects: on the one hand a *direct application* of already developed mathematics to real situations with mathematical content, on the other hand a "dressing up" of purely mathematical problems in the words of another subject or of everyday life; such *"word problems"* often give a distorted or falsified picture of reality (which is sometimes done deliberately for didactical purposes).

1.2. Mathematics and other subjects at school and university

At *school and university*, relations between mathematics and the real world, especially between mathematics and other subjects, have mostly played an important role. Here, the lines of development have not taken a straight but rather a "wavelike" course, i.e. there have been phases where extramathematical applications in mathematics instruction or mathematics in the teaching of other subjects were strongly taken into consideration and phases where mathematics was more isolated from other disciplines. In recent years a worldwide trend towards a stronger (re-)emphasizing of applications and links to other subjects as well as an extension of the range of application fields in school and university teaching of mathematics can be observed (cf. Burkhardt 1983, Niss 1987, and Blum/Niss 1989).

When dealing with relations between mathematics and other subjects at school or university, we can distinguish between different aspects (see Niss 1981 and Blum/Niss 1989): Firstly, mathematics instruction may essentially serve two different *purposes*:

(1) to provide learners with knowledge and abilities concerning mathematics as a subject,

(2) to provide learners with knowledge and abilities concerning other subjects to which mathematics is to offer some services.

Secondly, the *organizational framework* of mathematics instruction may take two different shapes:

(a) mathematics may be taught as a separate subject,

(b) mathematics may be taught as a part of and integrated within other subjects.

Thirdly, we have different *educational histories*:

(A) Mathematics in school offering general education, viz. at the primary, lower secondary, and upper secondary level,

(B) mathematics in vocational education,

(C) mathematics in university courses for future mathematicians or mathematics teachers,

(D) mathematics as a service subject in university courses for future scientists, engineers, economists etc.

Now the situation can be illustrated by the following matrix:

organization	purpose	
	(1)	(2)
(a)	examples: (A), (C)	examples: (D), (B), partly (A)
(b)	examples:	examples: (B), partly (D)
	integrated curricula	

In all cells of this matrix, *relations between mathematics and other subjects* may play a role. Also in (a1), examples taken from other subjects may be used for various purposes (see section 3.1). When dealing with (a2), (b1) and (b2), it seems quite natural to include applications from other subjects in mathematics instruction. However, one can sometimes find a "division of labour", both in (a2) and even — on a much smaller scale — in (b2) such that separate mathematics courses devoid of applications are given in order to teach once and for all the mathematical concepts, methods and results needed in the subject being served. In section 3.1 of this report, arguments will be given which will call this approach into question.

Possible relations shown in this matrix also include truly *integrated* curricula (second row), both in school and in university, whereby teaching and learning is taking place in an interdisciplinary and cross-subject way. There are, however, only very few materials and there is even less everyday teaching in this sense.

1.3. The topic of Theme Group T6

Theme Group T6 was dealing with all questions and problems concerning the *relationship between mathematics and other subjects at school and university*, embedded in the more general framework of relating mathematics with the real world as developed in section 1.1. The work of the group concentrated on the *relations* involved, especially on the *role of other subjects for mathematics instruction* and the *role of mathematics for other subjects*, at all levels of the educational system, with particular reference to *mathematics as a service subject*. With respect to the matrix constructed in section 1.2, all four cells were considered, provided that it was possible to distinguish segments of instruction with mathematics as an explicit object of attention. Only such instances of (b1) or (b2) with mathematics totally integrated within other subjects were excluded.

The topics of the group were divided into *two main areas*, I and II, one more "*theoretical*" and one more "*practical*". Area I comprised the following *theoretical and basic aspects*:

I.1. *Historical, epistemological, and methodological aspects* of the relationship between mathematics (instruction) and other subjects.

I.2. *Empirical investigations* into the learning and teaching of mathematics in connection with other subjects.

I.3. The *role of computers* in this field.

I.4. *Recent developments in applications in practice* and their relevance for mathematics instruction.

Area II comprised *examples, materials, and projects* linking mathematics and other subjects, for all educational levels:

II.1. For the *primary* level (5–10).

II.2. For the *junior secondary* level (10–16).

II.3. For the *senior secondary* level (16–19), including vocational education.

II.4. For the *tertiary* level (19$^+$), with special reference to mathematics as a service subject.

To each of these eight topics, an *organizer* had been assigned beforehand:

I.1: Ubiratan D'Ambrosio (Brazil),

I.2: Gabriele Kaiser-Messmer (FRG),

I.3: Rolf Biehler (FRG),

I.4: Dilip Sinha (India),

II.1: Alan Rogerson (Australia),

II.2: David Burghes (UK),

II.3: Rudolf Bkouche (France),

II.4: Dilip Sinha (India).

1.4. Survey of the work of Theme Group T6

In the *first session* W. Blum gave an introduction to the theme of the group and a survey of the programme. Then U. D'Ambrosio introduced the more theoretical area I of the group (see section 1.3), and A. Rogerson gave a short presentation on this area, dealing with some basic questions in an interdisciplinary curriculum project for schools, the "Mathematics in Society Project" (see Rogerson 1986). Then D. Burghes introduced the more practical area II, and Roger Jean (Canada) gave a short presentation on this area, concerning mathematics as a service subject for biologists (see Jean 1987).

The *second session* was devoted to area I and was divided up into four sub-groups according to the points I.1 to I.4 mentioned in section 1.3. After a short introduction given by the respective organizer, in each subgroup there were several short presentations as well as intense discussions. For details see chapter 2 of this report.

The *third session* was devoted to area II and was structured in the same way, according to the points II.1 to II.4. For details see chapter 3.

In the *fourth session* U. D'Ambrosio and D. Burghes summarized the activities of the various subgroups of the second and the third sessions, and this was followed by discussions. Then Jon Ogborn (UK) presented some general reflections on the relations of mathematics and other subjects, especially the sciences, exemplified by the use of computers in data analysis. Finally W. Blum looked back on the topic of the group as well as forwards towards essential activities and research areas for the future. For details see chapter 4.

Some selected papers presented during the sessions of Theme Group T6 will be published in the joint *Proceedings* of Group T6 and Theme Group T3 on "Problem Solving, Modelling and Applications" (Blum/Niss/Huntley 1989).

2. Area I: Theoretical and basic aspects

The following four sections of chapter 2 refer to the subgroups from the second session (see section 1.4). In order to make this report more concrete we will describe one interesting presentation in each subgroup in some more detail.

2.1. Historical, epistemological, and methodological aspects

In his *introductory remarks*, the subgroup organizer, U. D'Ambrosio, emphasized an historical approach to mathematics, recalling that at the outset mathematics was a part of the intellectual efforts of mankind to understand themselves, their environment and their relationship with nature and among one another, as well as to decipher the numerous mysteries posed by nature, by its phenomena and by the universe as a whole. He argued that mankind was driven into understanding and explaining reality, coping with it, managing it, and gaining from it. For this understanding certain techniques were developed, according to diverse cultural contexts. In the course of history, some of these techniques disappeared, some survived or became even stronger, among those being a mode of rational thinking called "mathematics", which includes measuring, counting, classifying, ordering, inferring etc. In summing up, D'Ambrosio stressed that we are therefore also facing an historico-epistemological problem when we discuss the relations of mathematics and other subjects, and interdisciplinarity.

Then three *invited speakers* presented their papers: Günter Ossimitz (Austria) on theoretical mathematical models in economic and management sciences, Jeff Evans (UK) on statistics and the problem of induction, and Shmuel Avital (Israel) on mathematics and cultural values.

As an *example*, we refer here to Ossimitz' contribution. The presenter pointed out that mathematics in economic and management sciences is mostly descriptive mathematics or elementary arithmetic. He emphasized that the fundamental act of mathematization in this field is measurement, through which qualitative stuff is transformed into quantitative structures. In his final thesis he argued that the relation of mathematics to economics is comparable to that of chemistry to medicine.

The *discussion* in this subgroup focussed on the idea of recovering the intimate relations between mathematics and other subjects by recovering the humanistic values of mathematics. Here, the idea of ethnomathematics was also brought forward.

2.2. Empirical investigations

In an *introductory survey* of the state-of-the-art concerning empirical research on the learning and teaching of mathematics in connection with other subjects, given by the organizer G. Kaiser-Messmer, it was pointed out that there are different strands of research, e.g. isolated investigations restricted to quantitative-statistical methods, applied problem solving research, research on curriculum projects based on a theoretical background and closely linked with classroom experiences, and personal reports on certain school or university courses.

The *invited speakers* in this subgroup were Barbara Binns and John Gillespie (UK) on experiences with the "Numeracy Through Problem Solving Project" at the University of Nottingham, Jan de Lange (Netherlands) on a curriculum project on mathematics for the life and social sciences for the upper secondary level, and Christopher Ormell (UK) on research on "application readiness" in mathematics of lower secondary pupils.

As an *example*, we choose J. de Lange's presentation. The author reported on the experiments that eventually led to the introduction of a new curriculum for upper secondary students aiming at a study of the social and life sciences in the Netherlands (cf. de Lange 1987). This curriculum uses the real world as a starting point for extracting mathematical concepts ("conceptual mathematization"). Teachers' and students' reactions were discussed which showed especially a need for many teachers to change their attitudes. In addition, the problem of process oriented assessment was mentioned.

The *discussion* accented three items in particular. Firstly, there is still a considerable gap between everyday school practice and the educational debate on applications to other subjects, but the gap has become reduced during the last few years. Secondly, most students and teachers respond positively to examples taken from other subjects, provided that they are challenging and fit in the syllabus. Thirdly, the ability to link mathematics with other subjects is not at all easy for students and demands special instructional phases.

2.3. The role of computers

For many years now computers have been bursting more and more into many areas of society, including the educational system, and also into mathematics instruction at school and university. The use of computers as a tool, as a means for doing numerical or algebraic calculations or for drawing, as an aid for creating new teaching methods, has implications also for the learning and teaching of mathematics in connection with other subjects (see Blum/Niss 1989).

For example, more complex applied problems with more realistic data become accessible earlier and more easily, or problems which are too demanding can be simulated numerically or graphically. As to goals, routine calculatory skills are becoming more and more devalued and abilities such as modelling, applying or experimenting are becoming revalued upwards. With regard to contents, new topics which are particularly close to applications in other subject areas can be treated more easily now, e.g. data analysis at the upper secondary level or dynamic systems at the tertiary level. Computers entail, however, also many kinds of problems and risks, e.g. the devaluation of routine skills will make mathematics instruction more demanding for all students and too demanding for some of them, for linking mathematics to other subjects is an ambitious activity, with or without computers. And teaching and learning may become even more remote from real life than before, because real life now may only enter the classroom via a computer; simulations may replace real experiments.

In his *introductory remarks* the subgroup organizer, R. Biehler, concentrated on the use of software, especially for modelling and for simulating systems. He reported on promising experiences with using such software in the classroom under certain favourable circumstances, and he also stressed the need for more and deeper empirical investigation in this field. Further, he accentuated the role of computers for deepening the student's understanding of the model conception of the relation between the real world and mathematics.

Four *invited speakers* presented recently developed software and reflections on their possibilities. Jon Ogborn's (UK) lecture focussed on his micro-computer modelling systems DMS and CMS for secondary schools. Two Hungarian Colleagues presented ideas for restructuring the mathematics and the science curriculum in Hungary by using computers for simulation and games, with the emphasis on statistical models. R. Biehler discussed how far computer supported analysis of real data can be helpful for developing an adequate concept of probability.

J. Ogborn *for example* showed that the use of computers gives hope to reducing some technical problems with mathematics, e.g. in postponing analytical methods in favour of discrete methods. Modelling tools may allow a more flexible change and extension of initial models and a numerical and graphical exploration of the consequences of models from different viewpoints. Examples from modelling growth, traffic flow, or atmospheric energy transfer processes were given. So computational modelling serves here as a link between mathematics and science.

In the *discussion*, criteria for appropriate modelling tools were addressed. The need for still using paper and pencil in many situations instead of computers was stressed unanimously. Eventually, the idea of using a "virtual computer" in the classroom was discussed, i.e. to choose approaches to topics and styles in learning and teaching which take into consideration the existence of powerful computers.

2.4 Recent developments in applications in practice

The *introductory talk* by the organizer, D. Sinha, centred essentially around the broad range of recent practical applications of mathematics. He mentioned examples from the physical sciences as well as from engineering, biology, ecology, psychology, communication, and linguistics. Explicitly he spoke about qualitative studies in developmental biology which have led to catastrophe theory and to bifurcation theory, about chaos theory resulting from atmospheric science studies and about fractals in connection with the oceanic sciences. Reference was made to the problem of nonlinearity in physical, social or biological phenomena. An essential aspect of this talk was the role and impact of the new technologies in the practical use of mathematics, e.g. in engineering, medicine or economics.

Furthermore, the speaker reflected on the educational relevance of these newer application areas. He discussed possibilities and strategies for making these areas accessible to learners at the tertiary or even at the secondary level. The idea of joint instruction by a mathematics and a non-mathematics teacher was called to mind and strongly recommended. Sinha regarded such collaborative and interactive ventures as a necessity. Again, he stressed the crucial role of computers, now as a tool for getting examples of recent applications percolated down to the instructional level. He finished his presentation by pleading for an inclusion of such new examples in mathematics curricula in order to cope with the demands of the changing socio-economic and cultural context.

There were three *invited speakers*, R. Jean on recent developments in bio-mathematics and H. Khare together with B. Bawerjee (India) on a survey of some recent examples in applied fields.

In the *discussion*, the participants consented to three recommendations. First-ly, some "leading examples" taken from those newer areas of applied mathematics should be worked out in detail. Secondly, strategies and modalities should be iden-tified to stimulate cooperation and collaboration between mathematics teachers at university and school and their colleagues in other subjects. Thirdly, mathematics teachers should be encouraged to keep uptodate with respect to recent develop-ments in applied fields and to incorporate newer examples into their teaching.

3. Area II: Examples, materials, and projects

The sections 2 to 5 of chapter 3 refer to the subgroups from the third session (see section 1.4). Again, for each subgroup one interesting presentation will be described in more detail. Section 1 deals with some common aspects of this chapter.

3.1. Arguments in favour of applications to other subjects in mathematics instruction

There are many arguments in favour of references to reality in mathematics instruction, especially of connections with other subject areas, mostly agreed among all participants. We briefly refer to five kinds (cf. e.g. Blum 1985 or Niss 1988).

"Pragmatic" arguments: Learners should be taught how to use mathematics to describe special situations taken from other subjects, to understand them better and to cope with them better. This can only be done by dealing with certain applied examples in mathematics instruction.

"Methodological" arguments: Learners should acquire "meta-knowledge" and general capabilities and strategies for applying mathematics. They should learn how to translate between the real world and mathematics, they should reflect on methods of application, and they should come to know possibilities and limitations of the application of mathematics, which includes a critical appreciation of the use or misuse of mathematics. All this can only be achieved by incorporating suitable applied examples into mathematics teaching.

"Formal" arguments: Learners should be taught general "formal" abilities (such as argumentation or problem solving) and attitudes (such as an openness towards problem situations), which can be done also (but not only) by means of examples taken from other subjects.

"Scientific theory" arguments: Learners require a balanced picture of mathe-matics as a total cultural and social phenomenon, to which *inter alia* references to other subject areas also belong.

"Learning psychology" arguments: Suitable applied examples (as well as suitable purely mathematical examples) can motivate or illustrate mathematical content, can serve in the structuring of larger mathematical subject ranges, can contribute towards better understanding and longer retaining of mathematical topics and can improve the learner's attitude towards mathematics.

These arguments are based on certain *educational aims* which are implicitly contained in the arguments: pragmatic, methodological, or formal aims, aims based on scientific theory or on cultural history. Such aims have different relevance according to different educational histories. From these aims, more *mathematics-oriented aims* can be inferred such as:

- learners should be taught how to handle mathematics in a well-founded and rational manner, particularly with regard to problems in other subjects;
- learners should acquire adequate basic ideas, related to the real world, and basic conceptions with regard to the essential mathematical concepts, methods and results.

There are also some arguments *against* references to reality in mathematics instruction, based on certain obstacles and barriers (see section 4.1). When weighing the arguments and counter-arguments against each other in group discussions, the result was a strong plea for including applications in mathematics instruction.

An important question in many discussions was: where to find *examples* for applications, suitable for teaching? Two very useful resources for materials and literature, relevant to the subject, are the survey articles by Pollak (1979) and by Bell (1983). Further, the extensive bibliography by Kaiser et al. (1982, with a supplement to appear in 1988) should be mentioned. Many references to current curriculum projects as well as to interesting individual contributions can be found in Blum/Niss (1989).

Many more examples for links between mathematics and other subjects were presented in the four subgroups during the third session. The range of application fields comprised non-traditional ones like architecture, art, biology, computing, environment, finance, language, music, and politics. Many of these examples incorporate the use of computers to a substantial degree, e.g. by simulations, spread sheets, or symbolic algebra. We are now going to report on those subgroups.

3.2. The primary level

Invited speakers were Morten Anker (USA) on architectural mathematics, Drora Booth (Australia) on spontaneous pattern painting and Piero del Sedime (Italy) on mathematics and social conditions. The organizer, A. Rogerson, gave an introduction to the topic of the subgroup by stressing the particular advantages of the primary level for linking mathematics with other subjects and especially for integrated curricula, for project work etc.. He also presented the "Mathematics in Society Project" in connection with integrated curricula.

An interesting *example* was M. Anker's presentation. He reported on experiences with children in an "architectural math lab for cubic city planning". Here, primary level pupils designed and built a "children's city" with houses, people, cars etc. in miniature. In doing so, they were inspired to use mathematics creatively as a tool for exploring, describing, and reconstructing their environment. In this project, mathematics was brought together with architecture, art, and social studies.

3.3. The lower secondary level

There were nine *invited speakers*: Emma Castelnuovo (Italy) gave examples of connections between elementary geometry and reality. Ikutaro Morikawa (Japan) showed some real world illustrations of geometric topics. Andrew Begg (New Zealand) discussed possibilities for combining mathematics with Maori language and culture. Maria-Cristina Zambujo (Portugal) gave examples for linking mathematics with biology, ecology, geography, history and languages within a project studying several aspects of a local river, by essentially using computers. Hans-Wolfgang Henn (FRG) presented reflections on and examples for analysing real data. Bruno Vitale (Italy) reflected on the exploration of the space of informatics and the realm of open mathematics. Laurie Aragon (USA) explained some teacher training materials in mathematics applications. David Hobbs (UK) presented the "Enterprising Mathematics" project, a contextual course for the 14–16 year old. Finally Kumiko Adachi (Japan) gave an example for the integration of mathematics with music, design, science and crafts.

From among the various interesting presentations we refer as an *example* to Henn's in more detail. The speaker considered two cases, the measurement of a single value and the investigation of the functional interrelationship of two measured values. His main aim was to show how students should be taught to handle numbers critically in such situations. After specifying his notion of exactness he gave examples of a reasonable calculation of mean values in applied situations and of adequate methods of linearizing given pairs of numbers resulting from measurements of real data, among those a dropping ball or the decay of beer foam.

3.4. The upper secondary level

Seven *invited speakers* gave short presentations: Paul Bungartz (FRG) showed how to use recent real applications in teaching probability. B. Chaudhuri (India) reported on a study on the interaction between languages in the teaching of mathematics and informatics. Solomon Garfunkel (USA) presented the "High School Mathematics and its Applications Project". John Goebel (USA) spoke about an American curriculum project for high schools, with emphasis on applications. Yvette Horain (France) talked about some teaching experiences at the upper secondary level. Bernard Parzysz (France) showed how he teaches solid geometry through shadow problems. Finally Mary Rouncefield (UK) explained a project on the use of statistics in other subjects such as biology, geography, psychology, sociology and economics.

As one of many interesting *examples* we consider Parzysz' talk. His starting point was the "vicious circle" which results from the fact that studying spatial geometry at school requires the drawing of plane projections and vice versa. To break this circle, the author uses shadows cast by an electric light bulb and by the sun. He presented concrete materials and examples used in the classroom, among others cubes, "Dürer's window" and boards.

3.5. The tertiary level

There were seven *invited speakers*: Michel Helfgott (Peru) presented his approach for teaching differential equations to students of science and engineering. Anthony Briginshaw (UK) reported on mathematical language as an information transfer mechanism. Ruth Hubbard (Australia) spoke about incorporating mathematical reading and study skills into mathematics service courses. Megan Clark (New Zealand) analysed factors affecting the flow of students into mathematics, science and technical training. Eric Muller (Canada) raised and discussed some important issues related to service courses in mathematics, starting from experiences in Canada. R. Jean broadened and concretized his presentation given at the first session (see section 1.4). Finally Arno Jaeger (FRG) reported on experiences with a new approach to teaching linear algebra and optimization for beginning students of business administration.

Jean's talk was one of several interesting *examples* for the teaching of mathematics as a service subject, which was the central topic of this subgroup (cf. also Howson et al. 1988 and Clements et al. 1988). Based on aims such as "to instill in students the ability to use the mathematical approach in biological situations", the author pleaded for the so-called "integrated method", where mathematics is taught through biological subject matter in contexts relevant to the undergraduate biology programmes. He gave some examples, taken from genetics or the growth of populations, from the theory of predation or from animal behaviour.

4. Problems and prospects

In many discussions throughout the work of the Theme Group, barriers and obstacles to the linking of mathematics with other subjects were identified. We will briefly refer to some of them in section 1. In section 2 we will enumerate some important activities for the future.

4.1. Obstacles to applications in mathematics teaching

In spite of all the good arguments in favour of applications to other subjects in mathematics teaching (cf. section 3.1), such relations often still do not play as important a role as one would wish in "mainstream" mathematics instruction at school and university. This is due to certain obstacles (well-known amongst mathematics educators for a long time), among others the following (cf. Blum/Niss 1989):

Obstacles from the point of view of instruction: Many mathematics teachers are afraid of not having enough time to deal with applications in addition to the compulsory mathematics material. Some teachers doubt whether relations to other subjects belong to mathematics instruction at all because they would disturb the clarity, the purity, the beauty and the universality of mathematics.

Obstacles from the learner's point of view: Mathematical routine calculations which can be solved by merely following some recipes are more popular with many students than applications, because applications make the mathematics lesson more demanding and less predictable.

Obstacles from the teacher's point of view: Applications also make instruction more demanding for teachers and more difficult to assess. Very often teachers simply do not know enough examples, or they do not have enough time to up-date examples, to adapt them to the actual class and to prepare them in detail.

Participants of the Theme Group agreed that the obstacles related especially to learners and teachers are really serious, but that in the light of the arguments given in section 3.1 mathematics teachers and educators should continue to make every effort to *overcome* these obstacles, especially by an adequate pre-service and in-service teacher education or by stimulating every kind of contact, or rather cooperation, between mathematics teachers at school and university and their colleagues in other subjects. And both teachers and educators should *insist* that *applications* to other subjects *become and remain an essential part* of mathematics instruction, even in the face of what has been mentioned, e.g. that instruction becomes in fact more demanding for students and teachers.

4.2 Future activities

The participants of the Theme Group agreed that the following, among other things, is necessary for all levels of instruction:

1) *To develop more* concrete *"local" examples* for relations of mathematics to other subjects, suitable for teaching. Such examples should be more suited to a motivation or an illustration of certain mathematical topics or to a description and a better understanding of special problem situations taken from other subjects.

2) *To develop more* concrete *"global" examples and project materials* for relations of mathematics to other subjects, suitable for teaching. Such examples should be more suited to developing general abilities such as translating between the real world and mathematics or developing adequate attitudes such as an openness towards problem situations.

3) *To devise more examples and conceptions for the use of computers* in mathematics instruction, with special respect to connections between mathematics and other subjects.

4) *To gain more experiences* regarding both successes and failures in the teaching and learning of mathematics in connection with other subjects, and *to establish* the broadest possible *opportunity to exchange* these experiences.

5) *To carry out more* controlled *empirical investigations* concerning mathematics instruction with respect to relations to other subjects. For instance: what could the actual effects of applications to other subjects be? How do learners react to that? What are the possibilities and risks of the use of computers in an "application-oriented" mathematics instruction?

6) *To connect* more closely *practical teaching experiences* on the one hand and *basic theoretical questions* (such as methodological and epistemological aspects concerning the interaction between mathematics and other subjects, or questions concerning the educational aims of instruction) on the other hand, in both directions.

7) *To embed all reflections and activities* in mathematics education concerning the relation of mathematics to other subjects *into a "theory of mathematics education"*.

All these activities are research activities in a broad sense. But certainly the most important thing to do is still:

8) *To intensify the efforts to integrate applications to other subjects into "standard" everyday mathematics teaching*, by means of curricula, of textbooks and materials for learners and teachers, by pre-service and especially by in-service teacher training.

References

Bell, M. S.: Materials Available Worldwide for Teaching Applications of Mathematics at School Level. *Proceedings of the Fourth International Congress on Mathematical Education* (Ed.: M. Zweng et al.), Boston 1983, 252–267.

Blum, W.: Anwendungsorientierter Mathematikunterricht in der didaktischen Diskussion.*Mathematische Semesterberichte* 32 (1985) 2, 195–232.

Blum, W. et al. (Ed.): *Applications and Modelling in Learning and Teaching Mathematics.* Chichester 1988.

Blum, W.–Niss, M.: Mathematical Problem Solving, Modelling, Applications, and Links to Other Subjects: State, Trends and Issues in Mathematics Instruction. In: Blum–Niss–Huntley (1989).

Blum, W.–Niss, M.–Huntley, I. (Ed.): *Applications, Modelling and Applied Problem Solving: Teaching Mathematics in a Real Context.* Chichester 1989.

Burkhardt, H. (Ed.): *An International Review of Applications in School Mathematics.* Ohio 1983.

Clements, R. R. et al. (Ed.): *Selected Papers on the Teaching of Mathematics as a Service Subject.* Berlin/Heidelberg/New York 1988.

Howson, A. G. et al. (Ed.): *Mathematics as a Service Subject.* Cambridge 1988.

Jaffee, A.: Ordering the Universe: The Role of Mathematics. In: *SIAM Review* 26(1984)4, 473–502.

Jean, R. (Ed.): *Une Approche Mathématique de la Biologie.* Montreal 1987.

Kaiser, G.–Blum, W.–Schober, M.: *Dokumentation ausgewählter Literatur zum anwendungsorientierten Mathematikunterricht.* Karlsruhe 1982 (Addendum 1988).

Kaiser-Messmer, G.: *Anwendungen im Mathematikunterricht (Vol. I/II).* Bad Salzdetfurth 1986.

DeLange, J.: *Mathematics — Insight and Meaning.* Utrecht 1987.

Niss, M.: Goals as a Reflection of the Needs of Society. In: *Studies in Mathematics Education, Vol. 2* (Ed.: R. Morris), Paris 1981.

Niss, M.: Applications and Modelling in the Mathematics Curriculum — State and Trends. *International Journal for Mathematical Education in Science and Technology* 18 (1987) 4, 487–505.

Niss, M.: Aims and Scope of Applications and Modelling in Mathematics Curricula. In: Blum et al. (1988).

Pollak, H. O.: The Interaction between Mathematics and Other School Subjects. In: *New Trends in Mathematics Teaching, Vol. IV* (Ed.: UNESCO), Paris 1979, 232–248.

Pollak, H. O.: Recent Applications of Mathematics and Their Relevance for Teaching. In: Blum et al. (1988).

Rogerson, A.: The Mathematics in Society Project: A New Conception of Mathematics. *International Journal for Mathematical Education in Science and Technology* 17 (1986) 5. 611-616

Supporting survey presentation:

Werner Blum (FRG) and *Mogens Niss* (Denmark) presented a joint survey lecture for Theme Groups T3 and T6. Selected papers of the speakers in these groups are planned to be published in a separate volume.

THEME GROUP 7: CURRICULUM TOWARDS THE YEAR 2000

Chief organizer: John Malone (Australia)
Hungarian coordinator: János Urbán
Panel: Hugh Burkhardt (UK), George Eshiwani (Kenya), Victor Firsov (USSR), William Higginson (Canada), Christine Keitel (FRG), Richard Phillips (UK)

Few educators who are involved daily with the mathematics curriculum of their homeland would claim that the syllabi with which they work are completely satisfactory. It is for this reason that curriculum change was chosen as a Principal Theme of this Congress just as it was at the last. Now, four years beyond ICME 5 and as we approach the last decade of the 20th Century, the mathematics education community is desperate to plan effective and efficient educational experiences that will prepare today's students for life in the 21st Century. The mathematics educator's role in this task has been made especially urgent and difficult because of a host of mathematically relevant cultural changes that have occurred which have altered how the future generation will learn, work and live.

This Theme Group was concerned with "what", "why" and "how" curriculum changes in mathematics may best be achieved. It was organized into four subgroups, each of which focussed on a different important aspect of the field. These subgroups themselves divided into smaller working groups for some of the sessions.

Each of these groups will contribute a report to a separate fuller publication on this Theme, which will also contain the full texts of some of the contributed papers and a Theme Survey. Details of this publication may be obtained from the Chief Organizer: John Malone, SMEC, Curtin University, Box U 1987, Perth 6001, Western Australia.

This summary of the work of the Theme Group begins with a brief Overview of the challenges facing mathematics curricula around the world. This is followed by reports of the discussions of each of the subgroups. Finally, there is a summary of the Survey Lecture.

Two of the subgroups focussed on the "what" and the "why" of curriculum change. The largest, with well over one hundred participants, took a broad view under the title

"The mathematics curriculum in the Year 2000 and the changing character of social demands on, and need for, mathematics"

Another subgroup explored one specific force for change:

"The impact of computer technology on the mathematics curriculum"

The other two subgroups focussed on the "how" of curriculum change — mechanisms for agreeing and for implementing change, and how they might be improved. This is an area that has not received much explicit attention in the past. Again, one subgroup tackled

"The dynamics of planned curriculum change"

It took a "systems" approach to the whole problem, while the last group explored one key aspect

"The teacher as a critical agent in curriculum change and implementation"

An Overview

Critics will argue that past attempts to predict future trends in mathematics education have been based, at best, on longitudinal studies and other research results that have merely "ploughed old ground" and offered curriculum innovators little help in operating through the current period of rapid change, during which the very definition of success is altering constantly. Is this a true representation of the inevitable relationship between educational research and curriculum development initiatives, or can a closer, more constructive relationship be established and sustained? What new developments in mathematics itself have, or will have, an influence on future mathematics curricula?

Because it seems that the main lesson offered by the history of mathematics education over the past 25 years is the confirmation that educational programs respond to cultural forces, and not vice versa, mathematics educators are compelled to look beyond their own profession to broader cultural and social concerns. Equally, we must strive to identify those technologies that will influence future cultures, and mathematics itself. How do we go about this? Does this mean that there is a need for a major overhaul of the content of the curriculum?

The increasing reliance of societies on information processing and transfer is changing the ways in which mathematics is used; particularly in response to the impact of the computer and the calculator, it would seem inevitable that the curriculum for all ages must change profoundly. How and how fast remains to be explored and established. How should these changes be introduced? Is it a case for the reform of syllabus content only, or are other changes needed in the pattern of teaching and learning to ensure that mathematics as a vehicle for observing, understanding and influencing life is available to all?

A review of the historical evidence makes it clear that there is no established method of planned curriculum change. Goals and outcomes are usually far apart, when the latter are looked at in detail. What are the distorting factors, the barriers that prevent change and the "levers" that promote it? What might we do about this? Can we learn anything from the study of local versus national control of the curriculum operating in various countries? Modifying any complex system in a systematic way must be associated with comprehensive and detailed feedback to ensure the full, effective cooperation of all those involved in implementing the change. Pressure and proven support are complementary essentials but how, in each situation, can they best be designed and deployed?

In particular, what is the role of the teacher as change-agent in the school curriculum? To what extent does the teacher drive the curriculum or does the reverse situation operate? What exemplars are there for us to study? Curriculum development which has as its goal the improvement of mathematics learning and teaching in schools, must be concerned with factors that constrain the opportunity to learn, including matters of teacher quality and preparation. How can we identify these factors and manipulate them to our advantage?

There are, of course, many other major issues that are influential on the direction, style and emphasis of the mathematics curriculum throughout the world over the next ten years or so. Provision of an appropriate mathematical education for the less academic secondary-school student has become a key issue in the face of relatively high youth unemployment and a desire by governments to encourage potential school-leavers to stay on at school as long as possible. The effects of the "knowledge explosion" on the content of the mathematics curriculum; reductions in the time allocations for mathematics, due to the changing nature of schooling; the potential of parents for promoting a child's mathematical education; the need for relevant research into mathematics learning and teaching to inform curriculum design and decision making; curriculum factors mitigating against females; reform of that vehicle for curriculum change — the textbook — so that it becomes readable, engaging and capable of being understood by students without constant teacher interpretation — all of these are issues deserving of consideration and discussion.

This Theme Group served as a focus for discussion of such issues.

The Mathematics Curriculum in the Year 2000 and the changing character of social demands on, and need for, Mathematics

Leader: Christine Keitel (West Berlin)

Presenters: Abraham Tesfai (Ethiopia), Luciana Bazzini (Italy), Lawrence Shirley (New Guinea), Yves Chevallard (France), Anthony Ralston (USA), George Schoemaker (Netherlands), Wu-Yi Hsiang (USA), Norma Presmeg (South Africa), Frances Curcio (USA), Hirotoshi Hirahata (Japan), Joseph Fishman (USA), Ann Chisko (USA), Bob Davis (USA)

There is growing pressure being applied to introduce various topics into the mathematics curriculum to help prepare students for life in the 21st Century. In several of the socialist countries we observe a swing to-and-fro between a "practical" and a "scientific" orientation of subject matter, with an attempt to use this swing in the cause of dialectical advancement. With respect to developing countries, it is obvious that a more fundamental understanding of the question of content is a crucial prerequisite for appropriate application. From the Western industrialized countries we know that an up-to-date review and a critical analysis of the following questions should provide a fresh approach to the complex matter of content:

Demands and Qualifications

What skills, knowledge, metaknowledge and qualifications in mathematics will be required in the next 10 to 15 years and beyond?

What are the associated needs and demands and who defines them?

How should we match needs? Should we teach directly what is needed? Do we need more or less mathematics?

Is there a common core of needs and responses for everybody?

Is there a common core of needs and responses independent of historical, social and cultural changes?

What can historical and empirical studies on the role and function of mathematics in specific areas of social life tell us?

Implicit Mathematics in Society: Mathematics "frozen" in techniques, tools, instruments, conventions in specific social fields, such as the economy.

What are the topics that link "implicit mathematics" in social life and explicit mathematical applications? Can they be developed and taught?

What are the relationships between implicit mathematics and explicit learning? How can explicit mathematics education prepare for situations in which no explicit mathematics is normally required?

This subgroup looked at these questions, and at the curriculum in a broad sense covering concepts and techniques, strategic and tactical skills for their deployment, contexts of application, learning activities and classroom roles. The aim was to make suggestions for future developments, with specific examples. The discussion in each session was built around a set of presentations.

The presentations covered three types of different approaches to the questions.

a) Suggestions and proposals for changes within the existing curricula: some new topics, new emphases and changes in the treatment of mathematical and mathematics application themes.

b) Descriptions and discussions of recently developed new curricula and experiences with them in different countries.

c) General discourses and debates on the purposes of secondary mathematics education and the consequences which should be drawn for mathematics education from the fact that mathematical practice has changed irreversably.

a) The contributions here gave examples and argumentations of what necessarily had to be added or changed in the curriculum: *Presmeg* emphasized the necessity to broaden the curriculum with multi-cultural starting points in order to show the cultural roots of mathematical instruments and tools for mutual understanding; *Fishman* suggested that controversial social (real) problems should be analysed within the maths classroom; new topics as a base for new thinking tools like graphing and statistics should be introduced as early as possible was proposed by *Curcio*, and *Chisko* made suggestions for activities in mathematics education which enables students to develop analytical skills and stress and emphases on theories and high order thinking, as well as critical and reflective positive attitudes towards mathematics.

b) *Bazzini* demonstrated the recent solution done in Italy to integrate the humanistic tradition and orientation with the demands of new technology without loosing the advantages of both. *Shoemaker* described the Dutch curriculum development as a down-to-top approach which is based on a realistic use of mathematics leading to theories and concepts for understanding reality. *Wu-Yi Hsiang* presented the carefully developed new national curriculum in mathematics for junior and senior high school of the Peoples Republic of China with a strong emphasis on the basic principles of mathematics as a way of thinking. *Shirley* described the difficulties and obstacles encountered in poor developing countries when introducing new, mostly borrowed curricula. *Hirahata* proposed a far-reaching curriculum reform using calculators and computers extensively to allow space for more professional and vocational mathematics education. *Ralston* reviewed one specific attempt — the work of the Curriculum Framework's Taskforce of the Mathematical Sciences Education Board of the US National Research Council; the impact of computer science as well as technology on curricula and teacher education was seen as profound.

c) The relationship between social needs or demands and mathematics curricula which have to react to these changing needs, was principally questioned by *Yves Chevallard*. He argued that the more widespread use of mathematics in society which is so often taken as an argument for teaching more explicit mathematics to more pupils has been totally misinterpreted as mathematics in use has become more and more implicit, crystallized and imbedded in all kinds of "objects" or organizational forms of our social life and do not demand explicitly doing mathematics by more than few people in society. *Davis* offered an analysis of the interrelationship between teaching and learning in the classroom and research in mathematics education and its impact on curriculum development. He pleaded to recognize that any curriculum design needs research activities not only on theoretically better grounded psychological micro-level but also on the more socially determined macro-level of teaching and learning mathematics. As the last speaker he also provided an appropriate conclusion to the subgroup's discussion.

Main Points of the Discussions

The common tenor of the discussions was the insight that we generally switched too quickly into new curricula without having analysed deeply what had happened before. Although we unanimously agreed, that we can do better in mathematics education, there were different opinions about the main purpose of mathematics education and the urgent need to follow some fashions in education.

The presentation of the two highly demanding maths curricula for secondary education, a Japanese proposal and the experimental curriculum of PRC contrasted strongly with an American proposal to push into technology as quickly as possible as a solution for all the problems we have so far in mathematics education. The last was criticized by some participants as too superficially grounded fashionable trend which neglects the more important basic principles of mathematics education as were mentioned "training the rational mind of the pupils", "emphasizing the human thinking capacity instead of fostering skills and techniques", "discovering the principle advantages of basic algebraic or geometric thinking as the most important base of any technology".

Especially the implicit character of mathematics in social practice has to be recognized and carefully analysed in discussing new curricula. With respect to the relation between technology and mathematics, there have to be drawn quite different conclusions from this argumentation.

The Impact of Computer Technology on the Mathematics Curriculum

Leader: Richard Phillips (UK)

There are few aspects of our societies that are not being influenced in some way by information technology. Computers and calculators are certainly a powerful force for curriculum change. This subgroup looked in some detail at the implications for the whole mathematics curriculum.

Calculators, computers and similar devices are important for mathematics not only for what they can do in the classroom. The impact of the calculator on everyday mathematics — in commerce, in industry and in the home — has been a quiet but devastatingly successful revolution which has altered the way that almost everybody works with numbers in the many countries where the devices are freely available.

Early fears that calculators would reduce people's understanding of number have proved unjustified and there is evidence that the reverse is true, if they are used well; by reducing the mental load of calculation, our minds seem to be left clearer, giving us a better chance to understand what we are doing.

There is a similar but more complicated situation with computers. Their everyday use at home and at work may not be as common or as obvious as for calculators, but the diversity of human activities that they support is impressive — the drawing office that no longer draws on paper; the clothes manufacturer who can now make optimum use of his cloth; the supermarket that automates its stocktaking from its check-out machines — these are but a few of the myriad of activities of a mathematical character which have been changed forever by computers. There are lesser forces at work as well, within mathematics itself as well as in its applications.

Whether we like it or not, the practice of mathematics has been profoundly changed by computers.

How will this affect the curriculum? Change there has really only just begun; though there are many exciting ideas and possibilities they have mostly still to be fully developed, let alone implemented. In the classroom there is no doubt that information technology can support and enrich the existing curriculum in a valuable way. That is not enough; our objectives will also need to change, enhancing what is important but discarding those goals that are no longer useful. It has also become apparent that the presence of these new devices in the classroom can help teachers change the way they work, supporting activities like mathematical discussion, small-group work and non-routine problem-solving where there was none before.

Many, indeed most, questions remain to be answered decisively, including:

How far will instruction in various mathematical skills, such as "heavy arithmetic" and graph-plotting, become obsolete and how far will it remain important for conceptual understanding?

Is computer science going to change or replace mathematics in certain areas? In what ways will the relative importance of different parts of mathematics change?

What will be the roles of new styles of thinking, such as the algorithmic approach?

The subgroup examined some of these questions and their possible implications for the curriculum. It thus focussed on one aspect of the brief of the first subgroup, and related it to the concerns of Theme Group 2.

As a group of about twenty people from 14 countries, we faced some difficult problems. We wanted to see things clearly and in context. We wanted to know and to understand what was happening in mathematics classrooms in different countries now, as well as looking into the future. We were concerned as much with what "should be" as with what "would be".

We began with an exemplar — *Schoenfeld* (USA) offered us a review of many different ways that computers and calculators can support work with graphs. The relatively simple business of plotting or interpreting a graph automatically gives scope for a wide range of classroom activities. These include games with functions, such as GREEN GLOBS, and graph interpretation work, such as EUREKA.

We then heard statements about what is happening in four countries; Man Sharma talked about the US situation, Fumiyuki Terada about Japan, Brendon Kelly about Canada, and Joe Watson about England. In the discussion that followed we attempted to make crude estimates of current acceptance of information technology in the classrooms of different countries. Among the questions we considered were:

Is the use of calculators in the classroom actively encouraged, tolerated in some lessons, or actually forbidden? For most countries it is clear that there is much more acceptance at secondary than at elementary school.

Is there work with LOGO turtle-geometry? Here the reverse was found, with much more happening at primary level.

Are long division algorithms still taught and practiced? Despite the calculator, this is very common in most countries; someone argued the case for teaching such algorithms at a higher level, perhaps alongside polynomial division.

Is there small-group problem-solving work around a computer?

Is there work where students experiment with very simple programs in BASIC, or a similar language?

Are logarithms still taught as a means of calculation?

One important point that emerged clearly from our discussion is that national policy statements, such as national curricula, make the differences between countries seem greater than they actually are. For example, one country has a national curriculum that virtually excludes calculators from all of its mathematics classrooms; nevertheless, it has a sizeable number of teachers who see their importance and who bring them into the curriculum. Another country with a much more progressive policy has many teachers who resist these changes.

It was perhaps unfortunate that our discussions tended to focus on the more prosperous countries, largely because of the balance of the group. Poorer countries are unlikely to have computers in schools, but the use of the four-function calculator is widespread across the world, and its implications for the mathematics curriculum are global and profound.

For our last two meetings we worked in two groups. One focussed on the new generation of powerful calculators with graphical or symbolic capabilities — such as the HP28S, Sharp EI 5200, Casio 7500 G. These hand-held devices are much cheaper than computers and yet share many of their features — one member had used a calculator graphic display to do "turtle geometry" and to display fractal patterns! However,there was a general feeling that the computer, with its larger screen and larger keyboard, justified its larger price tag.

The curriculum implications of these powerful calculators were discussed in some depth; the group concluded that:

They demand changes in how we treat most topics in mathematics, and in the order in which they are taught. For example, the exponential is now much more accessible, and may be tackled earlier.

The place of algebra must be re-examined. Routine manipulation becomes much less important, with other topics such as linear algebra and the modelling aspects more accessible and central.

Proof may be of greater importance. Generating examples and results from a machine rather than by hand calculation increases the need and the opportunity for more rigorous mathematical argument, once we have acquired a good intuitive understanding through investigation.

Real data becomes more manageable. For example the SD button on a calculator provides the opportunity for experimenting with the statistical concept of spread without the immediate need for an algebraic definition; equally, graphical displays of data provide a more powerful resource for building such understanding.

The discussions ranged widely over other topics as well — the roles of programming, spreadsheets, teacher education, and the value of simulations of situations such as running a hot-dog stand. The calculator group felt that the most pressing needs for meeting the challenge of the curriculum in 2000 are for teacher training and research. Some members were concerned that the teaching of pure mathematics should continue to be of prime importance. There was some concern that new technology would remove some rich mathematical experiences; others felt that it would create as many new ones as were lost.

At the end of this series of discussions, some light at least had been thrown on these difficult questions.

The Dynamics of Planned Curriculum Change

Leader: Hugh Burkhardt (UK) Chair: Jim Ridgway (UK)

In many countries it is not easy to achieve planned curriculum change. The outcomes are often very different from what is intended, in various ways. For example, many students fail to learn the planned new techniques, or cannot use them in practical or mathematical problems, only performing successfully in routine imitative exercises. Many teachers do not play the intended new roles in the classroom that would allow students to acquire more independence in doing and using mathematics. With many good projects, there is no large-scale impact at all. Because classrooms are not often carefully observed, these problems may only be noticed many years later.

How might we do better? How might we achieve more reliable progress? How might we, at least, know reliably and cost effectively what we are achieving? This was the challenge for this subgroup.

The sessions opened with two brief introductory presentations, by *Jeremy Kilpatrick* and *Hugh Burkhardt*. After this the members of the subgroup spent the four sessions in discussion of the issues in three working groups, led by *Marjorie Carss, John Mack* and *Jim Ridgway*. In all, about forty people from 15 different countries took part.

Jeremy Kilpatrick reviewed the growth of "curriculum development" as an activity, pointing out its relatively recent origins. Historically, the curriculum had been more or less fixed and unquestioned. Another feature of the curriculum in many countries is that the intended pattern is rarely reflected by reality. "Every school in France is supposed to do the same thing at the same time, but they don't; every school in England is supposed to devise its own curriculum, but they don't." Generally, the outcomes of curriculum development projects have been very different from their goals, many having little direct impact, though perhaps some secondary influence.

He outlined different approaches to curriculum change taken in different countries over the last 30 years or so, noting a trend towards more coercive, centrally directed methods of change. Among the issues he raised were the maintenance of an appropriate level of teacher autonomy, and the limitations of a "one size fits everybody" curriculum.

Hugh Burkhardt pointed to the absence of expertise in curriculum development, an activity requiring many complex and subtle skills for which there was still no profession — most of those involved in directing projects have never done so before and, exhausted from the effort, will never again. Only a very small proportion of the education budget is spent on systematic research and development — 0.03 percent in England. This seems out of scale with the problems we face, or the pace of change; about 5 percent is typical in other changing fields. There is a clear need to see the education "system" as a whole, when curriculum change is planned.

The introduction of change should be seen as an "engineering problem", in which the probable effects of the planned change on all the interacting elements in the system are considered, and the design is developed and tested empirically in all its aspects. Feedback of the right richness and detail is a crucial factor that is neglected. As in the best of engineering, fundamental research has a role to play in guiding design and in developing new methods.

The working groups then went on to discussion within the framework provided by the following "brief":

"Think about how you try to make curriculum changes in your country.

Write down advice to your government on methods of designing and implementing a curriculum change, analysing examples from the past including difficulties and systematic ways of finding better methods, and their cost implications.

Consider the following:

Issues: obstacles to change; how might they be overcome? past experiences, especially failures; how do we learn what is really going on?

Agenda: system description barriers and levers, developmental feedback, maintenance and progress"

The Diagram on the next page shows some of the elements which need to be considered:

The Nature of Change

Changes can be "political" or "educational". Examples of political changes include new policies on participation rates, an imposed change in curriculum emphasis, or in the pattern of funding; educational changes arise from groups claiming expertise in curriculum matters. Change can affect the whole system, including mathematics, or be directed specifically at mathematics curricula.

For any type of change, there is a need to consider the overall social interaction and the internal system effects. The first requires that there be good information systems, enabling the educators to communicate with the community. This is helped by working continually to raise the level of public awareness on educational issues generally. As to the second, there is a similar need to keep the profession informed and conscious of its role as an active body in influencing and implementing change, as well as in proposing or opposing it.

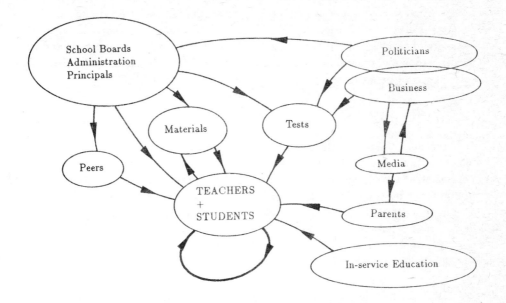

More specifically on the processes, there was a feeling that there is always a tension in the education system, arising from the perceived good and bad effects of a "steady state" curriculum versus an expectation of continual change.

Perceptions of Teachers

Change itself is often seen as an obstacle and rejected, especially by older teachers who are "effective" in the current system. Teachers resist change if they do not feel competent and confident — they "judge" the changes as misguided or see no need for change. The nature of a change and the impact that it is likely to have on them, and on other components of the system, are thus important. The functions and purposes of a change, and its likely effects must be spelled out for all mathematics teachers, as well as for their students and for the public.

Doing something new needs practice. It is important to create a climate in which teachers are sufficiently confident to experiment and develop their own skills; this confidence can only be based on successful experience of tractable challenges with effective support.

Teacher Education

INSET, In-Service Education and Training, appears to have achieved little in terms of large-scale impact on practice. There is a need to review existing methods and to experiment with, develop and test a variety of new approaches — candidates include whole-school INSET; day release; team teaching; advisory teachers; and many others, which were described for educational settings as diverse as Togo, Poland and Australia. INSET procedures must be integrated with other elements of teacher support and development such as teaching materials and time for discussion and reflection. The commitment of the whole school to the innovation was judged to be an important component of successful INSET.

Social Context

Teachers, parents, politicians and school administrators will all resist changes which go against their perceptions of the nature of mathematics, or of their role. Formal plans for curriculum change need to be supplemented by attempts to affect public attitudes towards mathematics. Around the world there has been some success in this direction through television, family and neighbourhood mathematics, and increased parental involvement.

Evaluation and Monitoring

Teachers find it difficult to evaluate themselves when they are coping with change; thus some kind of "trained observer" feedback is desirable, both in the short term and the long. It would be worthwhile to have this at classroom and school level, as well as at system levels. Each system could and should involve "developmental feedback" — formative evaluation procedures whose purpose is to provide improved information on the curriculum change process, in order to suggest modifications in design and/or implementation.

Lastly, in reviewing the kinds of change likely to achieve impact, positive or negative, the group underlined the important role of public assessment in influencing the curriculum. Since changes in assessment usually have an immediate impact on classrooms, those involved in the design of curriculum change should plan changes in the assessment procedures that will reinforce, even lead, the desired curriculum change. Equally, they should examine carefully existing assessment practices, and any proposed changes, in order to evaluate their effects on the curriculum.

Moral Responsibility

We make no apology for revisiting this theme. All change requires teachers to take risks in their professional lives. It is important that teachers are major contributors of ideas about the nature and goals of proposed changes, and that they receive support and feedback of proven effectiveness, appropriate to their needs.

Curriculum Development as a Research Area

Far more work needs to be done to study the mechanisms of innovation, if the perils of rejection, dilution and corruption of the innovation are to be avoided, and the whole educational community is to benefit.

The Teacher as a Critical Agent in Curriculum Change and Implementation

Leader: John Malone (Australia) Chair: Peter Brinkworth (Australia)
Presenters: Les Steffe (USA), Jan Thomas (Australia), Patricia Wilson (USA), Peter Reynolds (UK), Rafaela Guidi (Brazil), Peter Brinkworth (Australia), Madeleine Long (USA), Mike Pierce (UK), Yuang–Tswong Lue (People's Republic of China), Deborah Haimo (USA), John Hersee (UK)

In the corresponding subtheme discussions at the ICME held in Adelaide in 1984, attention focussed on teacher's and teacher adviser's roles in curriculum development and implementation. The debate concentrated on the need to ensure that teacher participation in curriculum change transcended nominal representation on curriculum development committees and that teachers should "participate in curriculum development at least at the level of action research" (p. 193). Advisers were encouraged to foster local networks of teachers, who would share experiences, and of schools which could share developmental tasks while simultaneously providing a clearinghouse through which to update information which would then be transmitted to the teaching body. In other words, the curriculum and the human resources associated with it were to provide opportunities for teachers in schools to turn theoretical insights into actual classroom practices so that the curriculum conducted in the classroom would be derived from the teacher's personal definition of learning.

Noticeable at this Congress was the call for innovation and the need to break from such a traditional view of the teacher's relatively minor role in curriculum change and innovation. A push for major reform was most apparent.

A need for radical change

When viewed from a historical perspective it seems apparent that the seeds of discontent and cries for educational reform were planted long ago by those who themselves were seekers of change — people like *Dewey, Price, James, Jefferson* and *Thoreau.* Now, towards the end of the 20th Century, circumstances appear to have caught up with us: Nations *are* at risk; students do not seem to be learning well and teachers receive little respect and possess little authority.

Strong evidence was presented by *Long* to support the assertion that never before has the demand for reform been stronger. The media of curriculum change, the platform reports (such as the Cockcroft Report in England and Wales, the Holmes Group and Carnegie Commision Reports in the United States) have in the recent past issued strong challenges to educators to question conventional expressions of the educational enterprise — for example, the concept of the lower school teacher as a generalist; the critical role of the teacher-training institution; the value of the traditional academic program for pre-service secondary teachers, and the value of teacher exposition as the sole vehicle of lesson delivery.

Subtheme participants described several approaches to break current mindsets and to design conditions for the full-scale emergence of the teacher as a professional educator. *Hersee* and *Price* suggested that the printed word is totally ineffective in communicating all other aspects of the curriculum besides content. Even the spoken word is of limited effectiveness due, at least in part, to the fact that the teacher is a product of his or her past and has been conditioned by it. Changes to a curriculum require teachers to change; to some degree it requires them to revise their ideas of what mathematics is. Bearing in mind the mass of past experience that teachers bring to their work, it is to be expected that they will translate a new curriculum into more familiar terms.

The need for teachers to be supported by their school system and by their peers as they take risks and proceed in the adventure of altering the status quo were two of three critical variables for ensuring productive teacher involvement in curriculum change identified for the subtheme group by *Wilson*. Based on a study of four curriculum development projects in the United States, her study indicated that the other critical variable related to a commitment of self by the teacher — the greater the teacher's investment of time, finance, effort, expertise and self-involvement, the more successful the implementation. It was of interest to note here that specific factors such as teacher content knowledge, the degree to which the existing curriculum was changed, renumeration of teachers, student population and the quality of materials and equipment, while appearing to be relevant in particular studies, did not generalize across the projects.

Brinkworth spoke of the "transition problem" — the difficulty of promoting the movement of teachers from one set of beliefs and practices to an alternative (better) one. The teachers' role in tackling this problem is to strive continuously to improve their practices, to exemplify and involve their colleagues in the process of improvement and to promote the transition from what is occurring to something better for students. In the end, however, successful teaching will be correlated with the teacher's personal belief that the methods he or she introduces *will* work.

Significant change was not only required of teachers. The curriculum and the methods used to implement it were also identified by the subtheme group as being ripe for change. One particular approach generated considerable debate and is the subject of the following section.

The meaning of curriculum in a constructivist's epistemology.

Many cognitive scientists now believe in a constructivist model of knowledge that attempts to answer the primary question of epistemology: how do we know what we know? The model can be summarized in the statement: Knowledge is constructed in the mind of the learner (Bodner, 1986; von Glasersfeld, 1984). Constructivist theory maintains that knowledge cannot be the result of passive receiving but originates as the product of a subject's activity.

Constructivist ideas were given scant recognition at the Adelaide congress (the term does not appear in the Proceedings) but achieved significant impact in Budapest. During the discussions of this subtheme group the point was stressed that, in an epistemology where the function of cognition is adaptive and serves in the students' organization of their experimental world, not the discovery of an objective ontological reality, the notion of a mathematics curriculum must be reformulated to include the mathematical knowledge of students and of teachers, mathematical learning and the mathematical environment. The notion of a *created* curriculum was discussed — a mathematics curriculum produced by the participants in ongoing mathematical teaching and learning. Herein lies one of the major differences between the teachers' role in curriculum development as perceived in Adelaide and that discussed in Budapest: The teacher has a more responsible task than that of simply implementing a curriculum imposed in the top–down manner mentioned earlier. A series of principles of curriculum design within a constructivist's perspective put forward by *Steffe* provided the bases for spirited discussion. Briefly, the generative power of children can be impressive when they are working in mathematical learning environments that are conductive to constructive activity. However, children's generative power has only begun to be charted, so mathematics teachers at all levels have an exciting choice between being participants in specifying the generative power of their students or taking what their students can learn as being already specified by an a priori curriculum.

What goes by the name of mathematics curriculum development in constructivism by necessity starts with the mathematical knowledge of teachers and students, and any person who purports to be a developer of curriculum must be first and foremost a mathematics teacher. This fundamental idea has powerful implications for curriculum developers as they strive to create abstracted mathematics curricula — a conceptual generalization that curriculum developers can abstract from a group of experiences for the purpose of categorizing and systematizing new experiences (von Glasersfeld and Steffe, 1987). By being embedded in ongoing mathematics teaching and learning, both as an actor and as a participant, a curriculum developer can isolate regularities in his or her experiential world.

Needless to say that the traditional homes of curriculum change in primary and secondary mathematics — Ministries of Education, Syllabus Committees consisting of tertiary educators, curriculum experts and teacher representatives — will need to be refurbished before the sentiments expressed above will gain general acceptance. It will not be easy to convince such groups that they should accept a model wherein the teacher, along with his or her students, are the primary actors in creating mathematics curricula in the context of ongoing teaching and learning. There will also be the problem of that group of teachers who will be unable to accept the challenge and responsibility which the constructivist epistemology places upon them. This is to be expected, and it was generally accepted by the subtheme group that the most promising scenario, in the short term at least, would consist of the co-existence of both forms of curriculum development — the "top–down" traditional model detailing broad topic areas but containing the flexibility and freedom for the constructivist teacher and his or her students to pursue the ideals of their epistemology. It would be a difficult structure to put in place, yet, nevertheless, one worth pursuing.

Other models for curriculum development

The constructivist model of curriculum development was but one of several discussed by the subtheme group. It, and at least one of the others, tend to view teaching and learning as a process of interaction and communication between student and teacher. Their goals are to determine what the student knows and to design appropriate experiences to capitalize on that knowledge. This implies that the focus of curriculum development needs to be on the teacher. It was agreed that curriculum development devoted mainly to the production of materials will not improve mathematics learning. Rather, teachers' views of the mathematics learning process will require "transition" to use Brinkworth's term, as an essential pre-requisite to the improvement of teaching practices.

Concerning other models discussed, recent changes as described in the United Kingdom, Australia, Brazil and the United States appear to hold great potential for curriculum development, because teachers and educators appear united in a desire to supplant an old system that is not working efficiently with something new. Undoubtedly the Cockcroft Report did much to shift the emphasis away from teacher exposition and the practice of skills and routines towards a focus on classroom processes, pupil learning and an enhanced role for the teacher. *Price* and *Reynolds* both referred to the situation in England where the examination system is being used as a model to bring about large-scale change *not* on syllabus content, but in teaching and learning methods and in advancement of classroom methods.

Thomas described a professional development model designed by the Victorian Ministry of Education in Australia which addresses the need for teachers to change if the challenges of mathematics for the year 2000 are to be met (Victoria: The next decade, 1987). The program calls for the devolution of curriculum development from the Ministry to educational regions and a process of teacher "transition" (Brinkworth) using a team of advisory teachers and extensive in-service activity to achieve its goals. The point was made that, despite the potential of this scheme, its success and indeed the future of mathematics is ultimately linked to the views of teaching in the community. The teaching of mathematics cannot be separated from teaching as a career, consequently, mathematics education esteem can be raised so that the profession is once again perceived as a valued career.

A change of Government in the mid-1980's provided the catalyst for change in the curriculum development model used in Brazil to that time. *Guidi* described how, at the instigation of a group of mathematics teachers, the way was made clear for the group to draw up curriculum proposals for Grades 1–8. In a concentric circle model, this core drew others in for support as the need arose and as the proposed program developed. It is estimated that between 1985 and 1988, over 4000 among Brazilia's 13000 specialist mathematics teachers have played some part in the development of the new program and are now well grounded in its theory and delivery. They in turn will assist in the process of "transition" with the rest.

The "teacher–leader" concept inherent in the Brazilian model is also being used in at least one major project in the United States. Long described how formal training is viewed as the most important and efficient way to develop professional teachers and provides the structure for the project FAME (Fellows for the Advancement of Mathematics Education) funded by the National Science Foundation. FAME's goal is to provide every elementary school in the USA with one well qualified "Mathematics Leader" to serve as the school mathematics specialist, spending half time in the classroom and the balance training other teachers, supervising and planning. The Mathematics Leader would be specially trained in a three-year program and would have a broad array of responsibilities including the development of a Mathematics Resource Team responsible for establishing sound mathematical practices within the school's larger community. The formal education the FAME teacher leader receives as part of the Mathematics Leader training is seen as a key element in the projects philosophy.

The above models call for a study in considerable depth of the incentives and influences that will encourage teachers actively to engage in change. This is another study that is worth pursuing.

♣ ♣ ♣ ♣ ♣

Supporting survey presentation:

Hiroshi Fujita (Japan) and *Geoffrey Howson* (UK)

The speakers were asked to discuss some key issues of importance related to the Theme. This summary is prepared by, and therefore the responsibility is of the Theme Organizers.

Geoffrey Howson outlined some of the central challenges facing the curriculum over the next decade. They included

Technology, particularly computers and calculators

Governmental attitudes and pressures

Demographic problems

Social context

Teacher supply

The crucial importance of technology is self-evident, though its detailed implications and their implementation will require a great deal of study and effort. The increase in government pressure, with increasing use of the criteria of industry and commerce, is a clear trend in Britain and in other countries. The changing social context of the curriculum is also clear, with the media playing an increasing role, though its implications are not.

While developing countries have long faced the problems of a population expanding faster than resources, developed nations are facing the reverse problem, which is proving a challenge; the number of 17–year olds will be considerably less by 1994 than in the recent past — by about 15 percent in France, Italy and Japan; by over 25 percent in the Netherlands, the UK and the USA; and by a staggering 45 percent in the FRG. Such falls create enormous problems, for society and for education. Teacher supply is just one such area, but a crucial one; shortages of skilled people will hit teaching even more severely than other richer fields.

Howson discussed a variety of views of mathematics and its pedagogy — the need for a systematic structured framework for study, the need for conjecturing to lead to explaining and justifying, the danger that mathematics for all may mean "no mathematics for all", the need constantly to review and redefine what school mathematics is and should be. Is the old danger that it is too close to university mathematics being replaced by becoming too close to engineering? Hardy said "Pure mathematics is on the whole distinctly more useful than applied." In modelling one can apply only the mathematics one knows.

The key problem of the 1990's is the design of curricula that are more effective without demanding too much of the weak teacher. Good teachers must lead, encouraging the weaker to develop and become stronger. Those of us outside schools can suggest and assist, but it is the teachers that must make the workable work, so that students leave school with an understanding of mathematics: its range, its nature, its practitioners and its methods. The talk made very clear the challenge that curriculum development faces in the 1990's.

Hiroshi Fujita described the current review of the school mathematics curriculum in Japan — a planning exercise leading to new textbooks and examinations. Particularly interesting was the introduction of a component of the curriculum focussed on "mathematical literacy" to complement that on "mathematical thinking"; so much effort in many countries has been focussed on encouraging proactive mathematical activity that it is interesting to see a major nation taking seriously the cultural, "spectator" aspect of mathematics. Should an informed awareness of fractals be included in the secondary curriculum, for example?

In the design of the curriculum itself, there is to be increasing emphasis on flexibility, and on the use of computers and calculators. High participation within two "streams" is to be supported by a core curriculum with options — remedial options, side options and advanced options.

A final word about the Japanese view of mathematical literacy. The emphasis here is on communication with mathematics, on a general global feeling for mathematics that can guide the selection of expert sources of advice if necessary, and on an understanding of the nature of mathematical reasoning. These are surely worthy objectives; the challenge of realizing them in each of our societies is formidable, but not beyond reach.

These talks will be published in full in the separate Theme publication, referred to at the beginning of this summary report.

MATHEMATICS, EDUCATION AND SOCIETY (MES)

Chief organizers: **Alan J. Bishop (UK), Peter Damerow (FRG), Paulus Gerdes (Mozambique), Christine Keitel (FRG)**

The main task of the scientific community concerned with mathematics education is to support the teaching and learning going on in the schools. However, increasingly the interrelation between mathematics education and educational policies has become a matter of worldwide consciousness and it is evident now that mathematics education has a serious political dimension. It was novel for an ICME congress to make this political dimension the issue of a whole day, i.e. the 5th day, with its special programme called, 'Mathematics, Education, and Society'.

The high number of contributions offered to the organizers indicates the increasing awareness of the relations between education in general as a universal human right and mathematics education in particular. On the day almost 90 contributors from over 40 countries were working together addressing issues of mathematics and mathematics education as human activities, which on the one hand belong universally to the core of the school curriculum all over the world yet on the other hand are always embedded in particular cultural settings. Both the historical dimension of the past, the present, and the future of mathematical education, as well as the scale from cultural and socio-political conditions in general to the social conditions of classroom activities and individual learning processes were fundamental dimensions of the programme schedule.

However, within this general framework of the 5th day programme the speakers as well as contributors were concerned with key problems of the social conditions and the outcome of mathematical activities.

For example, one important focus of the day was on the restricted opportunities to learn for children of certain minority groups in industrialized countries, for girls almost everywhere, and for the majority of the youth growing up in third world countries. There is a growing awareness of the importance of ethnomathematical activities as a means to overcome Eurocentrism and cultural oppression in mathematical learning.

A second focus was on the consequences of uncontrolled technological development which promotes specialization and suppresses pedagogical aims of education like general competence and individual autonomy. Creative uses of mathematics, meaningful learning related to social experience and social needs, and critical and multicultural perspectives in the designing of curricula and learning environments were vividly recommended by most of the contributors.

The outcome of the 5th day special programme is more an agenda for future activity than a balanced account of achievements and limitations of mathematics education under the present social and political conditions. This is the message which will be disseminated by means of a UNESCO publication containing the main arguments of all contributions to the 5th day programme of ICME 6.

The day was organized around four large themes. Timeslot 1 dealt with *Mathematics Education and Culture*. General presentations on the central themes helped to set the scene for the day. Timeslot 2 concerned *Society and Institutionalized Mathematics Education*. Timeslot 3 focussed on *Educational Institutions and the Individual Learner*. Within both timeslots several parallel sessions took place with panel presentations and discussions on specific topics. In timeslot 4 at the end of the 5th day program five hearings on *Mathematics Education in the Global Village* raised major questions society asks mathematics educators.

1. Mathematics Education and Culture

Mathematics education has a cultural history, a diverse present, and a cultural future. How can we represent it to ourselves? This was the leading question of the three opening sessions dealing with historical, actual and future perspectives.

1.1 Social history of mathematics education

This theme concerns some significant aspects of the history of mathematics education, which has contributed to the determination of today's situation. The first speaker, the Hungarian mathematics historian *Árpád Szabó*, showed in this presentation "Mathematics and Dialectics" that fundamental principles like mathematical defining, reasoning and proving were raised by the philosophy of the Eleatics; and that the system of mathematics as it is formulated in Euclid's Elements, is partly stimulated by the dialectics of Eleaticism, and is partly a further developed construction based on these dialectics. *Ahmed Djebbar* (Algeria) compared the contents of mathematics education of North Africa in the middle ages with its role in actual teaching. *John Fauvel* (UK) posed the question: "Should we bring back the mathematical practitioner?" referring to some important aspects of the social history of mathematics education in the Renaissance. In a case study of Britain he reported how British mathematics educators coped with didactic problems, how they had to struggle for acceptance of mathematics as a practical and academic subject, and in which debates about teaching practices they were involved. *Gert Schdubring* (FRG) discussed theoretical categories for investigations in the social history of mathematics education and some characteristic patterns. He tried to offer a more comprehensive framework for historical studies as well as for an understanding of the role of mathematics in both liberal education and vocational training.

1.2 Cultural diversity and conflicts in mathematics education

Cultural diversity is commonplace in education around the world, and mathematics educators have a particular interest in certain issues which are addressed in this theme. *Terezinha N. Carraher* (Brazil) in a paper read by David Carraher, described material embodiments of mathematical models found in use in everyday life in Brazil. *Lloyd Dawe* (Australia) discussed typical problems of mathematics teaching and learning in village schools of the South Pacific. *Murad Jurdak* (Lebanon) focussed on the roles of religion and language of instruction in mathematics in as far as the two factors operate as cultural carriers or cultural barriers in mathematics education. By contrasting two main and provoking theses, illustrated by the counter-examples of Saudi Arabia (rich) and Lebanon (poor) he explained the dilemma of balancing the needs of socio-economic development and the preservation of the ecology of the culture. The final speaker on theme 2, *Claudia Zaslavsky* (USA) emphasized the necessity for the expansion of the curriculum at any level to include culturally specific mathematical practices as social practices in daily life "ethnomathematics" and presented possibilities for doing so in the elementary and middle grades.

1.3 The cultural role of mathematics education in the future

Worldwide cultural domination and unequal distribution of opportunities to take part in education characterize the present situation. In theme 3, therefore, future perspectives of mathematics education under these conditions were explored. *Leone Burton* (UK) going beyond features already discussed on gender and mathematics education, focussed on another area which still requires investigation: this is the degree to which the discipline of mathematics together with its predominant pedagogy reflects an inaccurate, socially stereotyped and misguided view of the nature of mathematical activities. This she clarified by putting the question: "Mathematics as a cultural experience — whose culture?" *Desmond Broomes* (Barbados) described a strategy which seemed most promising for developing, implementing and evaluating an appropriate mathematics curriculum for rural communities: Community Involvement Strategy, in order to bring the curriculum closer to the activities of community life, to integrate educational institutions into the community, and to broaden the curriculum in a meaningful way so that it includes socio-economic, technical and practical knowledge and skills. *Kathryn Crawford* (Australia) argued that — as school mathematics is not socially neutral — students from minority groups need experiences which provide concept development in qualitative aspects of the contexts in which mathematics is used in the dominant technical culture, that is, they need to know what to reason about, instead of the overemphasis on instrumentalism offered by most teachers. *Philip J. Davis* (USA), took the line that mathematics education has to redefine its goals so as to create a citizenry with sufficient knowledge to provide social backpressure on the relevance and development of future mathematizations and to view applied mathematics as a "social contract". He argued for an "interpretative" approach to mathematics education.

2. Society and Institutionalized Mathematics Education

Mathematics education and culture, the overall theme of timeslot 1, was now examined at two different levels: in timeslot 2 at the level of institutionalized education embedded in the society, and in timeslot 3 culture was seen as a factor in the relationship between the individual process of learning mathematics and the social interactions framed by the institutions of the educational systems.

2.1 Mathematics as a cultural product

Mathematics is not culturally neutral. The relationships discussed here are between the cultural origins of mathematics and its future perspectives. By reflecting on the implications of the recently developed new and (mathematized) technologies and by combining arguments from the history and sociology of sciences with epistemological reflections, *Jens Høyrup* (Denmark), in his contribution "On Mathematics and War", asked the mathematicians (especially in their role as teachers) to take more responsibility for the social use of their work. He asked: "In which sense is it possible to understand scientific cognition as a neutral instrument, and in which not? What can mathematicians do in order to avoid the situation of their science and the scientific institutions in general staying or becoming virtual criminals against humanity?" These are questions which do concern all mathematics educators. *George Ghevarghese Joseph* (UK) deplored the fact that there has been a widespread Eurocentric bias in the production, dissemination and evaluation of scientific knowledge for centuries, and he seeks to destroy the existing Eurocentric paradimatic norms, not only in history, anthropology, and ethnology, but also in the history of mathematics and mathematics education. *Sam Ale* (Nigeria) described existing rural mathematics in Nigeria, which is oral and unwritten and generally serves as a tool for directly solving practical problems. He argued for taking this kind of rural and very meaningful mathematics as a base on which to build mathematics programmes for further instruction. *Leo Rogers* (UK) pleaded for examining the history of mathematics in its social context to identify possible conflicts in meaning and intention between "intended" and "actual" uses of mathematics, and to create a social epistemology of mathematics.

2.2 The image of mathematics in society

Mathematics is widely considered to be independent of social, institutional, and moral implications and consequences. This image was analyzed from different viewpoints. *Rudolf Bkouche* (France) discussed the recent reforms in France, their revisions and their social image; *Gilah Leder* (Australia) reported a case study, in which she tried to identify, quantify, and operationalize society's image of mathematics by examining the perceptions held about mathematics by a group of tertiary educators without special expertise in mathematics. *Stephen Lerman* (UK) showed that despite the apparent proliferation of philosophies of mathematics (as discussed by mathematicians, too) schools of thought fall into two streams, the absolutist and the fallibilist; i.e. the search for certainty of mathematical knowledge versus viewing mathematics as a social invention. Some implications for mathematics education were discussed. The last speaker on this topic, *Chandler Davis* (Canada) renewed the discussion of the responsibility of scientists (mathematicians) for the uses to which society puts the results of their research, and asked for "A Hippocratic Oath for Mathematicians".

2.3 Sociology of institutionalized mathematics

Mathematics and the way that mathematics is conceptualized, thought about, and learned, is dependent both on institutions and on political decisions. *Eduard Glas* (Netherlands) presented an historical example by discussing some cultural differences within the Paris Ecole Polytechnique 1754–1809. *Renate Tobies* (GDR) described the influence and the activities of Felix Klein in the Teaching Commission of the 2nd Chamber of the Prussian Parliament. *Emma Castelnuovo* (Italy) spoke about the teaching of geometry in Italian high schools during the last two centuries as a steady attempt to introduce geometry to a larger sector of society and to connect it more with reality. *Michael Price* (UK) offered some reflections on the role of associations in mathematics education and their relationship to and influence on other political and social pressure groups within the context of national educational systems.

2.4 The mathematics curriculum as a social issue

In the history of mathematics education, epistemological and socio-economic conditions of mathematics learning are strongly related. The issues surrounding this relationship were considered in this topic from different points of view. *Michael Otte* (FRG) in his presentation "Mathematics for all and the epistemological problems of mathematics education", argued that "the dichotomizations current in didactics are both false as far as practical, aesthetic, theoretical, philosophical, and many other kinds of experiences are concerned, and are equally essential for the development of the personality, — and productive, as far as the rationality of human learning and human knowledge is no 'flat world' but includes belief systems, convictions, dogmas, ties to contexts". *Neil Bibby* and *John Abraham* (UK) through considering some cases of controversy in the historical development of mathematics, analyzed the implications of studying controversies for mathematics education. *Bernard Charlot* (France) discussed social controversies in mathematics education by means of the introduction of the "modern mathematics program" in France.

2.5 Non-school alternatives for mathematics education

Expanding social needs for mathematical qualification, and within-school obstacles which prevent the development of a broader mathematical culture, call for new solutions to the problems of mathematics education. The panel members offered and discussed some possible alternatives. *John D. Volmink* (South Africa) discussed some examples of non-school mathematics education. Such examples are the more important under conditions of political systems that exclude systematically, for racial, political, religious or sexual reasons, certain groups of the population from adequate qualifications. *Sixto Romero Sanchez* (Spain) reported on the experiments with a very successful and socially acknowledged radioprogram to popularize mathematics developed and conducted by a group of mathematics educators at the Polytechnic University of Huelva. *Virginia Thompson* (USA) discussed the role of the already famous American "Family Math" - Project; and *Jeff Evans* (UK) analyzed what sort of statistical skills are needed by adults in nonspecialist work contexts and how these skills might be developed in more informal settings outside the educational system.

2.6 The mathematics demands of the economy

An important social constraint on mathematics education is given by the various demands of the economy with its challenges and its risks. *Guida M.C.P. de Abreu* and *David W. Carraher* (Brazil) showed that, although the general mathematical abilities of the Brazilian sugar cane workers vary greatly because of their poor schooling they use knowledge and techniques quite well to measure and calculate, activities which are cultural in nature, having their roots in the social traditions of the sugar cane communities. *Wang Chang Pei*, (China) gave examples of differences in mathematics education between rural and urban areas in China despite the national uniform curriculum. *Howard Russel* (Canada) presented the "Generic skills / Economic Development Project" as an attempt to match the demands of the economy. On the other hand, *Bernhelm Booss-Bavnbek* (Denmark) showed how a rapidly growing number of mathematical models and simulations turn out to be sufficient for the design of workable technological improvement, but many of them are not sufficiently transparent to provide the necessary means for their technological and social control. This places new and urgent demands on public education in mathematics for a democratic society.

2.7 Mathematics education under different cultural constraints

The conditions of mathematics education in industrialized and in developing countries are very different. The implications of serious cultural differences were discussed by the panel members: *Siaka Bamba* (Ivory Coast) spoke about some critical issues concerning mathematics education in the Ivory Coast; *Jens Naumann*, (FRG) reported a partnership-project with Senegalese villages involving Berlin teachers organized in the German United Nations Association, and practical aspects of basic mathematics teaching there; *Diane Rosenberg* (Argentina) presented her experiences in transformin the Dutch HEWET-Project into quite a different culture "HEWET–Argentina". *Munir Fasheh* contrasted the informal mathematical practice used socially in daily life and institutionalized formal mathematics, in order to analyze and criticize the role of "hegemonic education" implanted in developing countries.

2.8 Society as a source of ideas for mathematics teaching

Mathematics teaching, if it is to be relevant for the learner, has to take into account the broad range of sources offered by the specific culture of a society. *Brian Hudson* (UK) outlined the results of a project aimed at developing materials and approaches for the mathematics classroom from a global perspective and hence dealing with issues related to world development and the military technology/arms race. *Tadasu Kawaguchi* (Japan) showed mathematical thoughts being latent in various artistic activities like literary and painting work, theatre plays and movies. *Diana Schultz* (Australia) described the theoretical framework and practical strategies for developing and implementing a mathematics education program designed to make mathematics more meaningful to a larger majority of the school population (PRIMATE: real life integrated mathematics in teacher education). *Joop van Dormolen* (Netherlands) talked about the values of texts for learning mathematics for real life by showing examples of written materials and by reporting on how the values influenced the way the texts became what they are.

2.9 How autonomous is the mathematics teacher?

In any society the teacher is socially constrained and some of the consequences for mathematics educators were considered in this topic. *Diane Siemon* (Australia) described generally the social constraints surrounding teachers' work and pleaded for investigations of the belief-systems underlying the teaching and learning of mathematics. *Paul Ernest* (UK) asserted that all reforms depend on the teacher's system of beliefs, e.g. conceptions of the nature of mathematics and models of teaching and learning mathematics, and presented a model of mathematics teachers' knowledge, beliefs and attitudes which elucidates the theoretical relationships between the teacher's belief system, practice of teaching mathematics, and the learner's belief. *Kurt Kreith* (USA) reported about a teaching credential program for the recruitment and training of master teachers which has developed training and certification procedures that correspond, in appropriate ways, to those employed in other professions, such as medicine and law. *John Suffolk* (Zambia) spoke about the specific difficulties constraining mathematics teachers in developing countries.

2.10 Ethnomathematics and schools

Mathematical knowledge of a different kind from that which is usually dealt with in the school curriculum was considered in this topic. Sources for ethnomathematical ideas and their significance in schools were discussed by the panel member. *Gloria F. Gilmer* (USA) gave a survey report of research activities in ethnomathematics as reported to the newsletter of the International Study Group of Ethnomathematics. *Randall Souvviney* (USA) discussed the role of the Indigenous Mathematics Project in Papua New Guinea. *Eduardo S. Ferreira* (Brazil) showed by a lot examples from the history of mathematics and ethnomathematics that the genetic principle and ethno-mathematical methods are linked, and can be part of the same method of teaching mathematics in school.

2.11 Social needs and reforms in mathematics education

Developments in mathematics education are becoming more related to perceived needs in society, and in this session the speakers presented their views on this relationship from the perspective of the priorities for reform in their respective countries. *Cao Feiyu* (China) described how China tries to match the needs of social developments and socialist modernization in reforming mathematics education. He differentiated between strategies to suit long term and short term needs, common and different, social and individual needs. He reported some related investigations and possible measures to be undertaken. *Fidel Oteiza* (Chile) summarized a set of studies conducted to generate a framework for curriculum research and development in mathematics in Chile, in which opinions of professional mathematicians were collected, high school maths teachers, social scientists devoted to the study of the relations between education and labor, personal recruits and computer science instructors and specialists. *Teresa Smart* and *Zelda Isaacson* (UK) described a Higher Introductory Technology and Engineering Conversion Course, that recruited women only and was taught almost exclusively by women, in which they developed new materials and tried out new ways of working with mature women returners. *George Malaty* (Finland) set up the question to ICMI of what kind of reform is needed now in mathematics education and referred strongly to the Soviet example of continuous reform instead of following educational fashions which has been done for so long in Western countries.

3. Educational Institutions and the Individual Learner

3.1 Individual and social learning motivations

Mathematics is learned in a social setting and the significance of social motivations needs to be better understood by mathematics educators. *Anne-Nelly Perret-Clermont* and *Marie L. Schubauer-Leoni* (Switzerland) considered the learning of mathematics as originating from an intersubjective construction between the teacher and students and illustrate how the "didactic contract" controls all learning and teaching. They also addressed the social constructions of meanings of mathematical activities and concepts functions that the contract permits. *Timothy E. Erickson* (USA), in his paper read by Sherry Fraser, discussed certain attributes of cooperative exercises and cooperative materials, developed by the EQUALS project, and showed how they work together to help produce successful experiences for both students and teachers. *Albrecht Abele* (FRG) reported on an analysis of video-tape situations to show the interdependence of social learning and mathematics learning.

3.2 Cultural influences on learning

Children come to school as culture-bearers and the importance of this for mathematics learning was discussed in this topic. *Analucia D. Schliemann* and *Nadja M. Acioly* (Brazil) provided data on how everyday experience with numbers affect the way people solve problems and analyzed how the numbers and the arithmetical operations involved in the programme affect the efficiency and the strategy used to find the solutions. *Frederick Leung* (Hong Kong) discussed four characteristic of the Chinese culture that might influence mathematics learning: the Chinese stress on filial piety and respect for superiors, the Chinese stress on memorization and practice, the characteristics of the Chinese language and the high parental expectation of the Chinese. *Martin Hoffman* (USA) discussed the use of writing activities in mathematical curricula as a vehicle for student reflection and empowerment.

3.3 Are girls underprivileged around the world?

The issue of girls' underachievement in mathematics has been discussed for some years. In this topic special consideration was given to the social context surrounding the relation of girls to mathematics. *Gila Hanna* (Canada) reported on the results of the mathematics achievement of girls and boys in twenty countries, using the data collected by SIMS (Second International Mathematics Study), which show that in the majority of the countries girls are not underprivileged in relation to their achievement in mathematics in grade 8. This changes by grade 12, however, and she compared attitudes in the countries in which significant sex differences were observed to attitudes in countries in which such differences were non-existent to generate interesting hypotheses about the factors responsible for girls' inferior or superior achievement in mathematics. *Erika Kuendiger* (Canada) presented data on the attitude scale "Gender Stereotyping" of SIMS and discussed that in light of differences in the general learning environment. *Frank J. Swetz* (USA) reviewed findings from non-western societies which indicate that mathematical abilities and attitudes are culturally related and — as in the reported case of two Malaysian studies — that sex-related differences may contrast sharply with opinions popularly held in the West.

3.4 Societal determinants of learning

This topic concerns the influence which the childrens' society exerts on their mathematics learning. *Joan Bliss*, and *Haralabos N. Sakonidis* (UK) have investigated in a cross-cultural study — in England, Greece, and Spain — of children's ideas about what is really true in four curriculum subjects — mathematics, science, history, and religion, and reported on their findings. *Gustav-Adolf Lörcher* (FRG) described the specific problems in mathematics education which result from the fact that — both in developing in industrialized countries — many children learn mathematics in a foreign language, and tried to offer some ways to deal with them. *Ali Rejali* (Iran) presented a study of the causes for students not being interested in studying mathematics in Iran together with some suggestions for solving the problem. *Bernd Zimmermann* (FRG) demonstrated that "mathematics for all" and "mathematics for the gifted" are not only not exclusive alternatives, but also that the main issue "mathematics for — or better with — all" can be tackled by experiments and research in the special area of mathematically gifted students.

3.5 The social arena of the mathematics classroom

The special context for learning which is the "mathematics classroom" has particularly important social features which were explored in this topic. *Josette Adda* (France) showed how the taboos, the forbidden acts and compulsory rituals in the mathematics classroom create an artificial world as far from the mathematical "world of ideas" as from the external "world of everyday life" and discussed why this sociological context is in favour of pupils from high sociocultural level. *Tom Cooper* (Australia) analyzed primary mathematics teaching in terms of its role in the social and political function of schooling in reproducing the present society. He demonstrated that mathematics teaching is seen to have a formidable role in reproducing crucial ideologies, a role made more powerful because of the notion of objectivity that surrounds mathematics itself and masks hegemonic function. *Andrea L. Petitto* (USA) described the structure of classroom discourse and discussed the significance of this kind of discourse for the difficult and often gradual process of disambiguating arithmetic procedures and linking them to children's own intuitive understanding of number. *Terry Wood* (USA) analyzed the evolving regularities or patterns identified in whole class interactions as the teacher and children discussed solutions to arithmetic activities, focussing on the implicit, taken-for-granted obligations that the teacher and students feel under in particular situations, and on their expectations for each other.

3.6 Learning under difficult conditions

For some students, learning mathematics is specially difficult. Typically their problems are considered to be psychological, but in this topic there was a focus on the social nature of their problems. *Arthur Powell* and *Marilyn Frankenstein* (USA) explored the needs of "non-traditional" learners, "outsiders" of the educational system like women, ethnic and national minorities and other working class adults who are severely underprepared to meet successfully the challenges of the traditional curricula of higher education. They presented alternative approaches of "non-elitist" mathematics education which meet their intellectual needs in mathematics. *Françoise Cerquetti-Aberkane* (France) described the mathematics education program in special classes for children with difficulties and criticized certain parts of that "special pedagogy". *Nicolas C. Taylor* (South Africa) reported on some findings from case studies conducted among 13 year old students in Soweto into the methods employed in the solution of problems on equivalent fractions. His findings place a question mark over the utility of those psychological perspectives on learning which seek to separate the cognitive and affective domains, and those multi-cultural and ethno-mathematical considerations which see cultural aspects as providing a handle for lifting "underprivileged" learners into the mainstream.

3.7 Mathematics education in multi-cultural contexts

Within the mathematics classroom there can be serious cultural conflicts which will challenge both learners and teachers to negotiate satisfactory practice. *Raymond A. Zepp* (Macau) asserted that a new way of thinking about the relationship between language and mathematics is necessary, and described the new directions towards which recent research has begun to move. *Ina Kurth* (FRG) reflected on some special problems of children learning mathematics in a foreign language and illustrated her presentation with examples from the material particularly developed for such children. *Helen Watson* (Australia) in her paper read by Ken Clements, discussed her work with teachers who in their everyday practice are confronting the problems of teaching and learning mathematics in two worlds: Yoruba teachers in Nigeria in West-Africa, and Aboriginal-Australian teachers in the Tanami and Laynhapuy region of Northern Australia; together they have developed an understanding of the relation between conceptual systems, and have considered the perplexing problems of mathematical education in a bicultural situation. *Norma C. Presmeg* (South Africa) explored ways in which cultural continuity may be fostered in the context of mathematics education. She is well experienced in working with multi-cultural classroom situations and with adult students who might be in a similar situation to learners in developing countries around the world.

3.8 Ethnomathematical practices

How can the mathematics learning situation be structured so as to provide for the acceptance of the child's ethnomathematical knowledge? The panel members offered various examples of ethnomathematics: *Salimata Doumbia* (Ivory Coast) spoke about the mathematics in some traditional African games; *Sergio R. Nobre* (Brazil) described the mathematics involved in the most popular, but forbidden by law, lottery in Brazil. *Nigel Langdon* (Ghana) presented various ethnomathematical activities like mathematics of workmanship, craft and economy as cultural starting points for the learning of mathematics.

3.9 Social construction of mathematical meaning

Mathematics knowledge is both individually and socially constructed, and educators should give a greater priority to the latter than they have typically done. *Paul Cobb* (USA) outlined an analysis of two selected childrens' construction of arithmetical knowledge during a school year, focussing particularly on the major cognitive restructurings. *Erna Jackel* (USA) analyzed the social contexts that children mutually construct as they interact to complete mathematics tasks in small groups. *Stieg Mellin-Olsen* (Norway) described some of the project work of Norwegian teachers which develops creative uses of mathematics in social contexts within the classroom. *Jean-François Perret* (Switzerland) reported on empirical investigations about children's understanding of mathematical tasks set to pupils in schools and the discrepancies in meanings exposed between teachers and pupils. *Jan Waszkiewicz* (Poland) in his presentation on "The future cultural role of mathematics and its impacts on mathematics education" took an extreme antiposition to the concept of mathematics as social construction by renewing the platonist ideal of an universal mathematics independent of any cultural or social development and by relating social contexts exclusively to applications of mathematics.

4. Mathematics Education in the Global Village

In the five hearings of this timeslot questions were put, on behalf of society, by the Chair/Questioner to a panel. The Questioners focussed on the accountability of the international mathematics education community in relation to the particular hearing theme. The hearings were a summary and conclusion of the special programme of the day and concentrated the panels on very specific questions. Selected speakers from the special programme who had contributed to different themes and topics were questioned in these hearings in order to get their discursive or conflicting answers.

4.1 Hearing: Which and whose interests are served by mathematics education?

The two Questioners, *Christine Keite* and *Yves Chevallard* (France) raised the following questions amongst others in the hearing: Who defines qualifications provided by mathematics education, and what is the relationship between social needs and individual interests in changing social settings? Is the fact that mathematics, by its social use, has become more implicit and invisible although more widespread than technology, reflected in explicit mathematics teaching? What could be meant by democratization through education, and how can mathematics education contribute to building up democratic competencies while avoiding 'cultural imperialism'? Within the hearing the panellists brought out several particular points. *John Fauvel* (UK) raised the aspect of new technology and argued that there should be more teaching of judgement and wisdom rather than particular skills in order that people should not feel scared of this technology. *Kathryn Crawford* (Australia) agreed that what was needed was a greater confidence to deal with a highly mathematized society but also a better understanding of mathematical structures in relation to everyday life in order to be able to exert more democratic control. For *Ole Skovsmose* (Denmark) the link between interests and mathematics education was the idea of democratic competence in society, and in today's technology society he sees a need to increase people's competence in understanding and reflecting on mathematical models and applied mathematical knowledge. *Murad Jurdak* (Lebanon) felt that unlike language teaching mathematics teaching did not develop in relation to the learners' interests. Rather the interests of the state and of mathematics experts appeared to be of paramount influence. *Claudia Zaslavsky* (USA) showed how in the USA mathematics education was geared to the interests of an elite, through the control of the curriculum by national testing procedures, resulting in feelings of inadequacy amongst large numbers of people. Speakers from the floor interrupted the hearing by raising the more important points of general power imbalance between 'exploited' and industrialized countries, the dominance of interest groups through language, the patronizing stance of Euro–American educators and the plight of the oppressed black people in South Africa. There was a general wish that the issues which this hearing addressed should be debated much more widely within the congress.

4.2 Hearing: How does mathematics education relate to destructive technological developments?

The Enlightenment brought with it the optimistic idea of science as equivalent to progress. Today, faced with military research and enormous ecological problems, it seems that the time of scientific optimism is over. Questioner *Jens Naumann* (FRG) asked the panellists if there had been any qualitative or quantitative changes in recent decades in the scientific and technological developments calling for a much more prudent and cautious approach to the use and applications of scientific advances, and if so, what are the implications for mathematics education. *Jan Waszkiewicz* (Poland) voiced the opinion that, in the concrete case of Poland, mathematics and mathematics education do not contribute to any destruction. On the contrary, they make the citizens critical. In *Hugh Burkhardt*'s (UK) view it is not the scientist who is in the end responsible for the use of his/her discoveries. The social control belongs to all citizens. Mathematics education should educate all citizens to question decisions like the star-wars debate. What has shifted in the last decades is the scale of responsibility science has given to all citizens. *Jens Høyrup* (Denmark) argued that the level of understanding of the social implications of the work of mathematicians and scientists has deteriorated as they have become only elements in a segmented hierarchical system like bureaucracy. *Bernhelm Booß-Bavnbek* (Denmark) completed the reasoning by pointing out that as mathematicians are integrated in such a system they cannot individually control or assess their work. It is necessary to change the relation between human beings and their products, and mathematics education should contribute to this. It is not the scientific attitude in itself that should be blamed for destructive developments, but the 'military technical pollution' of scientific thinking. There are, today, unprecedented risks of technological applications in the civil and military field on the bases of models and simulations void of theoretical comprehension and insensitive to the limits of validity of existing empirical knowledge. Mathematics education should acknowledge this situation.

4.3 Hearing: Do mathematics educators know what they are doing?

Questioner *Jeremy Kilpatrick* (USA) asked the panellists for their view on this question and discussion started from this task. *Nicolas Balacheff* (France) raised the issues of who are the mathematics educators – mathematics teachers, teacher trainers, researchers; what kind of knowledge do mathematics educators need and how can research provide this 'professional' knowledge. *Marilyn Nickson* (UK) argued that it is from the teachers that everything stems. Mathematics educators in general can only be accountable in terms of what mathematics teachers do in the classroom. Action research and ethnographic research can do much to help mathematics educators improve their contact with classrooms. *João Ponte* (Portugal) presented the case that a mathematics educator is anyone professionally concerned with the intersection of mathematics and education. He therefore looks for a more professional engagement with relevant activities, such as curriculum development, assessment, teacher training and research. ICMI has for him a strong role to play in improving society generally. *Hans-Georg Steiner* (FRG) argued for non-simplistic approaches. He showed that much depended on what theoretical perspectives are adopted, as well as on the context in which the question was raised. All the panellists agreed that whatever micro-situation a mathematics educator worked in, there must be acceptance of some responsibility for the macro-situation as well.

4.4 Hearing: What are the challenges for ICMI in the next decade?

Questioner *David Wheeler* (Canada) offered the panellists the main question and three more specific ones based on the idea of ICMI as a scientific, international institution. What should ICMI's *scientific* priorities be in the 1990's? How could

ICMI improve the effectiveness of *international* cooperation in mathematics education? How should ICMI as an *institution* adapt to the challenges of the 1990's? *Aderemi Kuku* (Nigeria) argued that ICMI should continue to do what it does now, but also that it should concentrate on implementing the recommendations in its study series, that it should develop more themes such as the influence of first language on mathematics learning and increasing the role of women in mathematics, and that it should seek more funding for its important work. *Stieg Mellin-Olsen* (Norway) argued for much more open debate within ICMI. He deplored the subdividing of groups and issues into 'sections' which prevented debate, and he challenged the commission to get the debates out of the backrooms and corridors into the plenary context. *Gilah Leder* (Australia) agreed that in relation to its scientific work ICMI should promote more debate, but she also wished for more action and fewer words concerning international cooperation. She recommended ICMI to be more open to alternative approaches to mathematics education issues. *Bienvenido Nebres* (Philippines) felt that a major challenge for ICMI is how to keep discussion and involvement alive between the four-yearly congresses. He saw more regional conferences and other study groups (like PME, the International Group of Psychology and Mathematics Education) as ways to achieve this. He would also like to see ICMI as an international body making more contact with the key decision-makers in individual countries, who determine the kind of support mathematics education receives. *Edward Jacobsen* (UNESCO) proposed an ICMI study on how to provide *more* mathematics education with *less* resources, a challenge facing members from many countries. He wanted ICMI to improve its communicating role both internally and externally, through examining the status and role of National Representatives, through using the Bulletin in better ways and through increasing the international participation rate at the congress.

4.5 Hearing: What can we expect from ethnomathematics?

Questioner *Ubiratan D'Ambrosio* (Brazil) asked the panellists why each of them came to ethnomathematics. *Mary Harris* (UK) contrasted the low achievement of girls in mathematics at schools with their capacity to do complicated needlework, sewing, etc. The problem lies in who defines what mathematics is, in who defines which are the standards. In a male dominated society, womens' mathematics is not acknowledged. *Munir Fasheh* compared the monopoly of Western mathematics with that of Coca Cola: instead of drinking the pure water in their environment people drink Coca Cola thus reinforcing their economic dependency. Western Mathematics is the worst fundamentalism there is as it leaves no choice to believe in it or not. It justifies itself by argument of universalism and objectivity. What we need is meaningful mathematics. *Paulus Gerdes* (Mozambique) explained how racial and colonical ideology negated the capacity of Africans to do mathematics. In order to build up the economy of an independent Mozambique and to defend the country against South-African aggression, the people need to know mathematics in order to master as quickly as possible the necessary mathematics, and the formerly negated mathematical practices may serve as a starting point. *Patrick Scott* (USA) described the objectives and activities of the International Study Group on Ethnomathematics (ISGME), of which he is the secretary. The contributions of the panellists were interrupted by applause from the audience. The available time did not allow for either more contributions from the panellists or for questions from the floor.

♣ ♣ ♣ ♣ ♣

Releted event:

Meeting of the International Study Group on Ethnomathematics (ISGEm).

A meeting of the ISGEm took place on August 1st and it was attended by a large number of interested participants. Gloria Gilmer, from USA, who chairs the Executive Committee of ISGEm, reported on the activities of the group. She announced that the ISGEm will seek affiliation to ICMI during ICME–7. A calendar of future meetings was announced (Phoenix, Arizona, January 1989 and Orlando, Florida, April 1989), as well as plans for future actions. These acctions are geared towards a number of activities focused in four main directions:

1. Research on culturally diversified environments;
2. Curriculum development projects and classroom applications;
3. Out-of-school applications;
4. Conceptual and theoretical foundations.

Contributons of notices on activities related to Ethnomathematics were solicited for the *Newsletter of the ISGEm,* published regularly twice a year, under the editorship of Patrick Scott, School of Education, University of New Mexico, Albuquerque, NM, 87131, USA.

TOPIC AREAS

AND

INTERNATIONAL

STUDY GROUPS

TOPIC AREA 1: VIDEO AND FILM

Chief organizer: Michèle Emmer (Italy)
Hungarian coordinator: Klára Tompa

In recent years the role of new and old media in mathematics education has been rapidly increasing. The extensive use of personal computers, videotapes and videocassettes and now of videodiscs has become more and more important in schools and universities. Many of these technical media are used in an interactive way. Moreover the role of computer graphics is becoming more and more important in producing films and videos. In the future the same role will perhaps be played by the videodisc. The topic group was intended to discuss all these themes, starting from the work produced in different countries of the world.

The work of the group divided into the following subthemes:

1. Research in Producing Mathematical Movies and in Integrating them in the Classroom.

Due to the large number of participants the subtheme was organized in two sessions. The invited panelists for the two sessions were Jacques Courivaud (France), who gave a presentation on the "Integration of films in the Math classroom through examples"; Robert Reys (USA) "Video resources to help and improve Math teaching" and Yves Ardourel (France) "Production of mathematical movies".

Short presentations to the subtheme were given by Anik Demonget and Colette Pelé (France) "Films Analysis"; Peter Tannenbaum (USA) "Entry level Math project: teaching remedial Math through video" and Nitsa Movshovitz-Hadar (Israel) who presented videos from Israeli National Television. A presentation was given by Wendy Caughey (Australia) showing videos produced by the MCTP, the Mathematics Curriculum and Teaching Program from Monash University. Other presentations by Pasi Sahlberg (Finland) "Developing mathematical and scientific thinking through mathematical movies" and Joan J. Vas (USA) "A personalized system of instruction utilizing Math videos".

2. The Role of Media Centers and of National Institutions in producing and distributing Films and Videos: who is a Math-film-maker and who has to support him?

The panelists of the subtheme were Philippa Joy Surgey (UK) who gave a presentation on the Math videos produced by the BBC and Open University, showing several examples. The other panelist was Joel Schneider of the Children's Television Workshop (CTW) (USA) who gave a presentation on "Square One TV: Mathematics on Open Circuit Television". Square One TV is a daily, half-hour mathematics series for children of ages 8–12 years. It has been broadcast in the USA since January 1987. Part of the series was shown during the presentation.

3. Videodisc: a New Possibility for Math Education or just a New Expensive Tool for Old Ideas?

The panelists of the subtheme were Jean Delerue (France) who gave a presentation on "Mathématiques et interactivité: videodisque et banque d'images numeriques", showing in particular the use of a computer in connection with the videodisc; Images from the videodisc "Objectif Géométrie" were also shown during the presentation; Ruggero Ferro, (Italy) gave a presentation on "Communicating primitive abstract Math notions through interactive videodisc" showing the videodisc on elementary algebra and logic produced at the University of Padova.

A short presentation was given by Paul Eduard Laridon (South Africa) on "Videodisc inputs to learning mathematics".

4. Video and Film about Math Education Images of Children and Adults Learning and Teaching Mathematics.

The panelist of the subtheme was David Roseveare, producer of Math series for the BBC, UK. He gave a presentation on "Producing and using videorecording of classroom activity"; parts of videos produced by the BBC were shown.

Short presentations were given by Julien Clifford Smith (South Africa), by Klára Tompa on "Experiences in developing videoproductions for further training math teachers" and by Barbara Jaworski (UK) on "Video as a tool for teacher's professional development".

5. The Role of Films, Videoprojectors and Personal Computers: How to use them in an interactive way?

The panelist of the subtheme was Philippe Bernat (France). In his presentation he showed how images are essential to mathematics and how it is possible using microcomputers to realize in a short time interesting math images in animation. Examples were shown during the presentation.

A short presentation was given by Barbara Feluch (Poland) "Mathematics Lab with personal computers". She showed interactive programs for elementary schools in Poland.

6. Looking to Math Images. The Role of Math Films in Teaching Geometry; the role of animation; the optimal structuring of a Math film and videotape. Interdisciplinary movies: can they be useful to stimulate interest?

The panelist was Michele Emmer, chief organizer, who gave a presentation on "Interdisciplinary movies on Art and Mathematics: an example", showing several slides and the movie "Soap Bubbles" from the series 'Art and Mathematics'.

An important aspect of the structure of To 1 was the evening sessions. Three of them were organized.

1. The Aim Producing Math Videos and Films, presentation by Michele Emmer. In this session three movies were shown: *"Flatland"* by M.C. Escher, *"Symmetry and Space"* by M.C.Escher *"Geometries and Impossible Worlds"*, produced by Michele Emmer in the series 'Art and Mathematics'.

2. Producing Mathematical Movies by Computer Graphics, presentation by Michele Emmer. The two invited speakers, Thomas Banchoff and Heinz-Otto Peitigen, were not able to attend the congress.
   During the session the following films were presented: *"Hypercube"*, produced by Banchoff and Strauss; *"Hypersphere"* by Banchoff et al., *"Frontiers of Chaos"*, by Peitigen et al., *"Computers"* by Emmer, in the series 'Art and Mathematics'.

3. TV, Video and Film by National Institutions, presented by David Roseveare, of the BBC, and Joel Schneider, of the Children's Television Workshop. Videos produced by the BBC and in the series "Square One TV" were shown.

Other movies were shown by request in the video-bar and in other informal presentations, including a series of Hungarian films and the films of the Cité de Science et de l'Industrie de la Villette, Paris, France. Videos of the Israeli National Television were also shown.

A special evening event was dedicated to the memory of the mathematician George Pólya. A video was shown in which professor Pólya was discussing with students a geometrical problem. Title of the film: "Guessing and Proving".

All sessions, in particular the evening ones, were followed by interesting and stimulating discussions among the participants. There is no space to report in more detail on the presentations and discussions. In the *Proceedings of To 1 and Theme Group T2*, Computers and the teaching of mathematics, it will be possible to give more information.

TOPIC AREA 2: VISUALIZATION

Chief organizer: Claude Gaulin (Canada)
Hungarian coordinator: István Lénárt

Survey lecture:

Alan Bishop (UK): Survey lecture about Developments and Trends in the Area of Visualization in Mathematics

30 minute presentations in parallel sessions:

Judah Schwarz (USA): Using multiple representations to enhance learning, teaching and making mathematics

Françis Michel (Belgium): Visualization of affine transformations

Michele Emmer (Italy): Visualizing four dimensions

István Lénárt (Hungary): Visualizing spherical geometry

David Tall (UK): Recent developments in the use of the computer to visualize and symbolize calculus concepts

Theodore Eisenberg (Israel): Visualization in higher mathematics

Michael Swan (UK): The comprehension of graphs by pupils

Annie Bessot (France): Some aspects of the work of the French group "Espace, Géométrie, Graphisme Scientifique et Technique"

Michele Pellerey (Italy): The role of visualization in mathematical problem solving

Marion Walter (USA): Visualizing and problem solving in the classroom

Andrejs Dunkels (Sweden): Visualization of whole numbers with application to divisibility rules

Jan de Lange (Netherlands): Visualization, art and mathematics education

Ken Clements (USA): Assisting children to develop visual imagery in primary school mathematics

John Mason (UK): I for imagery and imagination M. Shaughnessy (USA): The "Mind's Eye in Mathematics" Project

A. Gutierrez – J.M. Fortuny – A. Jaime (Spain): Van Hiele Levels and Visualization
in three dimensions

Panel discussion

"Needed Research and Development about Visualization in Mathematics" with Norma Presmeg (South Africa) as chair and with panelists Ken Clements (Australia) and David Tall (UK).

TOPIC AREA 3: COMPETITIONS

Chief organizer: George Berzsenyi (USA)
Hungarian coordinator: István Reiman

Introduction

The rapidly growing importance of mathematical competitions was excellently reflected by the superb organization of this Topic Area at ICME 5 by *Peter O'Halloran* (Australia). The present organizer's major aim was to emulate the program staged in Adelaide, and to extend it to possible other areas not covered there. In view of the fact that Hungary is rightly recognized as the birthplace of modern mathematical competitions, and since Hungary continues to be a superpower in that field, ICME 6 provided the logical setting for such an extension. The resulting program, which will be summarized below, attests to this accomplishment.

Preliminaries

In order to ensure broad international representation, and to provide equal opportunities to everyone, in February 1986, the present organizer wrote to each of the national representatives of the International Commission on Mathematical Instruction (ICMI), requesting their assistance in shaping the program, and seeking recommendations for appropriate speakers from their countries. Several excellent contacts were established as a consequence of these appeals; appreciation is hereby expressed to those representatives who responded. Unfortunately, too many of these letters, and even repeated appeals were not answered — as a consequence, contact could not be established with the organizers of mathematical competitions in some countries.

A regular column in the Newsletter of the World Federation of National Mathematics Competitions (WFNMC) provided yet another avenue for the creation of initial contacts; the shaping of the program was also partially achieved through these "ICME 6 Update-s". Appreciation is hereby expressed to Warren Atkins

(Newsletter Editor) and to Peter O'Halloran (WFNMC President) for these opportunities. Nevertheless, to complete the program, it was still necessary to rely heavily on personal contacts established in a decade of varied involvements in the organization of many mathematical competitions.

Structuring

At ICME 5 there were four 90-minute sessions devoted to competitions. These sessions were reduced to 60-minute duration at ICME 6, and so it became necessary to utilize them very frugally. The possibility of parallel sessions didn't surface until the last few weeks prior to the Congress — by that time, only minor modifications were possible. Nevertheless, the resulting extra freedom was most welcome; it allowed for more leisurely presentations by the participants. However, even in this way, it was impossible to schedule time for questions and answers following the panel discussions.

The original intention was to address the following themes:

- The discovery of mathematical talents – first rounds of competition pyramids — contests for younger students — contests to popularize mathematics.
- The development of problem solving skills — year-round training via problem sections in journals, correspondence courses — other means of training and self-training.
- The selection of higher level contestants — second rounds of competition pyramids — intensified training sessions — the role of broadening mathematical strengths.
- Mathematical olympiads at the national, multinational, and international levels — directions for the future — reflections on and by former winners.

It was never intended to adhere strictly to these generic themes, but neither were they completely abandoned, in spite of appearances to the contrary. They served as guideposts in the selection of the talks, in their groupings into sessions, as well as in refocusing some of them. However, no attempt was made to create a one-to-one correspondence between the sessions and the themes; due to the busy schedule of the speakers and their differing times of arrival/departure, that would have been impossible. Most importantly, there were several other topics as well, which were equally deserving. These included the difficulties faced by developing countries, the need for involving teachers in the discovery and development of the mathematically talented, the creation of appropriate problems for competitions, various methods for evaluating competition results, and the growing need for continued dialogue about competitions throughout the world.

On Days 2 and 3, Sessions 1 and 2 were devoted to the International Mathematical Olympiads and to national summaries of mathematical competitions, respectively. On Day 6, Sessions 3A and 3B, and on Day 7, Sessions 4A and 4B featured other national summaries, along with discussions of various aspects of competitions. The duration of each presentation was, in general, 10 minutes — appreciation is hereby expressed to *John Hersee* (UK), *Agnes Wieschenberg* (USA),

Malcolm Brooks (Australia), *Alex Soifer* (USA), *John Webb* (South Africa), and
Pierre-Olivier Legrand (French Polynesia) for moderating the presentations and en-
forcing the stringent time limitations. Without their undivided attention it would
not have been possible to squeeze 33 lively presentations into the program!

Invited Presentations

In the listing below, Addresses 1–6 were given in Session 1, 7–10 in Session 2,
11–16 in Session 3A, 17–22 in Session 3B, 23–27 in Session 4A, and 28–33 in Session
4B. Extra time was allotted only to Addresses 1 and 7 — in the first case, the news
was still hot-off-the-press, while in the second case, it was desired to provide a
more complete survey. Nearly all of the talks were complemented by extensive
hand-outs — thanks are due to Elizabeth Fried and István Reiman (Hungary), for
storing/distributing these materials.

A written version of most of these presentations (along with those given in
Topic Area 10: Students of High Ability) is scheduled to appear later this year un-
der the auspices of the World Federation of National Mathematics Competitions.
Inquiries concerning it should be addressed to Peter O'Halloran (School of Infor-
mation Sciences, Canberra College of Advanced Education, P.O. Box 1, Belconnen,
A.C.T., Australia, 2616).

1. The Results of the 1988 IMO — R. Potts (Australia)
2. Past IMO Winners / Where are They Now? — M. Lehtinen (Finland)
3. Mathematics Contests: Time to Take Stock — E. Barbeau (Canada)
4. Reflections on the IMO — G. Jakovlev (Soviet Union)
5. Should the IMO Problems Reflect Progress in Education — C. Dechamps
   (France)
6. The Future of the IMO's — J. Hersee (UK)
7. A Kaleidoscope of Mathematical Competitions in Hungary — E. Fried and I.
   Reiman (Hungary)
8. The Canadian Mathematical Competition — R. Dunkley (Canada)
9. Gender Differences in the Australian Mathematics Competition — C. Annice,
   W. Atkins, G. Pollard and P. Taylor (Australia)
10. The American High School Mathematics Competition — W. Mientka (USA)
11. The Discovery of Mathematically Gifted Students — W. Engel (GDR)
12. Is Mathematics as a Career Field on the Decline? — N. Turner (USA)
13. Hungary's Role in the Development of Competitions — A. Wieschenberg
    (USA)
14. The First 25 Years of the Spanish Mathematical Olympiad — F. Bellot-Rosado
    (Spain)
15. How the Olympiads can Positively Influence the Official Spanish High School
    Program in Mathematics? — M. Conde (Spain)
16. The Canberra CAE Mathematics Day — M. Brooks (Australia)
17. The American Regions Mathematics League — A. Kalfus (USA)

18. The Discovery of Mathematical Talent in an Inner City Area — I. Brown (UK)
19. Competitions: Stimulating or Stultifying — A. Gardiner (UK)
20. Mathematical Competitions in Iran — A. Rejali, M. Tomanian and R. Zaare-Nehandi (Iran)
21. Statistical Evaluation of the Past Mathematical Competitions of Iran — A. Rejali (Iran)
22. Mathcounts USA — A. Soifer (USA)
23. A Method For Removing Guessing In Multiple Choice Examinations — G. Pollard (Australia)
24. Mathematical Competitions: Giving Marks and Comparison of Results — P. Kenderov and J. Tabov (Bulgaria)
25. The Prize Winning Solution at the 1988 IMO — J. Pelikán (Hungary)
26. Mathematical Competitions in Developing Countries — S. Ale and A. Sambo (Nigeria)
27. Mathematical Competitions in South Africa — J. Webb (South Africa)
28. Competitions in Czechoslovakia — A. Vrba and V. Burjan (Czechoslovakia)
29. The Polish–Austrian Mathematical Competitions — M. Brynski and A. Makowski (Poland)
30. Winter Mathematical Holidays in Bulgaria — E. Stojanova and L. Stojanov (Bulgaria)
31. Can a Local Problem Solving Contest Help to Develop Problem Solving Skills? — F. Grandsard and A. Schatteman (Belgium)
32. Mathematical Olympiads in Costa Rica — B. Montero (Costa Rica)
33. Mathematical Competitions in French Polynesia — P. Legrand (French Polynesia)

Concluding Remark

In spite of the fact that there were many other attractions at ICME 6, each of the sessions dealing with competitions was well-attended. Unfortunately, due to the parallel running of some sessions, not everyone could hear all of the talks. Moreover, even this way, no time was available for questions and further discussions. Consequently, it is recommended that more time should be devoted to this Topic Area at future Congresses. The constantly growing popularity of mathematical competitions necessitates a corresponding expansion in the time devoted to them.

Related Programs

Second Meeting of the World Federation of National Mathematical Competitions.

In view of the absence of WFNMC President Peter O'Halloran, the meeting was conducted by Vice-President Ron Dunkley. The Australian Mathematics

Foundation Ltd. pledged to continue its support of the WFNMC for 4 more years; Vice-President *Walter Mientka* indicated that the Committee on the American Mathematical Competitions will probably be willing to take over in 1992. The present officers of the WFNMC were reelected, with the addition of *George Berzsenyi* as Associate Editor.

Meeting of the Editors of Journals for Students.

Shmuel Avital (Israel) conducted the meeting, with emphasis on the need for publishing more complete and better motivated solutions intended for student readers.

A Documentary Film on the Australian Mathematics Competitions.

This excellent 17-minute film was first shown at ICME 5. The present showing of the film was well-received. It documents the history of the AMC, and it partially explains its unparalleled popularity: 1 out of every 4 students in Australia take the AMC.

Reception / Dinner / Folklore.

This event was hosted jointly by the organizing committees of the American, Australian and Canadian Mathematical Competitions, with Walter Mientka, Warren Atkins and Peter Taylor, and Ronald Dunkley representing these organizations, respectively. Special thanks are due to our physicist friend, Esther Toth, for her excellent arrangement of this relaxing and festive event.

TOPIC AREA 4: PROBLEMS OF HANDICAPPED STUDENTS

Chief organizers: **Olof Magne (Sweden) and Emmy Csocsán (Hungary)**

The participants worked in two groups. The short reports covered two basic themes:

1. Characteristics and guidelines for studying the processes of mathematics in different groups of handicapped children.

2. The use of electronics to help the process of learning mathematics for handicapped and retarded students.

Session 1:

Chair: Julia Csongor (USA)

1. The lecture of Olof Magne was read by his colleague Leif Hellström, who said that Magne had an operation on 4th July 1988, and therefore could not take part on ICME 6. Hellström talked about the contact between Sweden and Hungary in the field of mathematics teaching in special education and the positive role of the international seminar on "Teaching children with difficulties in mathematics" in Nyíregyháza in 1977.

Magne's lecture was about the psychology of remedial mathematics. Studies on remedial mathematics still represent a neglected area of research and development work. A pioneer was the Hungarian Paul Ranschburg who in 1904 wrote about arithmasthenia. A more modern term would be dysmathematica. This phenomenon may be treated as a psychological and neurological dysfunction or as a restriction of the scope of mathematical understanding. Three main models for remedial mathematics are suggested: (1) The behaviour deviation model; (2) The content deviation model, and (3) The factor-interplay model. Probably, all disabled persons are handicapped in their mathematical learning to some extent. Viewed from the content approach, major areas of mathematics often correspond to typical behaviour requisites. The inference is that remedial teaching in mathematics must emphasize balance between subject matter and individual learning reactions.

2. Françoise Cerquetti-Aberkane (France) spoke about handicapped or "maladjusted" school children. Her research work deals with children whose "handicaps" are of a social, cultural, emotional or even of a psychological or mental nature. These pupils were observed over a period of two years. After a general description of various problems the research group worked out methods to help the children to solve their difficulties.

3. Tarnai Ottóné (Hungary) talked about the problems of mathematics teaching of mentally handicapped pupils in the Hungarian school system. She has been dealing with the introduction of mathematics teaching in Nyíregyháza institutions for education of handicapped children since 1971. They have established effective methods on the basis of the handicapped pupils' conditions which can be built into the process of learning. They devised new books with instructions, suitable for teaching different groups.

4. Janza Károlyné (Hungary) spoke about the results and perspectives of teaching mathematics for mentally handicapped children. She talked about the research work in the National Institute for Pedagogy in Hungary, the methods of compensating development of intellectual functions, the improving of curriculum and the innovations in teaching mathematics for mentally handicapped children.

The reports were followed by discussion.

Session 2:

Chair: József Buday (Hungary)

1. Robert Dieschbourg (Luxemburg) discoussed LOGO, airtramp and socially handicapped children. A group of children have been introduced to the commands FD N, BK N, LT 90, RT 90. Movements of the turtle on the screen have been executed by the children on the airtramp and conversely. The command SETPOS(X,Y) enables access to the notion of ordered pairs, allows the illustration of their addition, preparing the notion of vector addition. Whenever possible, the integration of psychomotor and mathematical education was an aim of the study.

2. András Kertész (Hungary) showed different computer programs for teaching mathematics to handicapped children. He used a Commodore 64 computer.

Session 3:

Chair: Leif Hellström (Sweden)

1. Josette Adda (France) spoke about the Pygmalion-effect and socio-cultural "handicap". Mathematics teaching is neutral, neither for socio-cultural aspects nor for gender. Errors do not always lead to failure; there are many behaviours of teachers and students confronted with the same error. Stereotypes and prejudices influence teachers according to the status of parents, school attendance of elder brothers and sisters; teachers' behaviour depends on the labelling (stigmatization) of pupils by localization of school, type of section, previous results... The student

plays his/her role (according to family environment and his/her labelling). All this was presented through case studies more particularly from observations in "specialized classrooms" and during "orientation councils", in various schools in France.

2. Jenő Bődör (Hungary) talked about research and correction of dyscalculia at elementary school The aim of research was to investigate methods of correction that serve to eliminate dyscalculia. It aimed at complex and synthetic studying of developmental deficiencies in mathematical learning abilities. As causes of dyscalculia differed greatly, there were significant differences in symptoms and methods of correction.

3. Sarah L.B. Wildig (UK) discussed about hearing impaired children. Hearing impaired children experience many difficulties acquiring skills in mathematics compared to normally hearing children. This is due to a variety of reasons including the delayed acquisition of basic conservation concepts and restricted language skills. In England today between 80–85 % of hearing impaired children have access to the mainstream curriculum in mathematics. This in itself poses a number of problems due to the frequent mismatch in the age and the mathematics skill level of the mainstream hearing-impaired child. The lecture set out to highlight a number of these problems, and strategies employed in an attempt to overcome them.

4. Emmy Csocsán (Hungary) spoke about blind children's models of basic mathematical concepts. Blind children at the age 3–6 display various levels of knowledge and skills. The impressions established by manipulations have a decisive role in forming primary mathematical concepts. The impressions of the blind and those of the sighted are different from each other. This difference offers an explanation for the fact that the concept of natural numbers is not primarily formed from the concept of set, as is the case with sighted pupils. For them operations, i.e. functions, are more adequate. The whole process of cognition can be modeled with a system of paired objects. The process of forming the basis of set theory in the case of the blind can mostly be represented by the Neumann axiom system.

Session 4:

Chair: Robert Dieschbourg (Luxemburg)

1. Isydor Karwot (Poland) talked about the methods of teaching whole numbers for mentally and social handicapped children.

2. Helena Siwek (Poland) spoke about the problems of mathematics teaching for slightly mentally handicapped children in Krakow.

3. Klára Marton (Hungary) showed a computer for the physically handicapped. The Bliss system is a special communication method for non-verbal handicapped people. It consists of symbols and pictograms with which people can create sentences. Using personal computers makes communication easier and quicker. Klára Marton demonstrated the program in operation.

TOPIC AREA 5: COMPARATIVE EDUCATION

Chief organizer: Douglas Quadling (UK)
Hungarian coordinator: Tamás Varga

This meeting was dedicated to its Hungarian coordinator, the late Tamás Varga, in recognition of his outstanding contribution to international cooperation in mathematical education.

In the first three sessions the group studied three special topics from a comparative point of view: (1) the task of the mathematics teacher, (2) the management of curriculum change, (3) differentiation between pupils.

1. The Task of the Mathematics Teacher

N.C. Witman (USA) had observed classes in Japanese primary schools. She indicated the emphasis on instructing the class as a whole, information being imported through careful questioning, active pupil involvement and creative use of the textbook. Pupil participation is encouraged, e.g. by demonstration at the blackboard. There is a strong belief in the importance of the teacher's input, with 50–75% of class time devoted to developing a lesson: review of prerequisite knowledge, explaining new concepts, assessing comprehension, short opportunities for practice. Logical reasoning is stressed, and at times a pupil may challenge the teacher. Despite large classes, the atmosphere is non-threatening; this was attributed to the importance of the group in Japanese society. The class behaved as a large family, taking pleasure in the success of individuals but giving support to those less well endowed. (In discussion this was contrasted with experiences in some African countries, where conditions were apparently similar but negative pupil attitudes were common.

G. Gjone (Norway) outlined fundamental changes in years 1–9 in Norwegian schools. A centralized curriculum based on common textbooks and examinations is to give way to a large measure of local autonomy, with much of the responsibility for curriculum development (and perhaps also assessment) devolved to local groups of teachers. However, the tradition in teacher education is of unitary education in all school subjects; investigation and exploration play an important part in the work of the school. Also, in the early years teachers progress up the school with their pupils, thereby creating strong social bonds but requiring each teacher to have a comprehensive knowledge base. There is concern as to how to make proper provision in mathematics under such conditions; lack of individual expertise may had many schools to adopt open-plan, team-teaching arrangements. The question has yet to be resolved, how one can operate without subject specialists in a decentralized system. In Japan also the teachers are not mathematics specialists; but effective subject teaching is an important element in teacher training and in school expectations.

2. The Management of Curriculum Change

Y. L. Cheung (Hong Kong) contrasted older approaches, implemented only through textbooks, with current practice — less economical, but more effective — operated through government curriculum development committees (CDCs), which are able to identify and integrate resources. Teachers, the most effective agents of curriculum change, have to be involved not only in implementation and evaluation but also in planning and development; other agencies include teacher educators, mathematicians, textbook Writes, higher education, educational TV, examination authorities. Major changes require planning conferences, writing workshops, classroom trials, etc. and pose problems of liaison and timing: teacher training programmes must be changed, the concerns of existing teachers must be met — as well as those of employers. Gradual continuous adaptation — which can be more easily organized through the CDCs, and quickly introduced — is a better model for change.

V. Szetela (Canada) offered a specific example of major change, initiated in the early 1980s in response to current dissatisfactions, more critical attitudes and technological advances, and due for implementation at various levels 1987–90 — a time-scale necessitated by the iterative, consultative developmental model adopted. An initial intention to produce broad global objectives was gradually modified because of demands from teachers for more specific guidance. Some potential liaison problems have been eased through the participation of the British Columbia Association of Mathematics Teachers, intensive consultation with schools, contact with potential publishers, and well prepared in-service support. Points raised in discussion included the need to consult with other subject interests, and the extent to which curriculum change should be concerned with teaching style as well as content. A delegate from Uganda pointed out that in many countries the curriculum is dominated by requirements of external examinations over which teachers have no control, but which are the principal means of personal advancement in highly competitive societies.

3. Differentiation between pupils

Problems created by differences in vocational aspirations and personal capabilities attracted strongly contrasting solutions.

W. Headlam (Australia) spoke of the proposed Victoria Certificate of Education (VCE) for pupils in the final two years of schooling, based on principles of *access* and *success*. Qualification will be by satisfactorily completing 16 out of 24 one-semester studies (of which any number up to 8 may be in mathematics): The study has specified aims and objectives, broad content, work requirements and assessment structure, but the depth to which these are taken will depend on the background, abilities and vocational ambitions of individual pupils. By replacing book-based, fixed-content formal examinations with more open project and problem-solving tasks, the hope is to achieve maximum flexibility within a common system, and hence improvement in pupil motivation and attitude. There will, however, be difficulty in establishing standards for entry to higher education.

A. Coolsaet (Zaire) described a system of secondary education in which nationally determined curricula and examinations are used to achieve differentiation, by directing each pupil (after the two-year orientation cycle) to an appropriate course: short cycle (3 years, vocational studies) or long cycle (4 years, leading to a possibility of higher education, with choice from mathematical/physical, biological, technical, literary, economic or pedagogical sections). Mathematics is a crucial examination subject, both for initial placement and for re-allocation where necessary after the third and subsequent years. However, this means that many pupils experience failure in mathematics before they find their proper niche. Also, in the scientific/technical sections of the long cycle (in which the mathematics taught is partly modern/theoretical and partly classical/applicable) tension exists between those pupils whose sights are on higher education and those intending to take a job when they leave school.

4. Education and culture

The final session opened with a discussion on education and culture. Schools are a bridge between the home culture and the learned culture. In 19th century Europe, a variety of educational traditions emerged from the strengthening of national identities. When these were implanted in the quite different cultures of other continents, the situation was complicated by a language of learning different from the mother tongue, by multi-cultural groupings within a single school, and by a far wider culture gap between home and school. Nowadays, there are parallels to this in advanced countries which have admitted large numbers of immigrants. Mathematics could have a special role to play in bridging the gap, since its procedures — though not its applications — are relatively culture-free. (Evidence was quoted that second-generation immigrants in western countries are especially successful in mathematics.) However, mathematics as we know it developed largely through the medium of Indo-European languages; and whilst it seems that Chinese and Japanese pupils often find the subject congenial (possibly because their languages are relatively free of redundancy, have simple numeration systems and geometry-supportive forms of writing), attempts to convey mathematical concepts in many African and Polynesian languages have been unfruitful. The various international schools, which are well planned to throw light on such questions, have often avoided the issue by a policy of separation.

Political and social questions were also raised. Norway is moving away from centralization, the USSR is introducing more elective courses, the VCE will establish a broader base for tertiary education and employment, but in England and Wales there is to be more central control, less student choice, and proposals for broader sixth-form studies have been rejected. Do the reasons for these changes lie within or outside education? Japanese pupils have produced impressive results in SIMS, yet there is pressure for reform of mathematics teaching in that country. In schools in Zaire the pedagogical section is the destination of those who have failed (especially in mathematics). Recent government action in the UK implies low official regard for primary school teachers, yet in Japan they are highly respected and well paid.

These four trends — cultural, linguistic, political, social — could provide a suitable structure for continuing comparative studies within the framework of ICME.

Chairs: *J-P. Labrousse (France), D. Hill (USA), A. Hirst (UK)*

TOPIC AREA 6: PROBABILITY THEORY AND STATISTICS

*Chief organizer: **Kenneth J. Travers** (USA)*
*Hungarian coordinator: **Katalin Bognár***

Note: Lennart Råde (Sweden), who originally served as co-chief organizer, had to withdraw prior to the Congress.

Four sessions were devoted to this topic:

1. Ideas for teaching probability
   Contributors: Toshiji Kondo (Japan) and Michimasa Kobayashi (Japan)
   Organizer: Lennart Råde (Sweden)
2. Exploratory Data Analysis
   Contributors: Rolf Biehler (FRG) and Gail Burrill (USA)
   Organizer: Rolf Biehler (FRG)
3. Statistics at the primary school level (age: 5–11 years)
   Contributors: Andrejs Dunkels (Sweden), Lionel Mendoza-Pereira (Canada) and Julianna Szendrei (Hungary)
   Organizer: Lionel Mendoza-Pereira (Canada)
   Note: Benjamin Eshun (Ghana) was unable to attend the Congress.
4. Statistics at the secondary school level (age: 12–18 years)
   Contributors: Katalin Bognár (Hungary), Anne Hawkins (UK) and Albert Shulte (USA)
   Organizer: Gottfried Noether (USA)
   Note: The paper by Anne Hawkins was read by Gottfried Noether in her absence.

The major theme of this topic area was consonant with a primary thrust of the entire Congress, that the mathematics curriculum must be updated to reflect scientific and technological developments in order that school graduates will be better prepared for functional living in the 1990s and beyond. A significant component of the needed reform will be devoted to concepts and strategies that prepare pupils to deal intelligently with quantitative information (that is, with data). From the four presentations and ensuing discussions, *five identifiable topics* for deliberation emerged.

1. Exploratory Data Analysis (EDA)

EDA is an approach to organizing and analyzing data that has gained considerable attention in some countries (e.g. West Germany and the United States) in recent years. (An important reference for EDA is Tukey, J.W. (1947) *Exploratory Data Analysis*, Addison-Wesley, Reading, USA). The methods of EDA, which feature novel, easy-to-use techniques such as stem-and-leaf and box-and-whisker polost, have been successfully used by students at a variety of grade (age) and ability levels. Biehler (FRG) presented some key ideas of EDA and how these ideas may affect curricula and instructional practice. He provided examples of graphical representations of data, including ways of comparing several sets of data in one variable and of analyzing multivariate data. Mendoza-Pereira (Canada) and Dunkels (Sweden) outlined how stem-and-leaf plots have been used with children in lower primary grades. Such activities also provide practice in number ideas, such as ordinality and place value.

EDA-Based Projects

(i) Exploratory Data Analysis Teaching Experiment (EDATE). This Project at the University of London has developed teaching materials for a wide range of curricular levels and for students with varying ages and abilities. (Reported by Jon Ogborn, London Institute of Education, UK)

(ii) Quantitative Literacy (QL) Project. The National Council of Teachers of Mathematics and the American Statistical Association, both of the United States, appointed a committee several years ago to promote statistics education. This committee has produced several excellent publications and recently has developed the QL project to prepare teachers to deal with probability and statistics. QL has incorporated into its materials many ideas from exploratory data analysis (Reported by Gail Burrill (USA) and Albert Shulte (USA)).

2. Probability

Effective approaches to teaching probability make use of "hands-on", empirical methods. At the primary school level, Julianna Szendrei (Hungary) provided several examples of classroom experiments in probability that can be carried out using materials readily available to children — nut shells, bottle corks, etc.

For the upper secondary level, Toshiji Kondo (Japan) conducted a demonstration lesson in probability in which Congress members were participants. The lesson, which was, as promised, lively and enjoyable, included exploring the Buffon Needle Problem and generating a sampling distribution (using dice) to simulate radioactive decay.

3. Statistics

More and more, statistics is finding its way into the curricula of countries around the world. In order to obtain information on the teaching of statistics in the socialist countries, Katalin Bognár (Hungary) conducted a survey and received responses from Bulgaria, China, Czechoslovakia and the German Democratic Republic. Information about Hungary was available from an article prepared by Tibor Nemetz. It was generally found that in the primary grades a few ideas in statistics (e.g. measures of central tendency and dispersion) may be included in the textbook or syllabus for all children. At the secondary level, however, the pattern is likely to be that statistics is taught only to pupils who are in specialized courses leading to university study. The opinion was expressed by Dr.Bognár (and endorsed, it seemed, by many in attendance) that "statistics must not be considered a separate subject. Rather, probability and statistics must be integrated as in 'new mathematics' curricula. Statistics provides the empirical background for concepts in probability. On the basis of knowledge of probability, statistical inference can be made".

In the United Kingdom (reported by Anne Hawkins), there is soon to be released a new national curriculum. However, a recent "discussion document" about the new program makes no mention of statistics. On the other hand, there is at the upper secondary level a movement toward "half A-level" courses. This provision makes room for statistics, and is intended to address both applied and theoretical aspects of the topic.

4. Teacher Education

Implications for teacher education (both in-service and pre-service) were in evidence in each of the group sessions. As Biehler noted, it is important to resist the tendency to reduce the methods of EDA to mere techniques, rather than utilize the full investigative spirit of the EDA approach. Burrill put it another way. She observed, "Without guidance and direction, teachers will concentrate on the graph rather than the interpretation, on the calculation rather than what it says about the data". Kobayashi informed the group of the work of the Association of Mathematical Instruction, a national organization in Japan that is dedicated to the creation of enjoyable lessons.

Bognár reported, based on her survey of some socialist countries, on the existence of training programs for teachers of statistics and probability. "There is indication that teachers are more and more enthusiastic, and are producing their own material." The teaching of probability and statistics, it is clear, requires resourceful, energetic and imaginative individuals who are willing to try new approaches and deal with concepts that are, for many teachers, unfamiliar. Effective implementation of curricula dealing with these topics requires significant commitments of time and resources to teacher education programs.

5. Impediments to implementation

While there was evidence in each reporting country of significant numbers of enthusiasts who are doing excellent work in promoting probability and statistics in the schools, it was at the same time clear that in no country was the topic taught universally across all schools and all grade levels. This brief report concludes with a listing of some impediments to the teaching of this topic that need to be dealt with before widespread implementation will occur.

5.1 Unfamiliar content

For mathematics teachers, perhaps the majority, the content of probability and many statistics is not familiar. Pre-service preparation programs need this subject matter incorporated into their curricula. Extensive in-service training is needed for practicing teachers.

5.2 Novel instructional approaches

The effective teaching of statistics and probability requires a variety of approaches, including the collection and organizing of data, promoting discussion of results and assisting students in carrying out investigative projects, either as individuals or in groups. For many teachers, this requires considerable changes in instructional styles.

5.3 Availability of materials

The major projects that were reported upon have placed a priority on the preparation of instructional materials for use by teacher and student. Such materials may not be available in some school systems. However, as the presentation by Julianna Szendrei so clearly demonstrated, very good use may be made of objects that are part of the child's world (corks, marbles, etc.) and are available at little or no cost.

5.4 External examinations

— It is a fact in many teaching situations that examinations drive the content of instruction. For this reason, it was very encouraging to see in the United Kingdom an example of how modifications in the conventional examination process can facilitate curriculum change. There is a move in the UK toward a continuous assessment component of the syllabus that allows for students to undertake a statistical investigation (or, less often, a probabilistic project) as a part of the requirement for the school-leaving diploma (the General Certificate of Secondary Education). Such an option offers an excellent opportunity to pursue major goals of contemporary instruction in statistics and probability.

TOPIC AREA 7: PROOFS, JUSTIFICATION AND CONVICTION

Chief organizer: David Pimm (UK)
Hungarian coordinators: Mária Halmos and Viktor Scharnitzky

Proof is regularly offered as one (if not the) hallmark of mathematics, a characteristic which singles this discipline out from others, by its use as a 'guarantee' of the validity of its knowledge claims. Lakatos' view of proof highlights its social aspects, in particular the need for acceptance by a community of mathematicians, before something can stand as a proof. In Yu. Manin's thought-provoking claim, "a proof only becomes a proof after the social act of 'accepting it as a proof'."

The first session consisted of two talks and the second session took the form of five discussion groups.

The Teaching of Mathematical Proof: A Social and Cognitive Perspective

Lecture by Nicolas Balacheff (France)

The teaching of mathematical proof appears as a major problem for almost all countries. This problem has been generally conceived as the teaching of an abstract argument about a mathematical object. Balacheff reconsidered this problem in the light of results of his recent experimental research which documents the important roles played by both the social dimension of the teaching situation and by cognitive features associated with the learner.

On Proofs, Convincing, and Developing a Mathematical Point of View

Lecture by Alan Schoenfeld (USA)

Learning to accept proofs as meaningful arguments about mathematical objects is an important part of learning to think mathematically, but is far from the whole story. Understanding that proofs 'guarantee' that things must be a certain way is one instance (among many) of 'seeing the world from a mathematical point of view'. Schoenfeld's talk explored aspects of mathematical thinking, and ways in which classroom interactions could foster or hinder the development of such a perspective.

Formalism as Communication

Group-leader: Uri Leron (Israel)

Mathematical formalisms in general, and proofs in particular, are normally viewed as means for securing the validity of results. Suppose that we adopt an communication. What desirable mathematical styles follow?

Uri Leron started by claiming that proof is difficult for students; thus some math educators find ways to work around it (e.g. use intuition, visualization). Is teaching proof a hopeless task? Proofs involve the communication of results which are arrived at through discussion, negotiation and construction. Assuming correctness, we can ask how communicative is a proof? Yu. Manin writes "A good proof is one that makes you wise." So a good presentation of a proof makes the reader/listener wiser. Proofs are codes for ideas and professional mathematicians know how to decode them. Students may not know how to decode them, or they may mistake the code for the ideas.

While the introduction focused on the nature and presentation of formal proofs, the discussion went more in the direction of the understanding of proof, and the necessity of proof in the curriculum. Two issues arose: the development of the students felt need for proof and the development of their skills in the use of formalism.

Research on Proof

Group-leader: Gila Hanna (Canada)

What are some current aims and intents of proof research in mathematics education, and what should they be? How do we take into account the student's level of mathematical knowledge, of language and of understanding of the purpose of proof?

Gila Hanna spoke briefly about the common issues in current research on proof: demystification of formal proof, placing meaning above form and emphasis on proof as a debating forum. The idea of demystification of formal proof, and consequently reducing the importance it is given in the curriculum, comes from a desire to be more faithful to mathematics as practiced and not from pedagogical considerations. It will not make mathematics easier to teach; it will be more demanding of the teacher and students to insist on a demonstration of 'understanding' (whatever that means, than a memorization of a formal deduction. The feasibility of turning a classroom into a debating forum seemed attractive but its viability needs to be assessed. These issues were the subject of a number of research projects and results were given in short interventions by N. Balacheff (secondary), J. Volminck, P. Tinto and D. Chazan (all junior high), D. Alibert (tertiary) and E. Wittmann (across the curriculum).

Changing Notions of Proof

Group-leader: John Fauvel (UK)

The problem is no longer whether mathematical truth is time-dependent, but what consequences this has for mathematics education. Is there a pay-off for mathematics teachers in classrooms, or does the relativist historian's conclusion make life more difficult by subverting the claims which for some have justified mathematical pursuits? How can knowledge of past proof-activities help resolve this dilemma?

Dirac remarked, "I don't care about proofs, I want to know the truth". The great variations in the concept of proof across time and within and across cultures occupied much of the discussion. Thus it was clear from examples given by Indian and Chinese delegates that proof conceptions in those cultures had been very different from Europe's Greek legacy. Again, it was observed that (e.g.) topology and algebra have quite different standards of proof. It was argued, in a legal analogy, that proof is a special case of evidence. And since mistakes in long, complex proofs are normal, but do not apparently affect the proof of a result, proof should be seen as more like a living root system than a simple chain whose breakage destroys the system. A cheerfully realistic spirit permeated the group: "After all", one participant said, "if set theory turns out to be inconsistent, bridges won't fall down".

Research into Practice

Group-leader: John Mason (UK)

How can we use available research work about proof and proving — in particular, what we have heard in the plenaries — to affect our classroom practice? How might we work on bringing our pupils to a greater awareness of proof processes and the justification and value of proof as a social mathematical activity?

We attempted to reconstruct what Balacheff and Schoenfeld had outlined in the previous session, by recalling the technical terms they had used, talking in pairs to a neighbour about examples of these terms followed by a public airing of those examples and a discussion of the extent to which they were indeed examples. The title Research into Practice was construed by some as the research of practice, and by others as the translation of research findings into practice. One salient issue was the interpretation of the French word experience which means experiment rather than experience, and yet which carries the sense of the experience or state of experimenting.

Convincing Yourself and Others

Group-leader: Kaye Stacey (Australia)

What characteristics make mathematical arguments convincing to us and to students? Should we and, if so, how can we modify arguments and proofs used in teaching so that they are more convincing? Are there dangers in sacrificing attributes such as conciseness, the use of minimal assumptions and elegance to make proofs more convincing.

Convincing is an inner state, justification involves reasoning, while proof is peculiar to mathematics. This group was concerned with presenting proofs so that they were convincing to others, and not necessarily with traditional rigorous proof. One tension was between producing concise rigorous proofs and longer, less rigorous ones which convince pupils/students. Are these differences reflected in the atmosphere created in classrooms? Do pupils expect to be able to learn what is taught, and are they expected to reason, argue and justify?

TOPIC AREA 8: LANGUAGE AND MATHEMATICS

Chief organizer: Colette Laborde (France)
Hungarian coordinator: Zsuzsa Somfai

Four sessions were offered on this topic area, involving over thirty presenters from several countries in the world. During three sessions, the participants worked in subgroups. A plenary meeting was organized at the closing session in which the leaders of each subgroup reported on the work of their subgroup.

Language and mathematics is a theme of worldwide concern and the choice of the themes of the subgroups aimed both at taking into account the cultural and social differences arising in the various teaching contexts in the world and at elucidating regularities in the language processes involved in mathematics teaching and learning with regard to specific areas in mathematics (especially algebra). The following is a summary of important issues raised in each subgroup.

Subgroup 1: Natural Language and Symbolism

Leader: Pamela Matthews (USA)

Algebra was a question of high interest in the subgroup and gave rise to the discussion of different points of view. It is commonly assumed that mathematical rules are explicitly given, systematically derived, consciously deployed while rules of natural language syntax are recognized to be implicit, unconscious and automatic. *Kirshner* (USA) presented a research perspective viewing the symbolic elementary algebra according to the model of a natural language and giving a linguistically-based explanation of presence of linearity errors in the distributivity as unconscious experimentation from the learner with distributivity towards discovery of the maximal context in which it is legal to distribute. *Matthews* (USA) proposed to approach the learning of algebraic language by comparing the grammatical structures of natural language and the language of algebra. *Drouhard* (France) presented a teaching of algebra meant for adults and based on a syntactical analysis of algebra. In order to investigate the role of the language in pupils' understanding of algebra and the nature of the difficulties associated with algebraic symbolism, *Sakonidis* (UK) reported on an experimentation in which students were asked to expound on their interpretation of algebraic terminology and notation. Then their explication was used as a basis to develop further the students' understanding of the language.

Pimm (UK) used the case study of conic sections to explore certain mathematical processes of concept extension and change in terms of both the classical ethorical and semiotic concepts of metaphor and metonymy.

Parzysz (France) and *Penkov* (Bulgaria) approached the question of integrated activities in language and mathematics. Penkov presented an experimental teaching made in Bulgaria giving an integrated knowledge in the two areas and Parzysz reported on an in-service teacher training focusing on the linguistic problems encountered in the maths class.

Subgroup 2: Cognitive Aspects of Language in the Learning of Mathematics

Leader: John Conroy (Australia)

A theme occurring in several papers was to try to relate language development or language structures to the development of particular mathematical structures especially with regard to algebra.

Lowenthal (Belgium) considered the relationships between mathematical activity and language acquisition on a case study in which he used non-verbal communication devices to assist aphasic and normal children in the development of classification skills. *Cohors-Fresenborg* (FRG) discussed ways in which older students learn to use language with precision to develop axiomatization and formalization in mathematics.

Norman (USA), *Warrinier* (Belgium) and *Arzarello* (Italy) applied ideas of linguistics to ways in which students perform or develop competence in various areas of mathematics:

- the notion of linguistic transformation is applied to the investigation of the strategies that students employed in unitizing literal complexes (Norman);
- the triads of Pierce is applied to the student's acquisition of the concept of vector (Warrinier);
- the interaction between deep and surface structure of language is applied to the investigation of the student's strategies when solving word problems for multiplication (Arzarello).

Rachlin (USA) described the use of Vygotsky's notion of the zone of proximal development as a means of developing an algebra curriculum.

Another theme was the investigation of oral or written formulations of the students. *Relich* and *Conroy* (Australia) investigated the ability of students, recently introduced to multiplication to represent their ideas, orally, concretely and pictorially. *Gallo, Amoretti* and *Testa* (Italy) presented an account of students' ability to produce written and oral descriptions of geometric figures.

Subgroup 3: Social Aspects of Language in the Teaching of Mathematics

Leader: Hermann Maier (FRG)

The group focused on oral communication in the mathematics classroom. All presenters reported on observation of actual classroom activity as an important medium of getting data for later, mainly ethnographical, analysis. Two main themes were discussed: the verbal interaction between teacher and students, and the discussion among students.

Cuevas (USA) and *Dawe* (Australia) reported on an exploratory case study investigating the ways in which teachers use language functions in negotiating meaning in introductory algebra with respect to the context of American and Australian schools. *Maier* (FRG) analysed the processes of understanding for students taking part in classroom communication, especially in the question–answer sequences.

Bartolini-Bussi (Italy) carried out classroom experimentations in order to analyse how classroom discussions affect mathematical learning. *Pirie* (UK) analysed pupil–pupil discussion within a normal classroom setting in a longitudinal study and was especially concerned with the use of 'ordinary' language to discuss mathematics and with possible confusion arising from the use of mathematical language.

Subgroup 4: Cultural Aspects of Language in the Teaching of Mathematics

Leader: Kathryn Crawford (Australia)

The relationships between language, culture and teaching mathematics were explored from different perspectives: learning mathematics in a second language, the difficulties of children from minority (non-technical) groups to interpret mathematical information unless assumptions about western contexts for mathematical activity are made explicit, a cross-cultural study between France and Japan.

Kazadi wa Mashinda (Zaïre) discussed a study investigating the logico-linguistic problems of 14–17 year old Zairean students who are taught in French. *Crawford* (Australia) discussed the needs of students from non-technically oriented cultures and minority groups in multicultural societies for information about why, when, where and what as well as 'know how' about mathematical procedures and axioms.

Two presentations dealt with the language of mathematics in Japan. *Ozawa* (Japan) described the effects of cultural and political changes in Japan on written notations in mathematics. *Hosoi* (Japan) and *Denys* (France) made a comparison of mathematical geometric terms in French and in Japanese on two aspects: Kanji (ideograms) versus syllabic forms, the introduction of new geometric terms in textbooks from both countries.

Subgroup 5: *Communication with a Microcomputer*

Leader: Jacqueline Giard (Canada)

Microcomputers provide a new teaching and learning environment which requires a new way of communicating and the use of specific language.

The problem of the criteria on which the choice of appropriate languages may be based was discussed by *Giard* (Canada). *Rogalski* (France) analysed how the operative communication with the microcomputer is concerned with students' mental representations of the informatical device and what 'external' representations can be used to produce positive changes in the students. *Schwank* (FRG) investigated the cognitive processes underlying the behaviours of the students when they have to solve algorithmic problems. Some general problems introduced by the integration of computers in schools were raised by *Yanagimoto* (Japan).

The relationships between the language abilities of the children and the use of the microcomputer were approached by the presentations of *Yokochi* (Japan), *Dawe* (Australia), and *Sutherland* (UK). Yokochi reported on the changes in the means of expressions introduced by videos, microcomputers and to what extent it can affect the development of languages abilities. Dawe described a project which deals with the specific use of microcomputers for improving the mathematical thinking of children who suffer from a language handicap. Sutherland investigated, through peergroup discussion and through interactions with the computerfeedbacks, the relationships between the pupil's formal LOGO representation and their negotiation of a generalizable method in natural language.

Subgroup 6: *Textbooks in Mathematics Teaching*

Leader: Stefan Turnau (Poland)

Textbooks are not only meant for teachers but also for students. To what extent can the students understand mathematics textbooks? This question was considered from two points of view:

- the problem of the ambiguity of terms and symbols: should it be avoided or cultivated? (*Turnau*, Poland);
- the problem of measuring in an objective way to what extent a textbook is readable or understandable? (*Gagatsis, Patronis*, Greece).

TOPIC AREA 10: STUDENTS OF HIGH ABILITY

Chief organizer: Petar S. Kenderov (Bulgaria)
Hungarian coordinator: János Pataki

The work of this Topic Area consisted of two 90 minute sessions. The preliminary program contained 21 presentations three of which had to be replaced shortly before ICME 6 (see N°s 11,12,18 from the list of lectures below).

Although conditions with respect to work with High Ability Students (HAS) vary significantly from country to country, the main problems seem to be similar everywhere.

Here are some of the most important fears, questions and hopes, which were considered by the speakers:

- Can activities with HAS be organized in a way which does not contribute to a further separation and division of an already split society?

- How do we detect HAS? What is the appropriate age (if any) to look for high mathematical ability in students? What kinds of tests and/or competition problems are good for revealing non-standard thinking and high creative power? Can computers help?

- What can we offer to HAS? How can we teach them without damaging their connections with their natural environment (family, classmates, friends, etc.)?

- How can we complement and adapt the standard educational system, which is traditionally oriented to medium level students, so that something is done for HAS?

- What should and what could be done on a regional, national and international level in order to improve our work with HAS and to attract them to study mathematics? What should be the role of the Universities and other scientific and/or professional organizations in this direction? What should the literature for HAS look like?

– Will the HAS contribute substantially to the future technological and cultural development of his or her country?

While none of the above questions was completely and definitely answered, the talks in To10 did provide an interesting picture of the work with HAS worldwide. The participants (about 100) got new insights and new incentives for further work with HAS. The work of To10 was scheduled as follows:

Session 1

Chairman: John Hersee (UK)

1. John Hersee (UK): "Provisions for students of high ability in England"
2. Wolfgang Engel (GDR): "The furtherance of mathematically gifted students in the German Democratic Republic"
3. Horst Sewerin (FRG): "Mathematically talented children in Federal Republic of Germany"
4. Charleen M. Deridder (USA): "A study of selected factors to identify sixth grade students gifted in mathematics"
5. Mark Saul (USA): "Working with gifted students in Bronxville elementary and middle schools: The role of the elementary school teacher"
6. Ivan Tonov (Bulgaria): "Some problems in the detection of mathematically talented students"
7. Harvey B. Keynes (USA): "University of Minnesota talented youth mathematics project (UMTYMP)"
8. James J. Tattersall (USA): "Talented high school juniors and the governor's summer program in science and mathematics"
9. Antonin Vrba, Vladimir Burjan (Czechoslovakia), "Special classes for mathematically gifted pupils aged 14–18: system, curricula, textbooks, results, difficulties"
10. Tatiana Trushanina (USSR): "The Kolmogorov-school in Moscow"
11. A. Soifer (USA): "Centre for excellence in mathematics education — a new publishing house for talented students"
12. H. Siemon (Denmark): "On the work with two age groups students"

Session 2

Chairman: Mark Saul (USA)

13. A. Gardiner (UK): "High ability and technical proficiency: how are they related"
14. Bernd Zimmermann (FRG): "Mathematical investigation. Some case studies from a project to identify and foster mathematically gifted students"
15. John H. Webb (South Africa): "Identifying and stimulating students of high ability"
16. D.K. Sinha, Bharati Banerjee (India): "A case study of talented students in a bilingual setting"
17. Edward Lozansky (USA): "The international summer institute"

18. Gerhard König (FRG): "Gifted children and their education. A review of literature with special emphasis on mathematical education"
19. Betty K. Lichtenberg, Pamella Drummond (USA): "Enriching the curriculum through mathematics clubs in the secondary schools"
20. Gilah C. Leder (Australia): "Do teachers favour high achievers?"
21. Margaret C. Anderson (USA): "Empirical mathematical experimentation for middle school children"
22. Ian J. Putt and W.G. Patching (Australia): "Quantitative problem solving: a comparative study of quantitatively gifted and average students' mediating responses"

Some of the papers presented at To10 will be published (together with To3 materials) in a book by the World Federation of National Mathematics Competitions.

TOPIC AREA 11: MATHEMATICAL GAMES AND RECREATIONS

Chief organizer: David Singmaster (UK)
Hungarian coordinator: Tibor Szentiványi

Summary

This Topic Group seemed to be quite popular. The Lecture Room held about 200 people and was full each day, with some people standing. Supporting exhibitions of about 100 of David Singmaster's mathematical puzzles and of about 30 types of Zoltán Kovács' handmade educational materials in another room were attended by approximately 40 people each day and were open every day except Saturday.

Reports of Sessions

David Singmaster (UK): *Old and New Recreations in Mathematics*

In 1982, I began a project to locate the sources of problems in recreational mathematics. This was intended to produce a book of such sources. However, it soon became clear that the first stage must be the preparation of an annotated bibliography of the field. In 1984, I entered my material into a computer file titled *Sources in Recreational Mathematics*. A 1984 description of the project, more extensive than I can give here, has appeared as: Some early sources in recreational mathematics; in: C. Hay, ed.; *Mathematics from Manuscript to Print*; Oxford Univ. Press, 1988. pp. 195–208.

In 1988, I printed a Third Preliminary Edition, which covers about 290 topics on 192 pages, with an appendix of Queries on 23 pages. I am happy to send this to anyone interested, but if there is a great demand, I may have to charge for physical copies. I can also send it by electronic mail — mail requests to ZINGMAST@UK.AC.SBANK.VAX.

In looking through medieval works, I found a number of interesting illustrations of recreational problems. I showed slides of these in my lecture and discussed the history of the problems illustrated. Details of the topics and sources can be obtained from me.

Sándor Klein (Hungary): *Toys of prof. Dienes in Hungary*

The speaker described how he used the various toys of Prof. Dienes in his teaching and research into learning, using video of classroom situations.

Andreas Hinz (FRG): *The Tower of Hanoi*

The speaker demonstrated that even classical recreational problems can produce new results. He has studied and determined the average number of moves between a terminal position and an arbitrary position and between any two arbitrary positions. He has also studied 'illegal' positions and the number of moves required to legalize such a position. The problem with four or more pegs remains open.

David Singmaster (UK): *The Unreasonable Utility of Recreational Mathematics*

To begin with, I will attempt to define "recreational mathematics". Clearly it is mathematics which is fun. But almost any mathematician will say that he enjoys his work. So the idea of fun must be qualified in some way and I think the essential point is that it must be popular in some way. That is, recreational mathematics is fun mathematics which can be explained to the average person and will interest him.

Mathematicians tend to underestimate the public interest in mathematics. Yet somewhere approaching 200 million Rubik Cubes were sold in three years! More Rubik Cubes were sold in Hungary than there are people.

Another measure of the popularity of recreational mathematics is the number of books that appear in the field each year — perhaps 50 in English alone. Many newspapers and professional magazines run regular mathematical puzzles. Martin Gardner's columns were a major factor in the popularity of *Scientific American* and probably inspired more students to study mathematics than any other influence. (I am now contributing to the *Los Angeles Times* and the (London) *Daily Telegraph.*)

How is recreational mathematics useful? I claim it is unreasonably useful. I will first point out that it is of great pedagogic utility.

A. Recreational mathematics is a treasury of problems which make mathematics fun. These problems have been tested by generations. In medieval arithmetic texts, recreational questions are interspersed with more straightforward problems to provide breaks in the hard slog of learning. These problems are often based on reality, though with enough whimsey to make them appealing. They illustrate the idea that "Mathematics is all around you — you only have to look for it."

B. "A good problem is worth a thousand exercises." There is no greater learning experience than trying to solve a good problem. Recreational mathematics provides many such problems and almost every problem can be extended or amended.

C. Because of its long history, recreational mathematics is an ideal vehicle for communicating historical and multicultural aspects of mathematics.

Some Examples of Useful Recreational Mathematics

A. Probability and statistics and much of combinatorics have their roots in gambling problems.

B. Greek geometry was largely an intellectual exercise, but conic sections turned out to be just what Kepler and Newton needed.

C. Non-Euclidean geometry was developed long before Einstein considered it as a possible geometry for space.

D. The problem of the Seven Bridges of Königsberg, mazes, knight's tours, circuits on the dodecahedron were major sources of graph theory.

E. Recreational questions are a major source of problems in number theory, which have been a major source for modern algebra.

F. Another major impetus for algebra has been the solving of equations. The Babylonians already gave impractical problems where the area of a rectangle was added to the difference between the length and the width. Similar impractical problems led to cubics and their solution.

G. Even in analysis, the study of curves (e.g. the cycloid) had some recreational motivation.

Already one can see that perhaps half of all mathematics has some recreational sources. Now let us look at some recent examples.

H. Primality and factorization were traditionally innocuous pastimes, but now workers in this field get consulted on national security.

I. The Möbius strip has been patented several times as a single-sided conveyor belt and as a non-inductive resistor. Printer ribbons commonly have a twist so the printer can use both edges.

J. DNA molecules form into closed chains which may be knotted, or not knotted, providing a most unusual application for knot theory.

The Penrose Pieces

I will only sketch the ideas here, with some references.

My Polytechnic's coat of arms includes "the net of half a dodecahedron", i.e. a pentagon surrounded by five other pentagons. In 1973, I wrote to Roger Penrose on a Polytechnic letterhead which shows the half dodecahedron. Penrose had long been interested in tiling the plane with pieces that could not tile the plane periodically and the letterhead inspired him to try to fill the plane with pentagons and other related shapes. He soon found such a tiling with six kinds of shape and then reduced it to two shapes which could tile the plane in uncountably many ways, but in no periodic way. These have a generalized five-fold symmetry, and they are now called "quasicrystals". Penrose's "kites and darts" shapes were simplified further to "fat and thin rhombuses" and extended to three dimensions. These tilings have quasi-axes and quasi-planes, which can cause diffraction. Crystallographers determined the diffraction pattern which a hypothetical quasi-crystal would produce. In 1984, such diffraction patterns were discovered in rapidly cooled alloy and some 20 substances are now known to have quasi-crystalline forms. Indeed, examples were found about 30 years earlier but the diffraction patterns were discarded as being erroneous! Such materials may be harder or stronger than other forms of the alloys. So a mathematical recreation has led to the discovery of a new kind of matter on which we may be flying in the future! (See *Scientific American* for January 1979 and August 1986 for expositions of this topic.)

Binary Coding

Again, I will only sketch the ideas here.

Binary coding already occurs in ancient Egypt where multiplication was carried out by repeated doubling and then adding the appropriate terms.

Perhaps the next appearance of binary is in 'The Chinese Rings' Puzzle, which appears in first millenium China. Each move takes a ring onto or off a bar. For n rings, position corresponds to a binary n-tuple and the solution corresponds to part of a Hamiltonian circuit on the n-cube. (Hamilton circuits also appear as knight's tours in early works on chess as early as 1141.) This same circuit appears in the Tower of Hanoi (1983). This binary sequence is different the ordinary one.

Ordinary binary coding also appears in the arrangement of the I-Ching hexagrams which arose in China about 1060 and in the use of 1, 2, 4, 8, ... as a system of weights which is at least as old as Fibonacci (1202). In the 17th century, Harriot, Napier and Leibniz each discovered binary arithmetic and Bacon used a binary cipher. In 1953, Frank Gray of Bell Labs received a patent for the Chinese Rings binary coding as an error minimizing coding. Consequently, codes where adjacent code words differ in one place are called "Gray codes", although the Belgian telegraph engineer J. Emile Baudot had already used this idea in the 1870s!

Zoltán Kovács (Hungary): *GAMES AND REQUIRED MATERIALS*

The speaker described the variety of recreational materials that he had made for student use and invited the audience to see the exhibition of these materials.

TOPIC AREA 13: WOMEN AND MATHEMATICS

Chief organizer: Leone Burton (UK)
Hungarian coordinator: Julianna Csongor (USA)

Four one-hour sessions were devoted to a consideration of the relationship between gender, mathematics, classroom practice and the curriculum. A plenary session began the series with two addresses, one focussing on classroom practice (Gilah Leder, Australia) and the other on the curriculum (Heleen Verhage, The Netherlands). The two subsequent sessions provided opportunities for group discussion of these issues using prepared papers (listed at the end of this report) as a basis and focussing on identifying an agenda for future action. In a final plenary session, the results of the group discussions were summarized and synthesized (Leone Burton, UK). The four sessions were very well attended, underlining the increased recognition internationally of the importance of this topic. Participants welcomed the opportunity to improve their knowledge and compare the results of research and action projects in different countries.

Session 1: Classroom practice and the curriculum

Gilah Leder gave an overview of research on classroom pertaining to the relevance of teacher practices and beliefs as a partial explanation for gender differences in educational performance and participation in mathematics. She drew attention to the different techniques, qualitative and quantitative which have been used, and the different data thereby obtained. She underlined the persistence of results which demonstrate an imbalance between quantity and quality of teacher/pupil interactions when analysed by gender. The results of her study in Melbourne were then discussed. Teachers' interactions with the girls and boys in the mathematics classrooms were monitored using two different observation schedules. One focussed on teacher/child dyadic interactions using the Brophy and Good (1970) schedule. The other used Rowe's 1971 schedule to quantify time spent waiting for and interacting with students. The observations were made by video camera and a total of about 70 hours of classroom material was recorded. A mathematics test was administered to every student and teachers were also asked to rate their students' performance in class. Only at the grade 10 level did the boys achieve significantly better than the girls and, in general, students' test scores coincided with their teachers' expectations. However, across all grade levels, boys tended to interact more frequently than girls with their teachers in every category (discipline exchanges, work related exchanges, teacher initiated and pupil initiated, low and high cognitive questioning). "When all discipline exchanges were discounted, boys were still found to have more interactions than girls with their teachers." However, the data on wait-time intervals did not provide evidence of teacher bias in favour of one sex. Gilah Leder concluded that the different frequency of interaction patterns which have been previously reported in classroom studies seem still to persist despite the commitment of the teachers who participated in this study to provide equal learning opportunities for boys and girls.

Heleen Verhage described a major innovative project in mathematics education in The Netherlands and explained connections between the work of that project and the group Vrouwen en Wiskunde (Women and Mathematics). The project is addressing the mathematics education of all Dutch students between the ages of 12 and 16 years and a necessary part of the approach is that mathematics should be derived from the reality around the students and applicable to that reality. One detailed example was described, related to developing learning materials for the teaching of symmetry. A decision was taken to devise work on borders and tessellations, the former for 13–14 year olds, the latter for 15–16 year olds. Worksheets were devised by the project team but an idea of the teacher's, to let the class embroider borders, was incorporated into the lesson planning. The lesson was observed and some resulting transcript material of exchanges between teacher and boy and girl pupils was displayed. The transcript made clear that the girls were at an advantage in understanding both the context of the lesson and how to go about it. The teacher commented afterwards that a previous lesson had the opposite effect as it had been on brick patterns.

In discussing the above example, Heleen Verhage raised a number of unresolved questions. One such question refers to the use of contexts derived from women's traditional domains such as cooking or sewing. Is the use of such examples emancipatory, or does it confirm traditional roles? As the majority of curriculum developers in mathematics education have been male, chosen contexts have tended to reflect a male bias; how likely is the incorporation of non-male contexts and what important messages might be carried in this way? How should bias in the choice of contexts be controlled or assessed? A recent research project which dealt with aspects of emancipation in school textbooks was outlined. In particular, it aimed to confront stereotypes in language use and in illustrations but, in addition, attention was drawn to the need to offer opportunities to pupils to identify with the mathematics. In this respect, Heleen Verhage felt it important to differentiate the "static image" of mathematics as abstract, objective, well-defined and certain, from a "dynamic image" which is creative, relative and personal. Her conclusion was that the latter image is more likely to be attractive to women and girls and, for that reason if no other, should be incorporated into any new curriculum approach.

Session 4: Summary and synthesis — An agenda for action

It was generally agreed that classroom practice and the curriculum intertwine and that, although it was helpful to discriminate between them in order to have a sharp focus, the pupils' experience reflected their inter-dependence. The question of current interest in the performance of females and members of minority groups was raised. The answer offered warned of the link between economic pressure and demographic changes. In some countries, this is producing conditions which demand that the appropriate mathematical and technical skills be available from a broader range of prospective employees hence leading to interest in including those from previously disadvantaged groups. Discussion groups drew attention to different aspects of classroom practice and curriculum when identifying their items for inclusion in an agenda for action. Ten different subheadings covered all:

Classroom climate: Systematic work needs to be done on the effects of group work and discussion. In most classrooms the teacher is still a scarce resource, appropriation of which provides evidence of power and control. Can this role be changed, redefined, and what is the result of so doing? What strategies can teachers develop which do not reinforce attention-seeking behaviour by according it recognition?

Labelling: Is it necessary to label or even acknowledge certain contexts for aspects of mathematics as either male or female or can a mathematics which is inclusive of all groups be developed? What would it look like?

Appropriateness: How should decisions be made as to the appropriateness of content, of examples, and of presentation?

Curricular changes: The content of the curriculum is clearly undergoing considerable re-consideration in many countries in the light of both social experience and mathematical changes. We must be particularly aware of re-definition which leads to an easier curriculum (taken by girls) and a more difficult one (taken by boys).

Resources: In collecting examples, we should be actively seeking, collecting and sharing those which are gender inclusive, such as the embroidery example mentioned by Heleen Verhage, the investigation of the construction of a pair of socks and of an umbrella used by Mary Harris (UK) in the Maths in Work project, and the example given by Philip Davis during his presentation on the MES Day of the effects of removing gender bias from insurance calculations. One discussion group appealed "that we should go home and demand that we get textbooks which give equal representation to girls and boys, equal numbers in the pictures, the names used, etc. We should demand texts dealing with areas of everyday life with which girls are familiar".

Language: The use of language and its meanings in mathematics and mathematics classrooms is still under-researched from a gender perspective. An additional hazard to use and meaning is the metaphoric usage such that words like "hard" and "soft" become attached to, for example, pure science and social science which then carry gender messages.

Assessment: Despite some already established research results demonstrating gender differences in performance on multiple choice and essay type papers, many countries continue to submit all pupils to such examining styles without apparently taking account of existing research or undertaking their own. More needs to be done on different assessment strategies. To date, the only known studies from a gender perspective have been undertaken in Denmark (on examination of group work), in the UK (on extended investigations) and in The Netherlands (on allowing pupils to review and revise their assessments as part of the grading procedure).

Classroom organization: One extended study on single sex grouping has been undertaken in the UK. Further studies on different styles of organizing the classroom are urgently required and, in particular, on their effects on pupil behaviour, attainment and attitudes.

Cross-country studies: Why do some countries produce better results than others? What does better mean in this context? Information on different countries' strategies and performance needs to be made public internationally.

Teacher training: Extensive monitoring is necessary to establish effective strategies and the conditions under which they operate. Sensitivity and awareness training being undertaken in some countries needs monitoring. Classroom behaviour and the propensity for making changes also needs monitoring. Teacher training courses need to develop classroom strategies which can be modelled as gender sensitive methods of instruction for translation into schools.

The following list of research questions and action projects was produced.

Research questions:

1. What is the link between language and mathematics and its effects on the learning of mathematics?

2. How do the effects which were demonstrated in the opening plenary presentations develop?

3. How do effective girls develop their strategies?

4. Can we find out more about the interaction between social, personal and mathematical behaviours?

5. We should shift the focus from "the problem", i.e. girls, minorities, to the source and generate strategies and data by investigating gender stereotyping in males.

6. Is "equal" fair? Studies are required.

Action projects:

1. The development of information sources on role models is needed. One example given was a Directory of Women teaching Mathematics, Statistics and Computer Science in the Universities and Polytechnics of the United Kingdom (reference Leone Burton). Attention was also drawn to the listing of the names of the national representatives on the ICMI and participants were urged to contact their national representative and establish the procedures for more women taking on these roles.

2. Good examples particularly through cooperative ventures by researchers and teachers should be created and then publicized.

3. The agenda should be made specific and the difficulties associated with its implementation be recognized. The task is not an easy one!

4. More use should be made of what has been done so far, for example particularly in the USA where there has already been a considerable amount which could be adapted to different conditions rather than trying to develop new ideas (or reinventing the wheel).

5. A wider range of techniques should be used to support and encourage young women into mathematics. One example cited was that of competitions.

6. A final plea was made to improve the quality of our networking, in particular by identifying women in unrepresented countries who would be interested in becoming national coordinators of the International Organization of Women and Mathematics Education and, through them, extending access to information, research and activities.

Contributors:

The following papers formed the basis for the group discussions in Session 2 and 3:

Amit, M. – Moshovitz-Hadar, N.: Gender differences in achievements and in causal attribution of performance in high school mathematics.

Barnes, M. – Coupland, M.: Humanizing Calculus: A case study in curriculum development.

Becker, J.R.: Graduate Education in the Mathematical Sciences: factors influencing women and men.

Csongor, J.E.: Count on Women in Mathematics.

Fenaroli, G. – Furinghetti, F. – Garibaldi, A.C. – Somaglia, A.M.: Women in Mathematics: a checking test for appraisal and evaluation.

Flores, C.: Women in Mathematical Education in Nicaragua.

Hanna, R. – Kundiger, E. – Larouche, C.: Mathematical Achievement of Grade 12 Girls in Fifteen Countries.

Hubbard, R.: What Kind of Girls Choose a Mathematics Degree?

Isaacson, Z.: "They look at you in absolute horror": Women writing and talking about mathematics.

Kaur, B.: Girls and Mathematics: The case of GCE '0' level mathematics.

Marr, B. – Helme, S.: Women and Maths in Australia: A confidence building experience for teachers and students.

Milonas, E.: Do primary school girls do badly at mathematics?

Purser, P.M. – Wily, H.M.: Where have the mathematicians gone?

Rodgers, M.: Mathematics: Pleasure or Pain?

Rogers, P.: Some thoughts on power and pedagogy.

TOPIC AREA 15: THEORY OF MATHEMATICS EDUCATION(TME)

Chief organizer: Hans-Georg Steiner (FRG)
Hungarian coordinator: András Ambrus

TME was a Topic Area first at ICME 5 in 1984. It is a program pursued by an open international group heading towards a comprehensive approach to basic problems in the orientation, foundation, methodology and organization of mathematics education both as a discipline and as an interactive system comprising research, development, and practice.

Between the two congresses, three international TME-conferences were held whose proceedings are available: TME 1 as a post-congress conference about the overall TME-philosophy, Adelaide, Australia, August 1984; TME 2 on "Foundations and Methodology of the Discipline Mathematics Education (Didactics of Mathematics)", Bielefeld, FRG, July 1985; TME 3 on "Investigating and Bridging the Teaching–Learning Gap", Antwerp, Belgium, July 1988. Furthermore, the TME-group has organized working groups during international conferences of the International Study Group on Psychology of Mathematics Education (PME): at Nordwijkerhoud, The Netherlands, July 1985, at London, UK, July 1986, and at Montreal, Canada, July 1987, the overall theme being: "Relations between the Psychology of Mathematics Education (PME) and Theory of Mathematics Education (TME)" (see the reports in Zentralblatt für Didaktik der Mathematik: ZDM 1985/5 and 1989/1).

For Topic Area 15 at ICME 6 two themes of central importance for the analysis of the present state and future trends of mathematics education had been identified:

(A) How Does Recent Research Contribute to a Balanced Evaluation of the New Math Reform, and What is the Scientific Basis for New Orientations and Future Developments in Mathematics Teaching?

(B) Relations Between Research in Mathematics Education and Research in Science Education.

Of the two sessions with 90 minutes each, 135 minutes were used for contributions and discussions related to (A). The number of participants in (A) was about 200, in (B) about 80. The work related to (A) and (B) consisted of contributions from and discussions between the members of a panel, followed by questions and remarks from the audience as well as responses by panel members. The panel for (A) had the following members: T.A. Romberg (USA) (Presider), G. Brousseau (France), Y. Chevallard (France), T.J. Cooney (USA), A. Sierpinska (Poland), L.P. Steffe (USA), H-G. Steiner (FRG). The panel for (B) consisted of: J. Maass (Austria) (Presider), M. Artigue (France), P. Boero (Italy), D. Siemon (Australia), H-G. Steiner (FRG).

Concerning (A), H-G. Steiner started off with a short overview of the complexity of the theme and how it is related to TME. Some specific points of reference to research in further contributions were: the societal transformation of scientific knowledge into the knowledge taught in schools (Chevallard), the role of teachers and students in the constitution of meaning in classroom interactions and the didactical contract (Brousseau), teachers' professionality and the role of teachers' conceptions of mathematics and mathematics education (Cooney), epistemological obstacles in the teaching and learning of mathematics (Sierpinska), learning and teaching mathematics from a constructivist point of view (Steffe), needs for a research basis in formulating curriculum and evaluation standards for future school mathematics (Romberg).

From a TME point of view it is important to realize that the new math reform has been part of a comprehensive program in the late 50's and the 60's to promote societal changes on a scientific basis, and that in several countries, such as the FRG, the institutionalization of mathematics education as an academic field has also been part of this program. It is therefore no wonder that a sufficiently developed internationally shared research basis for the reform was almost non-existent. Emphasis was laid on content analysis and curriculum construction. The understanding of learning and the goals of teaching were roughly oriented according to Bruner's principles of the central role of the structure of the discipline and Piagetian conceptions of the overall importance of general operative structures and schemata in humans' cognitive development and activity. The reform movement and its scientific environment were surrounded by new developments in the social sciences, especially along the lines of socio-economic ideologies and theories. They did not however, penetrate deeply into mathematics education theory and research, but had some influence on the general ideology of the reform and related expectations.

In general, the failures of the new math reform were also partly caused by the deficiencies of overall societal programs for change. Social systems cannot be altered into intended directions by just applying scientific–technological projects. However, the recognition of the crucial failures and deficiencies had a profound impact on the self-concept of the social sciences and finally the sciences in general. Changes which involve human dimensions are more a matter of growth and often need catalysts rather than causal injections from the outside. The recent development of mathematics education as a discipline with an increasing degree of self-reflection and extending interdisciplinary cooperation is one of the fruitful reactions to and consequences of this insight.

Recent research in mathematics education and cognitive science has shown that some of the basic assumptions of the new math reform were inadequate. Learning and cognitive activities in mathematics are not primarily determined by the dominance of universal and coherent structural schemata but rather are highly domain-specific and depend on modes and means of representation and action, the human mind not being a homogeneous unity but rather a constructed and developing complex "society of minds". This has far reaching consequences for understanding and practicing learning and teaching of mathematics. In connection with a more empirically, oriented epistemology, research has also thrown new light on the nature of mathematical concepts, structures and theories in their relation to applications. Theoretical concepts in application-oriented theories — and many concepts of school mathematics are of this kind — cannot be defined once and forever and in a universal way, but need constraints to specific domains of applications and process of related interpretations. In this process the concepts are both founded and developed, but always have a certain degree of openness and fragility. Traditional as well as new mathematics teaching has neglected this by inadequate fixations, formalisms and overmethodizations. Furthermore, studies on classroom interactions have promoted the view that the classical didactical triangle consisting of the student, the teacher and the content has to be replaced by an understanding of the complexity of many interrelations that bind the three components together in an interactive system. Rather than seeing content as knowledge existing by itself, its meaning is to be understood as being constructed in negotiations taking place in the classroom processes. Another component of the fragility of knowledge in the educational context consists of the fragility of this interaction. It turns out that the logic of social interaction is often stronger than the subject matter logic, and failures of the new math reform have often been caused by this phenomenon. The fragility cannot be avoided but needs awareness and careful and constructive handling.

Apparently we now have a more developed and broader research basis for new orientations and curricular developments. However, there are still many underdeveloped domains, of which we mention some: lack of investigations on higher order learning and learning of more complex mathematical concepts and theories; the existing gap between research on learning and research on teaching representing two rather separated disciplines of scientific inquiry; continued and a greater variety of classroom interaction studies related to different age-groups of students and different topics from the curriculum; studies on the effect of computers as a kind of third component in the teaching–learning interaction.

An interesting debate about the quality and the applicability of existing research in mathematics education for future developments is presently taking place in the USA with respect to the NCTM-Standards for School Mathematics K-12, especially about the question of whether the authors of the "Standards" have referred sufficiently to research available and identified needs for research in order to fulfill the "Standards" (see contributions e.g. in "Mathematics Teacher" and "Journal for Research in Mathematics Education").

Concerning (B), after a short introduction by J. Maass, H-G. Steiner identified some aspects of the relation between research in mathematics education and research in science education and outlined some consequences for enhancing better future cooperation. P. Boero presented a model of how research between the two domains can be linked. M. Artigue reported about investigations related to the "didactical transposition" of concepts such as "differential" in the mathematics and the physics tradition in the teaching of mathematics as compared to the teaching of sciences.

At the level of relations between mathematics and science (especially physics) as school subjects, the following problems were indicated:

– coordination in content and time between syllabi, respecting the mutual roles of conceptual and procedural tools and fields of application taken by both fields;

– bridging (avoiding) conceptual and epistemological differences (discrepancies) related to concepts used in both domains like: variable, function, proportion, magnitude, differential, derivation, integral, vector, probability, space, etc.

– reflection of these (and other) points in the training of mathematics teachers and science teachers.

At the level of relations between mathematics education and science education the following topics of comparison were selected:

(I) Common research dimensions:

    (a) study of knowledge: foundation, structure, function, representation, means, development and growth of knowledge (epistemology, philosophy and history of mathematics and the sciences)

    (b) psychology: cognition, cognitive development, meta-cognition, social cognition, etc.

    (c) pre- and misconceptions among students and teachers

    (d) classroom interactions

(II) Metatheoretical comparisons:

    (a) use of systems approaches

    (b) the place of home grown vs. imported theories

    (c) relation of theory to practice

    (d) interdisciplinarity.

It was suggested that the TME group may take initiatives to encourage comparative studies between the two research domains and organize joint conferences.

TOPIC AREA 16: SPACES AND GEOMETRY

*Chief organizer: **Walter Bloom** (Australia)*
Hungarian coordinator: Vera Sztrókay

This topic area was held in four sessions, in parallel pairs. There were sixteen scheduled presenters, chosen from double that number who offered contributions, with each session attracting between 70 and 80 participants. The topic area was designed to focus on geometry in the current school syllabus in the light of changing attitudes of society and the concurrent development of high technology.

Session 1a: Spatial and visual aspects

Contributors: Walter R. Bloom (Australia), Robert Dixon (UK), John J. del Grande (Canada), Peter Bender (FRG).

The session was opened by Walter Bloom, who observed that geometry had featured highly in the first four International Congresses, and also in several meetings organized in the sixties and seventies. Such conferences attracted lively discussion on geometry, its role in promoting "logical thinking", and arguments on how best to teach the subject. However there has been a change in the presentation of mathematics at the school level, first through the introduction of the New Mathematics, and then because of the electronic computer and a growing interest in statistics. These, together with a general community call to teach "useful" and "relevant" mathematics, meant that geometry in many Western countries slipped from its central role in the secondary mathematics curriculum.

This was certainly the trend at the secondary school level in Australia and was reflected in the relatively minor role allocated to geometry at ICME 5 in Adelaide. The mathematics education literature also gave the impression of a swing away from geometry, with expositions on favourite approaches to its teaching giving way to corresponding articles on computing and statistics.

He indicated that parallel to this trend, there had been a growing awareness of the usefulness of geometry, not only in its role in day-to-day activities, but also as a vehicle to promote visualization in mathematics and many of the sciences.

Particularly relevant here was three-dimensional geometry. How could geography or astronomy be taught to a child with little idea of great circles on a sphere, crystal structure to a student who could not visualize a tetrahedron, architecture and engineering to a person for whom projections and cross-sections were foreign, and two-variable calculus to those who could not grasp the concepts of surfaces and solids of revolution. The omnipresence of geometry did not stop here; it arose in building solids, packing objects, making patterns, and estimating volumes, areas, and angles. These practical applications lend weight to the fundamental role geometry had to play in everyday life, something that was known as early as the time of the Ancient Egyptians.

The growing awareness of these attractions of geometry have been instrumental in its survival. There has also been a realization that the computer, which entered the classroom as a rival to the traditional mathematics syllabus, could in fact be a useful ally, especially in subjects like geometry where visualization is essential. This was the theme in both Sessions 1a and 2a.

Robert Dixon, the lead speaker for this session, was unable to attend the Congress. His paper, which was read by Walter Bloom, highlighted two important aspects of teaching geometry, namely through classical constructions and computer graphics. The emphasis here was on some of the interesting and visually attractive aspects that could be easily considered, designed both to appeal to students and to be useful.

John del Grande considered how spatial abilities such as

- Eye-motor coordination

- Figure-ground perception

- Perceptual constancy

- Position in space perception

- Perception of spatial relationships

- Visual discrimination

- Visual memory

could be enhanced and improved through an appropriate geometry programme. Geometry should be an integral part of elementary mathematics at the kindergarten level, relying on the child's intuitions about space and experiences of movement of objects in space.

Peter Bender put forward the thesis that thinking in terms of continuous movements and deformations could ease geometrical proof epistemologically. He indicated a psychological basis for continuity being a natural essence of mental spatial operations, and also a natural part of the notion of a (partially) rigid body, coming under the heading of congruence geometry. The above ideas would lead to the avoidance of problems in teaching the sophisticated concepts of geometric transformations.

Session 1b: Approaches to the teaching of geometry

Contributors: Noël Vigier (France), Frantisek Kuřina (Czechoslovakia), Stefan Turnau (Poland), István Lénárt (Hungary).

Noële Vigier discussed geometry for 15–19 year olds with the development of two themes considered by the inter-IREM (Institut de recherche sur l'enseignement des mathématiques) group, namely:

1. Space representation induced by plane geometry;
2. Problem solving in geometry.

These were presented in the context of the relatively recent modification of the geometry curriculum in France.

Frantisek Kuřina concentrated on two major problems of the teaching of geometry for 8–15 year olds:

1. Conceptual accuracy of knowledge and logical structure of the discipline;
2. Simplicity and vividness of knowledge, and practical applicability and exploitation of the concepts,

in the light of whether the teaching of mathematics is the conveying of a ready made mathematical theory, or the translation from the language of science to that of the student, and whether instruction should lean on the student's experience.

Stefan Turnau spoke about the integration of geometry into the secondary school mathematics curriculum in Poland, following the abandoning of the deductive approach, and emphasized the following five principles:

1. Phenomenological and formal geometry;
2. Plane and space geometry;
3. Synthetic and analytic methods;
4. Ancient and new problems, methods and approaches;
5. Geometry instruction and methodological education.

István Lénárt outlined his approach to the teaching of the axiomatic method in the elementary and secondary schools. This made use of the set of great circles on a sphere, with the relationship of common perpendicularity suitably interpreted as a binary operation.

Session 2a: The role of graphics and the microcomputer

Contributors: Judah L. Schwartz (USA), Uri Leron (Israel), William Wynne Willson (UK), Roger V. Jean (Canada)

Judah Schwartz gave the lead lecture in this session with a presentation of the computer program *Geometric Supposer*, and outlined how it could encourage the student to both discover geometric results and develop proofs.

Uri Leron highlighted three aspects of turtle geometry in which recent progress had been made:

1. Formalizing intuition and visualization;
2. Intrinsic representation of curves;

3. The relationship between turtle geometry and Euclidean geometry via an isomorphism of their associated groups.

William Wynne Willson saw the arrival of the microcomputer as a chance to rethink the geometry curriculum, with its role as a supreme drawing instrument, a superb maker of pictures. The relationship between the microcomputer and pencil and paper drawing was to be compared with that between the electronic calculator and pencil and paper calculation. The microcomputer was to provide motivation by encouraging exploration, allowing movement, and easing the study of transformations and the drawing of envelopes and representations of three-dimensional figures.

Roger Jean provided a source of material in the context of botany from which the ideas of the previous speakers could be tried, especially with regard to the use of the microcomputer. He referred to one of the great mysteries of botany, the explanation for the spiral patterns observed in plants and inside their buds, and expressed the hope that computer simulation of these would lead to an explanation of the physico-physiological mechanism that induced them.

Session 2b: Other geometries

Contributors: Emma Castelnuovo (Italy), Morten Anker (USA), Gerard Audibert (France), Françis Michel (Belgium).

Emma Castelnuovo gave the lead lecture in this session. Her talk emphasized the connection between (computer) drawings of a fractal and the theory concerning the curve's dimension. Based on the snowflake curve she indicated how the dimension of a fractal curve could be shown to lie between 1 and 2. Applications of fractals to chemistry, physics, and technology were also given.

Morten Anker showed how mathematics could become relevant and alive as children reconstructed their classroom and neighbourhood environment in miniature, and carried out the various measurements associated with these including lengths, areas, population counts and costings.

Gerard Audibert reported on the study of Perspective Cavaliere (the oblique projection on a plane) by the IREM at Montpellier, and indicated its importance in the representation of space.

Francis Michel traced a development of school geometry teaching, from the study of shadows and photography for young children, to the drawing of transformations of pictures and then computation within a coordinate system.

TOPIC AREA 17: INFORMATION AND DOCUMENTATION

Chief organizer: Gerhard König
Hungarian coordinator: Judit Szabó

1. Introduction

There has been a substantial increase in publications dealing with research in mathematical education in general and in particular on experiments in various countries, new pedagogical concepts and insights, topics, and teaching concepts. One of the features of the growth is the increasing number of conference proceedings, collections of papers, reports, etc. being published. Another aspect is the expansion of journals in this field in both number and page count. Journals are of great importance for everyone interested in national developments as well as for an international exchange of ideas. About 50 000 scientific journals serve worldwide as channels for scientific communication. Of these about 400 concentrate on mathematics education and/or computer education.

This ever increasing flood of information is a problem encountered in most fields of science: for example, some 120 000 books and papers on physics and engineering are published every year and some 40 000 on mathematics and its applications. It is well known that the production of what we may call scientific literature will continue to increase exponentially unless there are drastic changes in the practice of scientific research. Educational professionals like other scientists are thus faced with the problem of how to extract from a vast pool of potential information those items which they need for their own work.

The purpose of our topic area was to provide an insight into how to cope with this flood of information. We tried to give to the participants some information on the institutions and services which may help us to keep up to date with current progress in mathematics education: abstracting journals, on-line databases and information networks. In our discussions we thought about how to handle information and which information we need.

2. Presentations

Two papers were presented and a short oral communication.

In the first presentation "Fast Access to Literature in Mathematical Education" G. König from Zentralblatt für Didaktik der Mathematik (ZDM) gave an overview of abstracting services and on-line databases in the educational field. In particular, he introduced Zentralblatt für Didaktik der Mathematik (ZDM)/International Reviews on Mathematical Education. This unique and well established journal is an information and reference journal in mathematics education and computer education from pre-school level to teacher training and adult education. ZDM provides access to the worldwide literature published worldwide for educational professionals.

The second paper dealt with information services from the Charles University, Prague. In their paper "Information Services provided by the Sectional Information Centre, Pedagogical Faculty of Charles University" P. Libertova and I. Stransca reported on the yearly information bulletin "Didactics of Mathematics". This bulletin describes and reviews East European journal articles. In addition they discussed:

1. A system of scientific information in the Czechoslovak Socialist Republic.
2. The role and function of the Central Library-Sectional Information Centre, Pedagogical Faculty, Charles University, in the system of educational information with special reference to mathematical education.

In his short oral communication G. Schubring from the Institut für Didaktik der Mathematik (IDM) described the library of this well known research centre. The IDM library is a well equipped library for mathematics educators with an enormous amount of research literature and many volumes of journals in mathematics education, pedagogy and psychology.

3. Discussion

The discussion focussed on the following points:

3.1. Overview of journals

An overview of journals within our field, being mathematics and education, is a major demand by the developing countries. ZDM has prepared its "Source Journal Index" which gives a sound basis for a complete list of all relevant journals. UNESCO is also working on a revised version of a "Listing of Mathematics Education Journals" which is scheduled for publication in 1989. There is also a special need to differentiate between journals for mathematics educators, journals for mathematics teachers and journals for mathematics students. In particular an overview of journals for students at all levels would be appreciated.

3.2. Dissertations

In the USA it is possible to be kept informed on relevant dissertations through reviews in journals and other channels. In other countries there is no central information source (except France with an IREM catalogue). So far, there seems to be no need to catalogue the vast number of dissertations being published in the field of mathematical education as the dissertations vary in terms of quality.

3.3. Grey literature

Most up-to-date research is published as reports. The following questions have been raised. Which institutions publish which reprint and report series? Should it be necessary to collect all and to have a good hard copy collection of non-conventional literature at some local sites? There is no overview of report publishers but there are a few document centers. The German Göttingen University Library and the library of the Centrum voor Wiskunde en Informatica in The Netherlands collect reports on mathematics and mathematics education. American reports can be purchased in microfiche or paper copy from The ERIC Document Reproduction Service.

4. Concluding Remarks

Members of the group intend to continue, during the years to come, the work on some topics identified as important.

TOPIC AREA 18: SYSTEMATIC COOPERATION BETWEEN THEORY AND PRACTICE IN MATHEMATICS

Chief organizers: Bent Christiansen (Denmark),
Piet Verstappan (Netherlands)
Hungarian coordinator: András Ambrus

International cooperation on this theme was started at ICME 5 in Adelaide (1984) and continued at working conferences in Lochem (1986) and in London (1988). The work at — and between — these conferences has clearly demonstrated that theoretical ideas and teaching practice are mutually dependent aspects, but also that it is possible and fruitful to distinguish between influences and developments from theory to practice and from practice to theory. The two sessions at ICME 6 each dealt with one of these mutually related themes by means of: (1) a retrospective introduction; (2) a presentation looking to the future; and (3) a panel debate. The program was planned by the chief organizers in cooperation with Heinz Steinbring (FRG). The number of participants at the two sessions was around 80 and 60 respectively.

Session 1: Changes of Practice by Means of Theory

Bent Christiansen chaired the session and gave the introduction. To avoid misunderstandings of the theme of the day, he first commented upon the fact that theoretical constructions normally are initiated by practice and reflections on practice. Next, he emphasized the now widely accepted fact that direct mediation of theoretical ideas about the teaching/learning process and about the teachers' roles is normally inadequate and that a careful study of indirect approaches and means therefore is highly needed. Such a study is attempted in the SCTP-context by means of analyses and comparisons of examples of systematic cooperation between theorists (researchers) and teachers in the classroom. Details about the work at the first two SCTP-conferences are available in the reports [1] and [2], but Christiansen would on this occasion illustrate how the work of SCTP had developed over the years since 1984 by a brief account of major issues which had been subjected to analysis and debate at the three conferences.

At the *Adelaide Conference,* six cases of interaction between theorists (the speakers) and practitioners were presented. Major issues discussed at a final session were: (1) What is the role of didactical theory for the researcher in his/her cooperation with their students? (2) What are the potentials of such cooperative projects in teacher education and in curriculum development?

At the *Lochem Conference* a deeper analysis of the problem field was based upon a much enlarged number of examples of cooperation between members of the group and practicing teachers. A major aim was to analyse and discuss the themes: (1) Scientific knowledge. (2) Practitioners' knowledge and know-how. (3) Systematized relationships between (1) and (2). (4) Theory building concerning (3). The investigation took place in these interrelated perspectives: (a) Textual materials. (b) Teacher education. (3) Examples of systematic cooperation. (d) The reality of the classroom.

At the *London Conference* four levels of didactical decision making were considered: (1) Instantaneously in the classroom. (2) Reflective observation in the classroom. (3) Planning for future lessons. (4) Long-term planning (with colleagues) of an extended program. These levels provided one focus for debate. Another focus was: the methodology used to implement systematic cooperation between theory and practice or theorists and practitioners. And a third: the materials, which represent the basis of that cooperation.

The case studies of cooperation and interaction between members of SCTP and practicing teachers published in the reports [1] and [2] all include proposals of models under which practice is subjected to influence from theory. Christiansen described a few of these proposals and concluded his introduction by emphasizing that systematic cooperation in the SCTP-context must be based upon an organized set of frames for interaction between the didactician and the practicing teachers which provides conditions for, and demands, that both partners contribute to the investigation of some substantial object of common interest.

Tom Cooney (USA) presented his ideas on the theme "From theory to practice in future perspectives" in an indirect way, using the discussion of the related theme "Developing of a research agenda to support systematic cooperation between theory and practice" as a medium. He mentioned in his introductory statement that Burton in her contribution to the Lochem conference (see [2]) proposes four models reflecting aspects of cooperation between researcher and teacher: (1) researcher/teacher as a team; (2) teacher as researcher; (3) researcher/teacher role differentiated; and (4) researcher as participant observer. He found that each of these models would have a place in developing research agendas for promoting systematic cooperation between theory and practice.

Cooney next mentioned that he and Brown have argued that we might profit from considering what aspects of practice reside in and define theory and what elements of practice have theoretical overtones. Research in the USA shows that theory is inseparable from practice in the way it has evolved. It was Cooney's view that whatever theoretical perspectives we hold, they manifest themselves one way or the other in the context of teacher education, and he agreed with Bazzini (see [2]), who has argued that the single biggest obstacle to progress in mathematics education in most countries is a weakness in teacher education.

Turning to the subtheme: Methods and meanings, Cooney expressed his belief that a humanistic orientation towards both research and mathematics is becoming more widely accepted in mathematics education, and he proposed that a conception of mathematics as a human activity in which mathematical meanings are negotiated through interaction among students and between teachers and students would enhance mathematics education. However, understanding the process of meaning-making constitutes but one of the research agendas that evolves from a humanistic perspective.

Cooney concluded that in his judgement, research for the support of systematic cooperation between theory and practice should use a mode based on the realization that we as researchers and teachers are fully implicated in the research process. Such a mode permits examination of the personal meaning and the negotiation of meaning that teachers develop. However, it is not only a matter of learning what teachers know and do, but also what they experience and feel. Collection, interpretation, and analyses of data must become interactive processes informed by both researcher and individual informant. Such a line of research would have the potential to lead to a deeper understanding of teachers' meaning-making processes and thereby provide a basis for constructing teacher education programs that are responsive to teachers' beliefs, needs and theories. Any effort to accomplish systematic cooperation between theory and practice, between theoretician and practitioner, must ultimately be based on the human condition that one understands, communicates, and respects the other.

A *panel* was now formed by Michael Otte (FRG), Erich Wittmann (FRG) and the two speakers. Christiansen invited questions and comments from the panel and the floor.

Wittmann expressed his agreement with an earlier statement by Christiansen to the effect that the didactics of mathematics must identify ways and means for a genuine support of practice. Such a development had in Wittmann's opinion to build upon a systematic cooperation between theorists and practitioners. However, a researcher cannot collaborate with an "arbitrary" teacher. The needed development must rest on collaboration with committed and competent teachers. Hence, the best approach is to work with a fixed group of teachers for a long time and study this case in the SCTP-context.

Verstappen asked questions about the form and content of teachers' theoretical means. What theoretical ideas do teachers have, and how are they constructed?

Cooney pointed in this connection to the need for many different approaches. For example, important answers might be obtained from long term cooperation with groups of teachers.

Otte pointed to the difficulties involved in the establishing of cooperation between theorists and practitioners. Thus, at IDM Bielefeld, fourteen years had been used to obtain some real cooperation with practicing teachers. Returning to the question raised by Verstappen, he asked: What is theory? And he answered: Theory is characterized by reflection, and practice by action. But theory cannot guide practice like a map. Theory can be useful only as far as it becomes a part of the reality of the teacher. Otte argued that theory has no *direct* relation to reality or practice. Teachers use means such as techniques, but not theory which is a different kind of activity. So, theory has to be converted. Therefore, you need continuously organized cooperation, but there are institutional and political barriers. Our problem is to enlarge the range of the influence from theory.

In the debate with the floor, these views were doubted because *knowing how* contains *knowing what,* and also objections were made against a too strong separation between theory and practice. Cooney's position that reflection on one's own practice is fruitful was supported. And Christiansen emphasized the importance of the continuation of cooperation between the theorist and the teachers. In that way theory can be converted into practical means. It was also remarked that theoretical beliefs form a core of divergence. Cooney offered a variety of contexts in classroom-settings, emphasizing that meaning–making does not consist of imposing theoretical meanings by the theorist, but in teachers developing meaning in a dynamic process making the unfamiliar familiar.

There were several other contributions to the debate. At the end, Otte addressed this question to the audience: Do you think that mathematics is a theory? Several answers came from the panel and the floor: Mathematics also contains tools, creates organization, human solving problem activity, etc. Piet Verstappen stressed the dependence of the answers on the philosophical position.

Session 2: Ideological changes of research and development projects by practice

Piet Verstappen, who chaired the session, based his introduction upon the three SCTP-conferences. He gave examples of ideological changes of research and development projects by practice. He emphasized that the terms "Systematic cooperation between theory and practice" can easily lead to misunderstanding. What exactly is systematic, what is cooperation, what is theory and what is practice? Some think that theory arises from practice, which is the comprehensive process from which theoretical moments emerge and derive their function, and which as a process constitutes the unity of theory and practice. Others think that theory is a science in itself or a philosophy or an ideology or that it is identical to didactics. These terms have unfortunately been interpreted in different ways throughout the SCTP-conferences.

Verstappen expressed the opinion that — strictly speaking — theory and practice never cooperate; practice absorbs theory and in doing so changes itself. In that way practice exploits theory as a means to change, and accordingly theory can be an effective agent.

However, also in cases in which theory has been absorbed by practice, the theorist maintains his active reflection. Hence, an ideologically loaded practice of the theorists always exists. Accordingly, the question of the ideological changes of research and development projects by practice is meaningful in the sense that these projects may change teaching practice and lead to corresponding modifications of the theorists' views and practice.

Changes in teaching practice take place in several ways, varying from hetero-nomic to autonomic, from what has to be done to self-modification. Verstappen referred to cases presented at the SCTP-conferences in which changes took place due to teachers' self-modification. Here, a "mirror" has to be held up for the teacher. For example: interviews in which episodes are used (Brown/Cooney); lesson transcripts (v. Harten/Steinbring); or videotapes (Speelpenning). Showing alternatives takes the middle position between heteronomic and autonomic changing.

Verstappen proposed that what applies to practice analogically applies to practice of the theorists, and he described in that connection three phases. In a first phase, the researcher's theoretical knowledge is not explicitly expressed, but implicitly present in acting. Thus Zawadowski describes (see [2]) in his innovatory case study that in the beginning he had no theory, only practice. In a second phase, the theorists are not any more merging in practice, but detach themselves from it through knowledge of practice. In this second phase theory and ideology have to be worked out, and the researcher must be able to operate in such a way that he can discover the truly prevailing opinion of the teacher. A third — functional — phase corresponds to a manner of acting and thinking in which the theorist is neither detached from practice nor absorbed in it. In this phase the relation between practice and theory is central rather than participation and detachment.

In his closing statement, Verstappen raised the question about the future development. He referred here to the third phase, in which both the theorist and the practitioner are looking for a fitting ideology. Therefore they need a structured relationship and an awareness of all possible ideologies and their implications for education, students and society. They also have to be aware of how ideologies are hidden in education. That is the way we have to go, it is a long way, however, and usually our projects are not granted the time needed.

Alan Bell dealt with the theme "From practice to theory in future perspectives" under three headings: (1) Theorizing about practice. (2) Designing the system. (3) The future of the group.

As regards (1), his introductory example of a principle derived from practice was: "You should not give children direct answers to direct questions." He proposed that the theorist, in cooperation with teachers, should build upon what the teachers bring to the course, and recommended that the designing of a syllabus had to take place from the bottom up and not vice versa. Also, the theorist should be looking for what pupils actually do and can do in the classroom. Ideas of a theoretical nature cannot normally be transmitted by means of theory. It is more feasible to demonstrate through practice how theoretical ideas work.

As regards the subtheme (2), Bell indicated the many partners involved in the educational process. He mentioned that around 300 advisory teachers were contributing to the development of the educational system in the UK and found this to be one of the important means to obtain input from practice to development of theoretical ideas about education. However, the English examination system has a strong conservative influence, and the danger of having a national curriculum imposed was now to be faced. He believed that teachers' personal responsibility and involvement would decrease if a common national curriculum became a fact.

Concerning the subtheme (3), Bell saw the strength of SCTP in the study of a full variety of approaches to cooperative work with teachers. However, such a collection of approaches raises a demand for ways to evaluate their effectiveness. This should be a major question to be investigated by SCTP.

Bell mentioned four other issues for future SCTP-efforts: (i) the development of the educational system (the relations between the partners, the influence from the examination system); (ii) new types of teacher education; (iii) the ethical problem of change; and (iv) evaluation of in-service activities. He concluded his presentation with the assertion that there is to-day a strong trend away from specific offerings (in the form of ideas, principles and methods) towards helping schools where they are. SCTP would have to take this tendency into consideration in the planning of its future activity.

A *panel* was formed by Luciana Bazzini (Italy), Heinz Steinbring (FRG) and the two speakers. Verstappen invited questions and comments.

Bazzini found that analysis and comparison of different models of cooperation between researchers and teachers constituted the major potential of SCTP. In fact these models include in themselves a linkage function between theory and practice in the field of mathematics education. She pointed to two important future tasks. Firstly, to build a bridge between subject matter specialists and teachers and schools, and to investigate the obstacles (of institutional, socio-cultural and psychological types) of models for such a bridging and cooperation. Secondly to promote that a critical examination is made by universities and educational institutions concerning their contribution to the educational process in school.

Steinbring emphasized that information of teachers about research often is negatively influenced by an inappropriate underlying belief about the nature of this relationship.

P. Ernest (UK) found that Bell's suggestion that theory arises from practice could be misunderstood as meaning that theoretical constructs are useful only if they arise directly from practice. This would be reminiscent of a Baconian view of science: theories are induced from observation. But all modern philosophers of science deny this. In continuation, he proposed that mathematics teacher education can offer students broadening conception and conceptual tools for thinking derived from theory and not directly from experience. However, these constructs cannot be "given" to student teachers, but must be actively acquired by them.

J. *Weber* (France) pointed to the fact that student teachers for the primary school in France are subjected to influence from different parties such as the inspector, the adviser in pedagogy, the teacher trainer, the researcher in the field of the didactics of mathematics. These parties often have diverging views on how to improve mathematics teaching. Moreover, the student teacher is similarly disoriented during the components of practice belonging to teacher education. Weber asked whether it would be possible to ensure coherence between the different types of causes? And how?

More questions were brought up, due to the limited time available, the comments mostly consisted only in a recognition of the importance of the issues raised.

At the end of the session, Verstappen asked Steinbring to comment briefly upon a proposal (by Seeger and Steinbring) concerning the future work of SCTP which had been circulated to the participants. Steinbring mentioned that discussion at the previous conferences had shown that the members of SCTP generally accept that the problem of theory and practice partly consists in overcoming the broadcast or conveyance metaphor. Seeger and Steinbring propose that this agreement is a basis for future cooperation and that it seems especially important that the question about abandoning the broadcast metaphor be applied also to the relation of research and theory to practice.

The chairman then closed the second and final session by mentioning that the proposal of Seeger and Steinbring would be a major point for deliberations.

References:

[1] Christiansen, B. et al., 1985: Miniconference at ICME 5 on Systematic Co-operation between Theory and Practice in Mathematics Education. Topic Area Research and Teaching. Royal Danish School of Educational Studies, Department of Mathematics, Copenhagen.

[2] Verstappen, P.F.L. (ed.), 1988: Report on the Second Conference on Systematic Co-operation between Theory and Practice in Mathematics Education, Lochem 2–7 November 1986, vol. I–II. National Institute for Curriculum Development, Enschede.

INTERNATIONAL STUDY GROUP ON THE RELATIONS BETWEEN HISTORY AND PEDAGOGY OF MATHEMATICS (HPM)

Chief Organizer: Ubiratan D'Ambrosio (Brasil)
Hungarian Coordinator: László Filep

The scientific program of HPM was organized in four one-hour sessions, distributed in two symposia on *non-euclidean geometries and their adoption in the school systems* and *the evolution of algorithms for use in schools*, a panel on *history of mathematics in the teaching of mathematics* and a session of short communications.

The planning of the scientific program resulted from the recognition that teaching of geometry continues to be a major challenge to mathematics educators and that by bringing into the classroom the challenge that non-euclidean geometries represented throughout history we may introduce an element favourable to a revival of interest in geometry in the school system. The same with respect to algorithms, which have for years dominated school mathematics in the elementary levels and now may face a lessening of importance in view of the rapid introduction of calculators and computers. Major events in the history of mankind are closely related to major changes in algorithms and the very strong cultural roots of algorithms justify an historical overview of their presence in the school system. The theme for the panel was suggested by the very essence of this International Study Group. To reserve one session for short contributed papers, without bounds on their themes except to be relevant and of interest to both History and Pedagogy of Mathematics, was quite natural in organizing the program.

Let us now report in specifics of each of these activities.

The Symposium on *non-euclidean geometries and their adoption in the schools systems* had three speakers: Nikos Kastanis (Greece), Massouma Kazim (Qatar), and Tibor Wessely (Romania). The session was presided by Benedito Castrucci (Brazil). Nikos Kastanis spoke on *the concept of space before and after the non-*

euclidean geometries: an approach for didactic reasons. Drawing from an analysis of the architectural forms of the Parthenon in Athens, he stated that this was the first historical hint of the weakness of identifying or reducing the contemplation of space to the Euclidean model. The second hint is drawn from Albert Einstein's view that the conceptual system of Euclid does not include space, which is first introduced only by Descartes. Kastanis proposes the following periodization for the study of space:

1. from myths to logos,
2. from scholasticism to analytic thought,
3. the epistemological revolution (19th century),
4. the shift in scientific outlook (early 20th century),
5. structuralism.

Massouma Kazim presentation on *an educational unit in non-euclidean geometry for secondary schools* proposes a program based in the following steps:

1. stating the objectives of the unit,
2. developing contents as follows:
   a) how non-euclidean geometries have emerged from the geometry of Euclid through history,
   b) present euclidean logic system and non-euclidean systems as well,
   c) give some modern applications of non-euclidean geometries,
3. suggesting methods of teaching the unit, emphasizing the nature of proof.

Tibor Wessely spoke on *the Bolyais*, first introducing the main ideas of János Bolyai's Appendix and then telling about the life and careers of Farkas and János Bolyai. The lecture was illustrated by a collection of slides showing the places where the Bolyais have lived and worked, as well as monuments and museum pieces honouring them.

The Symposium on *the evolution of algorithms for use in schools* had only one formal presentation, by Lawrence Shirley (Nigeria), and was presided by Victor Katz (USA). In his intervention on *historical and ethnomathematical algorithms for classroom use* Lawrence Shirley suggested that mathematical techniques taken from historical and diverse cultural ambiances can be used in the classroom as alternative algorithms. This results in cognitive gains. Also, children feel a sense of closeness with their ancestors by bringing ethnomathematical practices into the classroom and learning the methods and approaches they have developed in their cultural context to deal with problems facing them. An extensive intervention by George Ghevarghese Joseph (UK), further exemplified the drawing from ethnomathematical sources by reporting on algorithms used by street prodigies on mental arithmetic in India. The algebraic explanation of the algorithms they use can be an important element in teaching algebra.

The panel on *history of mathematics in the teaching of mathematics* had four participants. Evelyne Barbin (France), spoke on *a case for the teaching of mathematics in a historical perspective.* Her presentation described, with several illustrative examples, the experience of the IREM's in France in teaching the history of mathematics based on the reading of old texts. Helena M. Pycior (USA), spoke on *mathematics teaching with history*, which reported on a project offering a course on History of Mathematics for teachers of grades 7 through 12 and on the preparation and testing, for this same course, of sample projects on history applied to mathematics instruction. Árpad Szabó (Hungary), spoke on *relations between history and pedagogy of mathematics*, claiming that historical viewpoints in the teaching of mathematics help also to deepen the understanding of mathematics itself, and giving examples supporting this claim. Hans Wüssing (DDR), spoke on *the teaching of history of mathematics*, emphasizing the importance of bringing the social context into the historical presentation and of discussing the social momentum in which advances in mathematics take place. The panel was presided by Ubiratan D'Ambrosio (Brazil).

The fourth time slot was devoted to short contributed papers and was also presided by U. D'Ambrosio. Ten minute presentations were given by

László Filep (Hungary) on *using the history of mathematics in teacher training*;

Ryosuke Nagaoka (Japan) on *mathematical education beyond the training of mathematical literacy*;

Zofia Golab-Meyer (Poland) on *about some difficulties in understanding mechanics notions in history and in the school*;

Rudolf Bkouche (France) on *ce que l'histoire des mathematiques peut apporter a l'enseignement: l'exemple de la geometrie*;

Robert Hayes (Australia) on *history as a way back to mathematics*;

Circe M. Silva da Silva (Brazil), on *Forschungsprojekt über Geschichte der Mathematik und Mathematikunterricht in Brasilien.*

The presentations drew in general much interest and discussion.

A satellite conference of HPM, organised by Florence Fasanelli, from USA, took place in Florence, Italy, from 20 through 22 July, 1988.

Full texts of all the presentations, including some papers presented at the satellite conference, will appear as a special publication of HPM.

♣ ♣ ♣ ♣ ♣

IOWME – International Organization of Women and Mathematics Education — see Topic Area 13.

♣ ♣ ♣ ♣ ♣

PME – Psychology of Mathematics Education — No description of activities has been submitted.

PROJECTS

Hungarian coordinator: Gábor J. Székely

Organizers:	Institutions:
A. Ahmed	West Sussex Inst. of Higher Education (UK)
J. Albert	The Weizmann Inst. of Science, Rehovot (Israel)
C. Alsina	Spanish National Exhibition (Spain)
G. Arrowsmith	Centre for Math. Education, Open Univ., Milton Keynes (UK)
D. N. Burghes	Centre for Innovation in Math. Teaching, Univ. Exeter (UK)
H. Burkhardt	Univ. of Nottingham, Shell Centre for Math. Education (UK)
P. Costello	Swinburne Inst. of Techn. (Australia)
D. G. Crighton	Mathematical Instruction Subcommittee of the Royal Society (UK)
P. Drake	Univ. London, Institute of Education (UK)
J. Goodwin	Homerton College, Cambridge (UK)
M. Halmos	Hungarian Subcommission of ICMI (Hungary)
L. C. Hart	Georgia State Univ., Atlanta (USA)
P. L. Hennequin	French Subcommission of ICMI, Aubiere, Clermont-Ferrand (France)
D. C. Johnson	King's College, London (UK)
N. Kreinberg	EQUALS, Univ. of California, Berkeley (USA)
C. Little	Univ. of Southampton (UK)
F. Lowenthal	Belgian Subcommission of ICMI (Belgium)
B. Penkov	Bulgarian National Exhibition (Bulgaria)
Júlia Szendrei	OPI: National Pedagogical Institute (Hungary)

N. Thomas   Problem Solving Interst. Group of MERGA,
                             Mitchell College, Bathurst (Australia)
V. Villani   Italian Mathematical Union (Italy)
P. E. Walker   Kent County Council, Springfield, Maidstone (UK)

THE INTERNATIONAL COMMISSION ON MATHEMATICAL INSTRUCTION

Geoffrey Howson – Secretary of ICMI

When speaking at the closing ceremony of ICME 5, four years ago in Adelaide, I tried to dispel some of the confusion associated with the two sets of initials ICMI and ICME. Alas, I note that even in the Editorial of ICMI Bulletin No. 24, which all of you will have received, the wrong set of initials was used on one occasion. Perhaps, then, there is still a job of work to be done.

The International Commission has existed in some form or other and with two periods of hibernation, coinciding with the two great world wars, for eighty years. For the last thirty or so it has formed a sub-commission of the International Mathematical Union and, indeed, its officers are appointed by the General Assembly of that body. For many decades the Commission contributed in a major way to the International Congresses of Mathematicians, before the decision was taken to hold our own congresses. The first congress, ICME 1, was held in Lyons in 1969. This is our sixth, the first to be held in a Socialist country, and, I am pleased to say, the best attended ever. For many people ICMI is, then, closely, perhaps entirely, linked with ICME and the two sets of initials are considered to be interchangeable. However, ICMI is more than the congresses and I should like briefly to describe some of its other activities.

First, the ICMI studies. At Adelaide I outlined plans. These by now are achievements. Since 1984 studies have been held on

The influence of computers and informatics on mathematics and its teaching.

School mathematics in the 1990s.

Mathematics as a service subject.

I do hope you will read the publications which came out of these studies. Brief descriptions are given in Bulletin 24. Translations into other languages have appeared or are being prepared.

Next year should see the publication of a further book in the series. This time written by members of the Psychology of Mathematics Education group and entitled "Cognition and Mathematics Education". Also, as readers of Bulletin 24 will know, there will be an international meeting in September, 1989 in Leeds, England on "The Popularization of Mathematics". Further studies are being planned.

Mention of the Psychology of Mathematics Education Group which, of course, holds its own annual conferences prompts a mention of the two other international study groups affiliated to ICMI, that on the relations between the History and Pedagogy of Mathematics and that on Women and Mathematics Education. All three groups are very active in the periods between congresses. Again details of their officers are given in Bulletin 24.

An ICMI sub-committee which also has important work between congresses is that responsible for fixing the site of the annual International Mathematical Olympiads — competitions which are attracting rapidly increasing numbers of entrants. In 1989 the Olympiad will be held in the Federal Republic of Germany, in 1990 in the People's Republic of China, in 1991 in Sweden, in 1992 in the German Democratic Republic, in 1993 in Turkey — I could continue.

ICMI also attempts to work regionally though IACME, the Inter American Council on Mathematical Education, and SEAMS, the South East Asia Mathematical Society. Both of these held international conferences on mathematics education in 1987. In 1990 a regional meeting will be held in Beijing, People's Republic of China and SEAMS will meet in Brunei. IACME will meet next in 1991 in Miami, USA.

ICMI has been able to make small grants in recent years, not only to these regional bodies, but also to further the work of the African Mathematical Union.

When speaking of the grants which we in ICMI can make, I should also add that our own work, and our ability to help others, depends almost entirely on the generous grants which we in turn receive from IMU, UNESCO and the International Council of Scientific Unions, ICSU. We are most grateful for this help.

There is a clear need for regional meetings of the type mentioned. More specifically, focussed activities can also take place at a national level. As one example, I should like to mention our Spanish colleagues who prepared a national response to the discussion document "School Mathematics in the 1990s" resulting in a commercially published book.

Here much depends upon the National Representative to ICMI and even more on National Sub-Commissions. If ICMI is to function effectively then National Representatives have an important part to play in feeding information and ideas from the member countries to the ICMI Executive Committee and to the International Program Committees for the congress and vice-versa.

There is much work for all mathematics educators to do and many problems to be tackled. Let us, then make full use of international cooperation and take advantage of the worldwide network which ICMI provides.

I hope that you have all enjoyed this Congress and that already you are beginning to make plans to be at Quebec in 1992 for ICME 7. For many people, attending an ICME means making a great personal financial sacrifice. I am pleased to report that ICMI has succeeded in giving financial assistance to more people from developing countries to attend Budapest than it did Adelaide. Next time we must try to do better still. One letter of thanks, however, I thought most telling and touching. It was from a school teacher in a developing country who said that thanks to the help provided by ICMI and to the gifts of some of his former pupils he would be able to attend Budapest. It is good that school teachers should make the effort to travel thousands of miles to attend an ICME and even better that they should be helped to do this by grateful former students.

Now, however, is a time when we as Congress attenders can show our gratitude to all those Hungarian friends who made this Congress possible. There are too many to name all of them here. As we were reminded in the Prime Minister's address to the Congress, there have been many economic problems in Hungary these last two years — problems which have made the planning of this large and enormously demanding Congress even more difficult. We are grateful to all our Hungarian colleagues for the enormous efforts they have made to make this meeting so successful.

I must, however, make special mention of four people who have made particularly significant contributions to the Congress:

Professor Ákos Csaszar, Chair of the IPC and of the János Bolyai Mathematical Society, our hosts;

Professor János Szendrei, Chair of the Hungarian Organizating Committee (HOC);

Dr. Tibor Nemetz, Secretary of the IPC and the HOC;

Mrs. Cecilia Szabados, Vice-secretary of the Janos Bolyai Mathematical Society.

Professor Császár, the Congress unites in thanking you and all your Hungarian colleagues for the planning, work and hospitality which have combined to make this so memorable a Congress.